WITHDRAWAL

Teaching and Learning Proof Across the Grades

In recent years there has been increased interest in the nature and role of proof in mathematics education, with many mathematics educators advocating that proof should be a central part of the mathematics education of students at all grade levels. This important new collection provides that much needed forum for mathematics educators to articulate a connected K–16 "story" of proof. Such a story includes understanding how the forms of proof, including the nature of argumentation and justification as well as what counts as proof, evolve chronologically and cognitively, and how curricula and instruction can support the development of students' understanding of proof. Collectively, these studies inform educators and researchers at different grade levels about the teaching and learning of proof at each level and, thus, help advance the design of further empirical and theoretical work in this area. By building and expanding on existing research and by allowing a variety of voices from the field to be heard, *Teaching and Learning Proof Across the Grades* not only highlights the main ideas that have recently emerged on proof research, but also defines an agenda for future study.

Despina A. Stylianou is an Associate Professor of Mathematics Education at the City College of the City University of New York.

Maria L. Blanton is a Senior Executive Research Associate at the James J. Kaput Center for Research and Innovation in Mathematics Education and an Associate Professor of Mathematics Education at the University of Massachusetts, Dartmouth.

Eric J. Knuth is an Associate Professor of Mathematics Education at the University of Wisconsin-Madison.

STUDIES IN MATHEMATICAL THINKING AND LEARNING
Alan H. Schoenfeld, Series Editor

Romberg/Shafer *The Impact of Reform Instruction on Student Mathematics Achievement: An Example of a Summative Evaluation of a Standards-Based Curriculum*

Sarama/Clements *Early Childhood Mathematics Education Research: Learning Trajectories for Young Children*

Schliemann/Carraher/Brizuela (Eds.) *Bringing out the Algebraic Character of Arithmetic: From Children's Ideas to Classroom Practice*

Schoenfeld (Ed.) *Mathematical Thinking and Problem Solving*

Senk/Thompson (Eds.) *Standards-Based School Mathematics Curricula: What are They? What do Students Learn?*

Solomon *Mathematical Literacy: Developing Identities of Inclusion*

Sophian *The Origins of Mathematical Knowledge in Childhood*

Sternberg/Ben-Zeev (Eds.) *The Nature of Mathematical Thinking*

Stylianou/Blanton/Knuth (Eds.) *Teaching and Learning Proof Across the Grades: A K–16 Perspective*

Watson *Statistical Literacy at School: Growth and Goals*

Watson/Mason *Mathematics as a Constructive Activity: Learners Generating Examples*

Wilcox/Lanier (Eds.) *Using Assessment to Reshape Mathematics Teaching: A Casebook for Teachers and Teacher Educators, Curriculum and Staff Development Specialists*

Wood/Nelson/Warfield (Eds.) *Beyond Classical Pedagogy: Teaching Elementary School Mathematics*

Zaskis/Campbell (Eds.) *Number Theory in Mathematics Education: Perspectives and Prospects*

Teaching and Learning Proof Across the Grades
A K–16 Perspective

Edited by
**Despina A. Stylianou, Maria L. Blanton,
and Eric J. Knuth**

Routledge
Taylor & Francis Group
NEW YORK AND LONDON

NCTM

NATIONAL COUNCIL OF
TEACHERS OF MATHEMATICS

First published 2009
by Routledge
270 Madison Ave, New York, NY 10016

Simultaneously published in the UK
by Routledge
2 Park Square, Milton Park, Abingdon, Oxon OX14 4RN

Routledge is an imprint of the Taylor & Francis Group, an informa business

© 2009 Taylor & Francis

Typeset in Minion by
Swales & Willis Ltd, Exeter, Devon
Printed and bound in the United States of America on acid-free paper by
Edwards Brothers, Inc.

Library of Congress Cataloging-in-Publication Data
Stylianou, Despina A.
   Teaching and learning proof across the grades : a K–16 perspective/
Despina A. Stylianou, Maria L. Blanton, Eric J. Knuth.
      p. cm.—(Studies in mathematical thinking and learning)
   Includes bibliographical references and index.
   1. Logic, Symbolic and mathematical—Study and teaching.   2. Proof
theory—Study and teaching.   3. Mathematical analysis—Study and teaching.
4. Problem solving—Study and teaching.   I. Blanton, Maria L.   II. Knuth,
Eric J.   III. Title.
   QA8.7.S79 2008
   511.3′6071—dc22   2008031609

ISBN 10: 0–415–98984–1 (hbk)
ISBN 10: 0–203–88200–8 (ebk)

ISBN 13: 978–0–415–98984–8 (hbk)
ISBN 13: 978–0–203–88200–9 (ebk)
NCTM Stock Number: 13356

We wish to thank our families as this book is proof of their patience and support

Contents

Series Editor's Foreword
The Soul of Mathematics

ALAN H. SCHOENFELD

If problem solving is the "heart of mathematics,"[1] then proof is its soul. Problem solving and proving go together like . . . well, like heart and soul; like love and marriage; feel free to pick your favorite pairing. This dualism of exploration and confirmation, the lifeblood of mathematics, is the everyday work of every mathematician. Yet, the vitality of proving as an essential part of *doing* mathematics—of exploring, conjecturing, and subjecting one's conjectures to rigorous testing—is little understood outside the mathematical community.[2]

At core, mathematics is a discipline of sense making. Various mathematicians have described this in different ways. For example, Lynn Steen (1988) called mathematics the "science of patterns," meaning that the mathematician perceives regularities and then has a responsibility for determining whether that perception reflects something inevitable—"It *has to* happen, and here's why." George Pólya, in the 1965 film *Let us Teach Guessing*, and in his books *Mathematical Discovery* (1962, 1965/1981) and *Patterns of Plausible Reasoning* (1954b) uses very different language, but the point is the same: "Here is a mathematical object. What can I say about it? Here is a class of similar objects. What can I say about them? Oh, that's interesting: These two things share the same property. I wonder if this third one does. Hmm, I wonder if it's true in general. Now, what's going on here? If I can see why that's happening, then I'll understand it."

In what follows, I take two examples and play them out as cases in point. They are distillations of conversations that students in my problem-solving course and I have had through the years.

Example 1: What can you say about the sum of the coefficients of $(x+1)^n$? My students work this problem early in the course, but by the time they encounter it they have been introduced to the general heuristic, "I you don't know how to solve a problem, try a series of examples and look for a pattern." Working in groups, they determine quickly that the sum of the coefficients of $(x+1)^n$ for $n = 0, 1, 2,$ and 3, respectively, is 1, 2, 4, and 8. From this pattern, they guess that the sum in the nth case will be 2^n. I ask the class if they're convinced that the sum will be 2^n, and they say yes. I ask them *why* the sum should be 2^n, and they look at me as though I've come from outer space. So, I suggest that someone come to the board and perform the multiplication for $(x+1)^4$, which will be $(x+1)$ multiplied by $(x+1)^3$, which many of them have already computed.

The computation looks like this:

$$x^3 + 3x^2 + 3x + 1$$
$$\underline{x + 1}$$
$$x^4 + 3x^3 + 3x^2 + x$$
$$+\quad \underline{x^3 + 3x^2 + 3x + 1}$$
$$x^4 + 4x^3 + 6x^2 + 4x + 1$$

I ask the students to think about this computation in terms of the sum of the coefficients of each polynomial. After a minute or so a student says: "Look, when you multiply $(x+1)^3$ by x you write all the same coefficients of $(x+1)^3$ down, but each one is next to one higher power of x. Then, when you multiply $(x+1)^3$ by 1 you just write down all the coefficients of $(x+1)^3$ again. When you add those two lines, you wind up with two copies of each of the coefficients you had before, so multiplying by $(x+1)$ doubles the sum of the coefficients." The class agrees. I ask if there's anything special about the exponent 3, since we had multiplied $(x+1)^3$ by $(x+1)$. The answer, of course, is no. We now know that the sum of the coefficients of $(x+1)^4$ is 2^4. The sum of the coefficients of $(x+1)^5$ will be twice that, or 2^5; and so on. Because this is a mathematics class we go on to formalize what we have just reasoned as an inductive proof (it's good practice to do so, of course.) However, by then the proof is a formality—the students understand *why* the coefficients of $(x+1)^n$ must be 2^n, so the proof is a corollary to their understanding.

Once the class is comfortable with the doubling argument we pursue our understandings a bit. We note that multiplying *any* polynomial by $(x+1)$ doubles the sum of the coefficients of the original polynomial; multiplying by $(x+2)$ triples the sum of the coefficients; multiplying by $(ax+b)$ multiplies the sum of coefficients by $(a+b)$, and so on. It doesn't take long before the class realizes that for any two polynomials $P(x)$ and $Q(x)$, the sum of the coefficients of $P(x)Q(x)$ is the product of the sums of the coefficients of $P(x)$ and $Q(x)$ respectively. This is admittedly a relatively minor result, but it was a true discovery when we found it.

To sum up this episode in brief, problem solving and proving acted as partners in generating new mathematics. That is how mathematics gets done.

A day or two after the class has done Example 1 and some other problems whose answer is 2^n (e.g., "How many subsets does a set of n elements have?"), I give the students the following problem.

Example 2: Suppose you pick n points on the boundary of a circle. You then draw all the lines segments that connect pairs of those points. If the points have been chosen so that no three of the segments intersect at the same point (that is, the circle is divided into the maximum possible number of regions), into how many regions is the circle divided?

If you are not familiar with this problem, play with it for a while before reading on.

Straightforward diagrams show that for $n = 1, 2, 3, 4$, and 5 points, the corresponding number of regions is 1, 2, 4, 8, and 16. My students, having found this pattern, are convinced that it, like Example 1, must result in a power of 2 for each n (in this case, 2^{n-1}). By now they know that they'll have to prove it. So, as in the past, I ask them to look at what happens when they move from $n = 5$ to $n = 6$. They draw six points on the boundary, connect them, and find . . . 31 regions! This must be a mistake, they say, and they try again—and still get 31 regions. This time the familiar pattern, 1, 2, 4, 8, 16, . . . doesn't continue.

The class works through the problem and ultimately shows that the number of regions, given n points on the circumference, is

$$f(n) = 1 + \frac{(n)(n-1)}{2} + \frac{(n)(n-1)(n-2)(n-3)}{1 \cdot 2 \cdot 3 \cdot 4}.$$

Interestingly, as the reader can verify, $f(1) = 1$; $f(2) = 2$; $f(3) = 4$; and $f(4) = 8$, nicely following the exponential pattern. But, $f(6) = 31$.

There is an important lesson to this problem. Patterns can be deceiving, and there are many such patterns in mathematical folklore. There is a simple formula that gives prime numbers for each of the inputs $n = 1, 2, 3,$ . . ., 40, but which fails to yield a prime for $n = 41$. There is a formula $F(n)$ with the property that: $F(1)$ is not perfect square; $F(2)$ is not a perfect square; neither is $F(3)$, $F(4)$, and so on, for more than a billion examples. But, there is a ridiculously large number N for which $F(N)$ *is* a perfect square. As all mathematicians know, you can't trust a pattern to hold forever. Patterns can be suggestive, they can be seductive. They can turn out to be true—but they can also deceive you. A million, a billion, a trillion positive examples aren't enough. Something is only true for sure when you can prove it's true.

Together, Examples 1 and 2 illustrate the essence of proof from a mathematician's point of view. As Example 2 shows, proofs are necessary: Unless you have one, you run the risk of being fooled by a false pattern. But as Example 1 shows, proofs are hardly "mere" confirmation, verifying one's intuitions. For the mathematician, proof is a way to figure out how things work. If you think something might work, you try to prove it—you try to work out *why* it should be the way you think it is. Sometimes you succeed, sometimes you don't—but even in failure you may come to a better understanding of the phenomenon you're trying to make sense of. That is, *proving* is a form of mathematical sense making.

There is a huge difference between a focus on proving as sense making and on proof as the final product of the proving process. Freudenthal (1983), for example, referred to formal mathematical arguments as the "fossilized remains" of the reasoning process. Millions of students have "proved" that the base angles of an isosceles triangle are equal, but the proof they typically (re)produce is both dead and deadly. A proof is typically taught as something to be memorized, and proving comes to be understood by students as a ritual that confirms what they already know to be true, rather than as a means of developing understanding. There is ample evidence to support this claim, but the example that sticks in my mind comes from two college students to whom I'd given a geometric proof problem. In talking about the problem, they produced a beautiful argument, a line of reasoning that every mathematician would find compelling and correct. But then they stopped, describing their brief and lucid conversation to that point as being "totally unmathematical";[3] and then they proceeded to spend a large amount of time trying to produce a standard two-column geometry proof. Ultimately they did so, but the point is that they were wasting their time doing so. The substance, which they had right, was the mathematical argument. The form, the two-column proof, was a straitjacket that they—and millions of students, and their teachers—had taken as a proxy for mathematical argument, without understanding that what counts is the correctness and completeness of the idea rather than the form in which it is expressed. These students, their fellow students, and their teachers, had mistaken a fossil for the real thing. Those who delay "proof" until high school, and who insist on formalities instead of robust reasoning, are dissociating proof from sense making.

This book is about the real thing. It is about students (including young students) coming to grips with mathematical reasoning—students coming to realize that there are ways to put mathematical claims together so that, when you are done, things *must be* the way you say they are. As you will see, sense making (and thus proving) can start early: Elementary school students can make arguments that not only make sense, but that contain the seeds of what mathematicians would accept as formal proofs. This book shows students coming to realize that proving is a way of thinking, so that when you really understand why something works (and *that* is the business of mathematics!), the proof often follows as a byproduct of that understanding. In the manuscript that follows you will find elementary, secondary, and college students grappling with proof. You will find patterns of errors, which are useful to know; you will find patterns of reasoning that surprise you, demonstrating how, when given the chance, children can think logically and productively.

From my perspective, one of the best descriptions of the proof process comes from Mason, Burton, and Stacey (1982). They describe the phases of proving thus:

Convince yourself;

Convince a friend;

Convince an enemy,

I take this to mean the following: First, you play with something until you think you see some regularities and begin to believe that they might always hold. You begin to make an argument suggesting so, an argument that suggests plausibility to a friendly auditor. But will the pattern always be true, or will it, like Example 2, turn out to trick you? That's where the enemy comes in. The question is, can you make your argument airtight? Will it resist all counterattack? When you have an argument that started with your intuition, but that now withstands attack and shows that what you first intuited *must* be true, then you have a proof. That's what my college-level students produced in Example 1. But, the real sense-making process can and should start much earlier.

In the pages that follow, you will find frameworks for examining proof, examples that show that students can experience reasoning and sense making early in the curriculum, and examples of what happens when they don't. I hope that as I did, you will emerge from the manuscript with a richer sense than when you began of what mathematical proof and proving in schools can and should be.

Notes

1. See Halmos, 1980.
2. It is worth noting that the roots of "prove" come from "to probe," meaning "to test the validity of." Thus the maxim "the exception proves the rule" is not an oxymoron, but a correct statement: "The exception probes (the truth of) the rule" and finds it wanting.
3. See Chapter 10 of Schoenfeld, 1985.

Preface

The nature and role of proof in mathematics education has received increased attention, with many mathematics educators advocating that proof should be a central part of the mathematics education of students at all grade levels (e.g., NCTM, 2000). Such attention has given rise to an emerging body of research on proof—research that includes a focus on understanding the forms of mathematical thinking required for understanding proof, on the development of students' proving competencies, and on understanding curricular and pedagogical interventions that can enrich the nature of students' experiences with, and subsequent understandings of, proof.

As this research gained momentum and expanded in volume and scope, it became apparent that the field would benefit from a collective body of work that would not only highlight important findings regarding the teaching and learning of proof at different grade levels, but also define an agenda for future research. In an effort to address these needs, a working conference was organized in Providence, Rhode Island, in September 2004 to provide a forum for scholars to begin to articulate a connected K–16 "story" of proof. The present volume is partly an outgrowth of that meeting. Several of the chapters that appear in this volume are revised versions of presentations given at that meeting, and others reflect conversations that originated in that setting and subsequent presentations at various conferences. And while this volume moves us forward in our collective thinking about proof, we still have much to learn if we want to realize the vision of teaching and learning of proof for all students across the grades.

Structure and Organization

This volume is structured to capture issues in the teaching and learning of proof as they play out across the grades. The introduction chapter sets the stage with a look at the historical roots of the teaching and learning of proof in school mathematics, particularly in the United States. It provides a brief review of the literature and orients the reader towards some of the questions that are currently central in the field. Section I attends to broader theoretical and conceptual issues that apply to the teaching and learning of proof and that underlie the work in the subsequent three sections. Sections II–IV present research corresponding to the three grade bands in which we typically separate grades K–16: Section II addresses the teaching and learning of proof in the elementary grades (K–5),

Section III addresses the secondary grades (6–12), and Section IV addresses college and beyond. And while issues of teaching and learning have often been viewed, studied, or reported separately in the field, we found that teaching and learning were interwoven in the work reported here in ways that inform one another and, thus make such separation both difficult and artificial. Several chapters in this volume report on studies of the learning of proof that takes place in classrooms, where, inevitably, the teacher's actions are also investigated as a tool mediating learning. Alternatively, other chapters examine teachers' actions but use student learning as a lens to understand practice. Finally, the concluding chapter reviews the volume through a framework of questions identified by Harel and Sowder (2007) and synthesizes the contributions made by this volume toward answering these questions. Each section is preceded by an introduction summarizing the chapters in that section and highlighting the main themes that emerge.

Intended Audiences of the Book

Because the book encompasses diverse points of view that address different aspects of the teaching and learning of proof, readers might approach it using a variety of lenses and for different purposes. The theoretical discussions that are prominent in Section I might be of interest to researchers, teacher educators, practitioners, or curriculum developers looking for broader discussions on proof, while the more empirically based classroom research found in subsequent sections might be more relevant for those whose work and interests intersects with the issues of teaching and learning proof in particular grade domains. And while the volume is intended to tell a story across the grades, different chapters can be viewed independently to inform the study of proof at different levels. Our intent is for this volume to appeal to and inform the work of a diverse audience in fields of research, teaching, and educational development and policy.

Acknowledgments

We are indebted to a number of people and organizations for making the publication of this volume possible. We are grateful to the National Science Foundation (grants REC-0337703 and REC-0092746) for supporting the editors' research and the writing of their respective chapters, for supporting the conference, which led to the preparation of the volume itself, and for subsequently supporting the editing of this volume. We are also grateful for public and private funding that supported some of the individual research projects reported on in this volume.

We would also like to acknowledge the assistance of our editors Naomi Silverman and Catherine Bernard. Their advocacy for the publication of the book, as well as their technical assistance throughout its development, was essential to our work. We are particularly grateful for the support of our series editor Alan Schoenfeld. Alan provided invaluable feedback, advice and encouragement from the conceptualization of this project to its final print.

List of Contributors

Lara Alcock	Loughborough University, U.K.
Nicolas Balacheff	Laboratoire d'Informatique de Grenoble, CNRS, France
Kristen N. Bieda	Michigan State University, U.S.A.
Maria L. Blanton	University of Massachusetts, Dartmouth, U.S.A.
Catherine Brach	University of Michigan, U.S.A.
Daniel Chazan	University of Maryland, U.S.A.
Chialing Chen	University of Michigan, U.S.A.
Jeffrey M. Choppin	University of Rochester, U.S.A.
M. Manuela David	Universidade Federal de Minas Gerais, Brazil
Catherine Twomey Fosnot	City College of the City University of New York, U.S.A.
Evan Fuller	University of California, San Diego, U.S.A.
Gloriana González	University of Michigan, U.S.A.
Maria Hamlin	University of Wisconsin, Milwaukee
Guershon Harel	University of California, San Diego, U.S.A.
Aiso Heinze	Universität München, Germany
Patricio Herbst	University of Michigan, U.S.A.
Reuben Hersh	University of New Mexico, U.S.A.
Celia Hoyles	University of London, U.K.
Bill Jacob	University of California, Santa Barbara, U.S.A.
Eric J. Knuth	University of Wisconsin-Madison, U.S.A.
Dietmar Küchemann	King's College, University of London, U.K.
H. Michael Lueke	University of Maryland, U.S.A.
Carolyn A. Maher	Rutgers University, U.S.A.
Tami S. Martin	Illinois State University, U.S.A.
Anne K. Morris	University of Delaware, U.S.A.
Kay McClain	Arizona State University, U.S.A.
Sharon M. Soucy McCrone	University of New Hampshire, U.S.A.
Talli Nachlieli	Levinsky College of Education, Israel
Stephanie Ryan Nichols	University of Texas at Austin, U.S.A.
Kurt Oehler	University of Texas at Austin, U.S.A.
David A. Reid	Acadia University, Canada

Kristina Reiss	Universität München, Germany
Deborah Schifter	Education Development Center, U.S.A.
Annie Selden	New Mexico State University, U.S.A.
John Selden	New Mexico State University, U.S.A.
Edward A. Silver	University of Michigan, U.S.A.
Jennifer Christian Smith	University of Texas at Austin, U.S.A.
Larry Sowder	San Diego State University, U.S.A.
Gabriel J. Stylianides	University of Pittsburgh, U.S.A.
Despina A. Stylianou	City College of the City University of New York, U.S.A.
Keith Weber	Rutgers University, U.S.A.
Michael Weiss	University of Michigan, U.S.A.
Sera Yoo	University of Texas at Austin, U.S.A.
Vicki Zack	St. George's Elementary School, Montreal, Canada

Introduction

Building a Proof Story Across the Grades

Mathematical reasoning at all ages is predicated on one's ability to develop sound and convincing arguments. While the formality and form of these arguments, or proofs, will vary across grades, the need for students to be able to develop, understand, and interpret arguments appropriate to their level of experience, is widely accepted.[1] Indeed, recent reform documents have called for the learning of proof to become a central goal of teaching and learning mathematics across all grades (MAA Committee on the Undergraduate Program in Mathematics, 2000; NCTM, 2000; RAND Mathematics Study Panel, 2002). The *Principles and Standards for School Mathematics* [*PSSM*] (NCTM, 2000) has emphasized that proof should be a part of all precollege students' mathematical experiences in order to deepen and extend learning and to democratize access to these ideas to a broader population of students. It recommends that all students learn "to recognize reasoning and proof as fundamental aspects of mathematics, make and investigate mathematical conjectures, develop and evaluate mathematical arguments and proofs, and select and use various types of reasoning and methods of proof" (p. 56). Similarly, the Committee on the Undergraduate Program in Mathematics, appointed by the MAA to identify major issues and develop tentative recommendations for an undergraduate mathematics *Curriculum Guide* (MAA, 2000), has emphasized that proof should be a part of all students' mathematical experiences at college level and that students "understand and appreciate the core of mathematical culture: the value and validity of careful reasoning, precise definition and close argument" (p. 6).

The task of incorporating proof at all grades is situated within the larger need for a connected K–16 "story" of teaching and learning proof. Such a story would include understanding how the forms of proof, including the nature of argumentation and justification as well as what counts as proof, evolve in students' thinking across grades and how curriculum and instruction support this evolution.

In this introductory chapter, we begin by providing background on the current effort to bring teaching and learning proof to the forefront of mathematics education across grades K–16. We provide a brief account of the role of proof in school mathematics historically and of the evolution in the research base on teaching and learning of proof. We then describe the contributions this volume makes to

1

this field by highlighting some of the central themes addressed by the chapters as well as the questions and issues discussed in the chapters.

The Teaching of Proof in the US—A Brief Historical View

The origin of proof in western mathematics can be traced to the 6th century BCE[2] when the Greeks developed the notion that mathematical facts, in particular geometric facts, must be established deductively, not empirically. As a result, the Greeks radically transformed the empirical mathematics of the Babylonians and Egyptians into the "demonstrative" science that we know today (for a more detailed discussion on the origin of proof see Arsac, 2007). The *Elements* of Euclid, produced at around the 3rd century BCE, consists of 465 propositions comprising the geometric, number theory and, in a way, algebraic knowledge of that era. It set the standard of rigor and "dominated the mathematics of the western world until the late 19th century and in essence is still intact in our days" (Harel & Sowder, 2007). Consequently, the teaching of deductive reasoning became a characteristic of formal mathematics training. Early examples of teaching proof are found in descriptions of the mystic Pythagorean school, but also in the curriculum proposed by Plato in the *Republic* as part of the education of future guardians of the state (see Fowler, 1999). Some believe that Plato's Academy welcomed students with the inscription "Let no one unskilled in geometry enter" ("Ἀγεωμέτρητος μηδεὶς εἰσίτω")—attesting to the importance of training in rational reasoning for the citizens of the Athenian democratic society.

The teaching of proof as a fundamental part of all school mathematics is not an original proposition in American schools either. Proof has been taught primarily as part of high school geometry for more than a century. As Herbst (2002b; Chapter 15 in this volume) explains, one can trace the imperative that students should learn "the art of proving" in a geometry course to the late 19th-century report of the Committee of Ten (Eliot et al., 1969). The Committee of Ten, a group of educational leaders formed to examine the role of the high school curriculum and its relationship to college admissions, concluded that part of the value of teaching mathematics would be to "train to the mind's powers of conceiving, judging, and reasoning; [...] in formal geometry we have the best possible arena for training in deductive reasoning" (Hill, 1895, p. 354, as reported in Herbst, 2002b).

This choice of the Committee of Ten was not an arbitrary one; it has its roots in the historical view of the role of proof in mathematics. Davis and Hersh (1981) point out that it was natural for geometry to be taught according to the Euclidean tradition of proof (established for over two millennia) while other areas of mathematics such as algebra and arithmetic were not associated with proof until the 19th century. The choice of geometry as the context in which students would develop proof skills was, then, a natural one, and it is not surprising that "this custom [...] has had an enduring influence on how Americans think about mathematical proof" (Herbst, 2002b). Furthermore, the two-column format that was introduced as an instructional tool for proof as well as the custom of "proving" already accepted statements led to the widely held popular belief that proof is a tool for validating obvious statements—an exercise in format and logic rather than an integral part of doing mathematics.

Proof maintained its stature as a topic that characterized the teaching of high school geometry until the "new math" initiative in the 1960s attempted to incorporate formal proof in all areas of school mathematics. This attempt was unsuccessful, and, as Hanna (1995) notes, the demise of the new math movement brought a gradual decline in the role and stature of proof in mathematics curricula to the point of almost complete extinction; proof was relegated once again to a peripheral role in school mathematics, finding its home in geometry courses for college-bound students. New trends for more "student-centered" instruction often further marginalized the role of proof. While instruction in which students engage in classroom argumentation should not be antithetical to the teaching of proof, some interpretations of these approaches viewed instruction on proof as an authoritarian approach to the teaching and learning of mathematics. But, as Hanna (1995) argued pointedly, proof is not "authoritarian"—aiming to impose a body of knowledge on students: "A proof is a transparent argument, in which all the information used and all the rules of reasoning are clearly displayed and open to criticism [. . .]. Proof conveys to students the message that they can reason for themselves, that they do not need to bow down to authority. Thus the use of proof in the classroom is actually anti-authoritarian" (p. 46).

The decreasing presence of proof in school mathematics had not gone unnoticed and, in fact, has been a target of criticism by mathematics educators and mathematicians. Wu (1996) argued, for example, that the scarcity of proof outside of geometry is a misrepresentation of the nature of proof in mathematics. He stated that the absence of proof is "a glaring defect in the present-day mathematics education in high school, namely, the fact that outside geometry there are essentially no proofs. Even as anomalies in education go, this is certainly more anomalous than others inasmuch as it presents a totally falsified picture of mathematics itself" (p. 228). Greeno (1994) suggested that the absence may be due to misconceptions on the nature of proof (or a narrow view of the nature of proof) and recommended that this be rectified by revisiting the "epistemological significance of proof in mathematics" (p. 270) and, hence, reconsidering the function of proof in doing mathematics. Indeed, proof can have several critical functions within mathematics beyond the broadly accepted role as the ultimate method of verification. Balacheff (1988), deVilliers (1990), and Hanna (1990) pointed to proof as a tool for explanation, communication, verification, systematization, and discovery in mathematics.

Recent reform efforts in the United States are calling for substantial changes in both school mathematics curricula and teachers' instructional practices with respect to proof. In contrast to the status of proof in the previous national standards document (NCTM, 1989), its position has been significantly elevated in the most recent document (NCTM, 2000). Not only has proof been upgraded to an actual standard in this latter document, but it has also received a much more prominent role throughout the *entire* school mathematics curriculum and is expected to be a part of the mathematics education of *all* students. These efforts not only attempt to re-establish proof as an inseparable aspect of doing and understanding mathematics (as opposed to just proof as a tool for establishing the validity of statements), but also as an overall theme that may provide coherence to the (oftentimes) compartmentalized learning of mathematics from kindergarten to college, and as a tool to democratize access to powerful mathematics for *all* students.

However, as Herbst (2002b) argued, changes in the way proof is viewed by the broader public (including students and teachers) will require more than minor adjustments or calls for reform. The introduction of a narrow view of proof as a verification tool of obvious statements, done in a specific "two-column" format took a systemic force to be established: that is, the custom developed through a system of decisions and the development and coordination of necessary curricular and instructional resources. A change of this view to a broader view of proof as a learning and problem-solving tool integral to the doing of mathematics across the grades will probably require a similar type of force.

One important catalyst for change is the development of a strong research base that informs our understanding of the teaching and learning of proof. We summarize later the results of research conducted over the last several decades and use this to situate the work of this volume and point to ways in which this volume supplements and extends that work.

Challenges to the Teaching and Learning of Proof

Students' difficulties with proof are well documented (e.g., Balacheff, 1988; Bell, 1976; Chazan, 1993; Coe & Ruthven, 1994; Porteous, 1990; Senk, 1985; Usiskin, 1987; Williams, 1980) and can be attributed to several factors. Broadly speaking, because proof has not traditionally been used in instruction in a way that creates "personal meaning or explanatory power for students' (Schoenfeld, 1994, p. 75; see also Harel & Sowder, 1998), many students perceive the activity of proving as an extraneous task, not as a tool for thinking more deeply about mathematics (Alibert, 1988). Moreover, proving a statement for which the proof already exists or is intuitively obvious leads to the perception that proof activity is not a process of discovery (Harel & Sowder, 1998; Schoenfeld, 1994; Wheeler, 1990).

More specifically, however, a number of empirical studies, particularly at the secondary school level, that have addressed the difficulties students face when attempting to either read or write proofs have identified several areas as problematic. First, secondary school students struggle with what constitutes a proof and with understanding the power of a generalized argument as covering all possible cases. In separate studies, students were seen to rely predominantly on empirical arguments (Healy & Hoyles, 2000), could not adequately justify their solutions to mathematical tasks (Bell, 1976), or did not understand that a statement could be refuted by a counterexample (Galbraith, 1981). Senk (1985) and Usiskin (1987) both found that less than one-third of their respective high school geometry students were able to construct mathematical proofs. Elsewhere, Chazan (1993) found that students interpreted deductive proofs as verifications of single cases that could be refuted by counterexamples, and Fischbein and Kedem (1982) found that students who had expressed their agreement with a generalized argument still needed to verify the claim by checking single cases. Similarly, Porteous (1986) found that the majority of students studied checked particular cases using empirical arguments rather than appealing to a general argument that had already been established. That students do not see that "even a single practical check is superfluous" (Fischbein, 1982, p. 17) suggests a discrepancy between what they might acknowledge publicly as constituting a proof and their internal belief about what they perceive as a more convincing argument. Indeed, many students seem to see "proofs" based on checking a few cases as the most convincing form of argument (Bell, 1976). While this could derive from students' need to make sense of ideas by "trying them" in specific scenarios, it more likely stems from a lack of understanding of the power of generality (Williams, 1980).

Other areas that have been shown to be problematic for students concern the logic and methods of proof (Bittinger, 1969; Duval, 1991) and the problem-solving skills necessary to construct arguments (Schoenfeld, 1985). Given students' difficulties with proof, a number of studies based on the assumption that students' difficulty with proof is closely related to low problem-solving performance attempted to provide instruction in problem-solving heuristics as a way to improve proof writing. Nonetheless, results have shown that heuristic training often has little effect on students' ability to do proof.

Mathematical language and symbolism also play a role in students' ability to construct proofs (Laborde, 1990; Rin, 1983). Students' lack of understanding of mathematical language and notation inhibit their ability to understand and use definitions that play an important role in the proving process. Finally, students' capacity to construct proofs is predicated on, and thus can be inhibited by, their understanding of the concepts involved in a particular statement to be proved (Hart, 1994; Moore, 1994; Tall & Vinner, 1981; Vinner & Dreyfus, 1989).

It is important to point out that the vast majority of the work that took place on the learning of proof was situated at the high school level. Little is known about elementary school and even middle school students' understanding of proof. This gap in the existing research base is also largely due to the fact that proof has been traditionally viewed as a high school geometry topic. The current attempt in mathematics education to reposition proof at the center of mathematics teaching and learning across the grades calls for a systematic study of the learning of proof at different grade levels.

Another important note is that much of the work on proof has focused on the *learning* of proof and less is known about the *teaching* of proof. It is generally recognized that teachers' content knowledge is an important determinant of their classroom practices (Borko & Putnam, 1996); accordingly, there have been a number of studies that have focused on teachers' knowledge of proof (e.g., Jones, 1997; Knuth, 2002a, 2002b; Martin & Harel, 1989; Simon & Blume, 1996). Although the aforementioned studies inform our understanding of the teaching of proof (or at the very least, of factors that influence the teaching of proof), very few studies have focused on the teaching of proof in the context of teachers' day-to-day instructional practices. Herbst's work (2002a, 2002b) has advanced our understanding of mathematics classrooms as organizations and how proof and proving as a mathematical task is shaped by the practices in classrooms. Herbst takes students' performance and possible difficulties in proving as symptomatic of the existence of instructional patterns that need to be understood in terms of the history and culture of mathematics teaching in a society (particularly American society) and suggests that changes in students' understanding of proof can only result from changes in the didactical contexts in which it develops.

This brief review of the current research base points to several issues that informed the development of this volume. First, this literature has helped the community understand the complexity of teaching and learning proof. At the same time, however, this review points to several areas where research is still needed: little work has been done in addressing proof at different grade levels besides high school, and, even less in understanding connections in this research base on proof across the grades. And, while understanding students' difficulties in proof has been very helpful in pointing to the need to incorporate proof across the grades, we still know little about the kinds of instruction that facilitate the development of students' understanding of proof.

Contributions to this Volume: Current Perspectives on Teaching and Learning Proof

While the studies described above have provided a significant contribution to our understanding of (primarily high school) students' and teachers' understandings of proof and some of the impediments to that understanding, current ways of thinking within the mathematics education community have adopted a broader view of student learning and the teaching that happens in mathematics classrooms, one that looks beyond the difficulties and deficits that students bring to learning and emphasizes the ways in which students come to know (Steffe & Kieren, 1994).

This view is particularly important as we broaden our perspective of looking at proof from being a topic taught in specific advanced mathematics courses, to being a tool for studying and learning mathematics at all levels. Although the authors who contributed to this volume take different approaches to the nature of proof and its role in the classroom, one common theme across the chapters is a broad view of proof with respect to its nature and role and an emphasis on proof's organic integration in doing K–16 mathematics. The volume's authors examine instructional situations in which proof is a learning tool across topics and grades that arises organically in the study of school mathematics (see, for example, Chapters 4 and 7; Schifter and Maher in their work illustrate how proof and the act of proving can arise naturally out of rich problem-solving tasks within the current curricula). A longitudinal approach to teaching proof, which reflects current perspectives on how to organize school mathematics (NCTM, 2000), is not unique to proof. As Kaput (2008) wrote concerning algebra and its historical treatment as an insular topic for high school grades:

> [T]he kind of narrow view of algebra that has dominated school mathematics for many years in many countries as primarily syntactically-guided, symbolic-logic manipulations done to validate already accepted statements not only grossly understates the multiple functions of algebra historically as mathematics, it is also an inadequate foundation for reconsideration of algebra's place in school mathematics. (p. 8)

This volume adopts a similar view regarding the role of *proof* in school mathematics. That is, the depth and power proof has historically provided mathematics as a field of study will only be realized in school mathematics as it is developed in students' thinking as a connected theme across all grades.

The Emerging Meaning of Proof Across the Grades: From Argumentation to Formal Proof

What constitutes proof? The answer to this question will likely depend on the person being asked. As Balacheff (2002) observes, individuals and even mathematicians and mathematics education scholars conceptualize the nature of proof in different ways and, while these differences may be acceptable: "We may benefit from being able to witness our convergences and turn our differences into research questions." As Section I discusses in more detail, and as this volume bears out through empirical data, differences in the conceptualization of proof become more apparent as one looks at proof across the grades—proof indeed manifests itself in different forms across different grade levels. As students acquire more advanced notational and linguistic systems of expression and as their ability to reason deductively within these systems develops, proof takes on a more formalized meaning in the classroom.

In the elementary grades, proof is viewed more informally as reasoning and argumentation. Section II explores the intuitive notions of proof that young children hold and the classroom environments that can support and build on their innate tendency to ask for explanations and to share their justifications. What constitutes a proof at this level does not take the rigid format and symbolism that has come to be associated with proof, but allows children to engage in logical arguments that are accepted in their own learning community. Fosnot and Jacob (Chapter 6), for example, examine the development in children's deductive reasoning using processes similar to those used by mathematicians—reflecting on arguments, reorganizing ideas, and building justifications—while attempting to convince their peers of mathematical claims they make. However, the context for these arguments is an open number line and the symbols are whole numbers. Similarly, Schifter (Chapter 4) explores how children build arguments to establish certainty, while Morris (Chapter 5) considers how children can be prepared for a more formal study of proof by first studying and developing arguments about relationships among quantities.

At the middle school level students are still exploring proof as argumentation as McClain (Chapter 13) illustrates, while at the same time they are exposed to some extent to symbolic notations, although not as much as one would expect, as discussed by Knuth and his colleagues (Chapter 9). At the high school level proof gradually takes on a more formalized meaning. Students are introduced to specific two-column formats for writing proofs (see, for example, McCrone & Martin, Chapter 12; Herbst et al., Chapter 15) and instruction places more importance on the use of deductive reasoning. In college, students who major in mathematics-related areas and, hence, study formal mathematics are expected to develop an ability to use more rigorous arguments that obey certain syntactic rules of logic (see, for example, Weber & Alcock, Chapter 19; Selden & Selden, Chapter 20).

One may argue that early experiences with proof should be structured to prepare students for more formal studies of proof in the higher grades. Stylianides (2007) argues that there should be continuity in how the notion of proof is conceptualized in different grade levels so that students have coherent experiences with proof in school. Given the different role and meaning that proof takes in this volume, it is reasonable to question how proof can occur as a connecting theme across the grades as students' understanding of proof emerges.

In response to this, we argue that proof in the early grades does not necessarily aim to prepare students for the use of symbolic logic that is often at the core of proof in advanced courses. Proof across the grades aims to instill an appreciation for building reasonable, logical arguments while using mathematical tools that are within the reach of the users. Indeed, the chapters in Section II illustrate how the elementary grades' mathematics curriculum can develop argumentation as a habit of mind and, particularly, deductive reasoning; while students are not introduced to a formalized system of logic, the underpinnings of that logic can be developed via careful argumentation and the building of new concepts on accepted and validated previous knowledge in the classroom. As Sections III and IV bear out, this habit of argumentation can be further developed in high school and beyond while attending to the syntactic and formal aspects of proof.

Invariants in Student Reasoning with Proof Across the Grades

While the chapters across different sections of this volume examine how proof and argumentation vary across the grades, they also illustrate that students across the grades currently approach proof using some similar strategies. Studies conducted across grades K–16 and presented in this volume illustrate students' wide use of empirical arguments. Indeed, as this volume bears out, students choose empirical arguments to prove their conjectures both as a way to convince themselves and to present their reasoning to their teachers and peers.

The use of empirical arguments as proofs by students of all grade levels is hardly a new finding; it has been extensively documented in several studies in the past quarter of a century (e.g. Bell, 1976; Coe & Ruthven, 1994; Healy & Hoyles, 2000; Porteous, 1986; Stylianou & Blanton, 2006). It is common enough to be viewed as a specific and broadly used "proof scheme" (Harel & Sowder, 1998), or form of proof (Balacheff, 1988). Scholars and practitioners agree that it is important for students to develop more general arguments as they mature mathematically; as Küchemann and Hoyles (Chapter 10, p. 171) note: "A major challenge in mathematics education is to develop students' abilities to reason mathematically, that is to make inferences and deductions from a basis of mathematical structures."

The authors of this volume work to understand what underlies students' broad use of empirical arguments at each grade level. In the elementary grades, Maher (Chapter 7) and Morris (Chapter 5) identify investigations rooted in quantitative measures as a first step towards building generalizations. In middle school, Knuth, Choppin, and Bieda (Chapter 9) examine the curricula used in schools and question whether textbooks and instruction model deductive reasoning for students. Küchemann and Hoyles (Chapter 10), in a longitudinal study spanning middle and early high school grades (grades 7 through 9), found empirical reasoning to be widely used as well. However, Küchemann and Hoyles note that empirical reasoning can vary from inappropriate pattern spotting to more advanced or sophisticated uses such as a basis to check the validity of a structural argument. While chapters in Section IV (college level) did not specifically address the use of empirical reasoning, it is implicit in the premise of studies whose aim is to develop instructional means to move students beyond empirical arguments (e.g., Blanton et al. (Chapter 17); Smith et al. (Chapter 18)).

The persistent use of empirical arguments across students of all ages points to the need for a better understanding of why it persists in students' thinking and how curricular and instructional choices either contribute to this or can challenge this approach in students' thinking. The latter can help us understand what types of instructional tool may help students confront the limitations of empirical arguments and move towards the use of deductive reasoning.

Representation in doing Proof

Ball and Bass (2003) classify the use of justification or proof as one of several "mathematical practices" that should permeate the teaching of mathematics across the grades. A second such "mathematical practice" that has been emphasized in current calls for reform, and, in particular, by the *Principles and Standards for School Mathematics* (NCTM, 2000) is "representation"—choices we make for expressing and depicting mathematical ideas and the ways in which we put them to use. Representation and justification have both been elevated to be two of the *process standards* in this document and are expected to be at the center of the mathematics education of all students.

As this volume indicates, representation is closely linked with the teaching and learning of proof across the grades. In Section II, Schifter (Chapter 4) and Morris (Chapter 5), for example, explore how children use representations as a tool to reduce abstraction and build deductive arguments. Schifter examines children's use of representations of story contexts not only as tools to aid them in specific arguments but also as tools to "establish general claims in the elementary classroom." In that sense, appropriate choice of representation can be instrumental in facilitating the development of early proof.

As students advance in their study of mathematics, representation takes on a different role; the format in which the proof is presented is as important as the mathematical concepts on which the proof is built. Chapters in Section IV observe that learning to prove at the college level requires students to work within a new *representation system* in which proofs are composed in a special genre based on formal rules of logic. Selden and Selden (Chapter 20) and Weber and Alcock (Chapter 19) both discuss the genre of proof and, while they acknowledge the challenges that it may present to students, offer insights into teaching practices that develop facility in this representation system. Weber and Alcock, in particular, frame their discussion around the argument that learning to prove in college classrooms requires students to work within a new representation system. Because this process can be simultaneously limiting and empowering, it is important to understand both the challenges and affordances that students face within the (new) representation system of mathematical proof. The volume opens up several questions with respect to the relationship between representation and proof as "mathematical practices," their teaching and learning and their development in students across the grades.

Multiple Research Paradigms on the Study of Proof

Earlier we discussed some of the literature on the teaching and learning of proof, work that focused primarily on cognitive issues of individual students' understanding of proof. More recently, however, scholars in mathematics education (e.g., Alibert & Thomas, 1991; Schoenfeld, 1999; Tall, 1991) have identified the need to research the development of students' concepts of proof and the factors that influence it in ways that pay "particular attention to studies emphasizing the nature of proof as *an activity with a social character*" (Alibert & Thomas, 1991, p. 216, emphasis added). Indeed, when proof is viewed as a communication tool in mathematics communities, using a more social perspective can be a fruitful approach.

Several of the chapters in this volume are bringing to the forefront some of the social issues associated with proof teaching and learning. In particular, in the early grades we see a greater emphasis on the development of classroom communities that allow students to communicate their reasoning and build norms and representations that provide the necessary structures for proof to have a central presence. Schifter (Chapter 4), Morris (Chapter 5), and Reid and Zack (Chapter 8) all examine the learning of proof within particular classroom cultures in which it is the norm for students to build and communicate arguments that subsequently are owned by the community and advance the mathematical understanding of these communities. But, more interestingly, similar emphasis is also characteristic of some of the chapters that examine the teaching and learning of proof in high school and college. Blanton et al. (Chapter 17) and Smith et al. (Chapter 18) examine the learning of proof that takes place in college classrooms in which mathematical meaning is socially negotiated. Whether these researchers are looking at early experiences on proof at the elementary and middle school level or the advanced learning of proof at high school and the college levels, they use a social lens to explore similar questions about the nature and role of proof in classrooms.

At the same time, several chapters pursue further cognitive research paradigms in order to explore questions related to individual student learning of proof and its development over time. Maher (Chapter 7) at the elementary school level, Knuth and his colleagues (Chapter 9), and Küchemann and Hoyles (Chapter 10) at the middle school level, as well as McCrone and Martin (Chapter 12) at the high school level examine individual student development on proof. Similarly, at the college level, Weber and Alcock use a cognitive lens to examine the learning of proof as situated within a specific representation system and the implications that this may have on individual student learning. Each of these chapters uses a cognitive paradigm as a means to understand the development of student proving competencies as well as the instruction that occurs at a particular level and curricula designed to foster the development of those competencies.

The two dominant paradigms in the field, cognitive and social, underlie the design of several of the studies in this volume and provide lenses for interpreting their findings. Yet, in recent years, we have witnessed a growing tendency to study social interaction as complementary to the study of cognitive factors when characterizing mathematical growth. Current understanding of our ways of knowing recognizes a symbiosis in the social and cognitive domains of learning (see, e.g., Bauersfeld, 1994; Cobb & Bauersfeld, 1995; Cobb & Yackel, 1996). This symbiosis also characterizes some of the chapters in this volume, in which the researchers are attempting to find lenses to examine student learning and the teaching that facilitates that learning using both cognitive and social perspectives and to find ways in which the two perspectives inform each other. In their work, Blanton and her colleagues, as well as Smith and her colleagues employ a sociocultural lens to understand the development of proving competencies, yet they also measure individual student growth and the effect of a particular type of instruction on students' abilities to prove.

It is important to note that our intention is to build on and extend existing research by allowing different perspectives to be heard. While we aimed to build a coherent picture of the existing research, we did not search for a single point of view; rather, we acknowledge the existence of different views and methodological perspectives and used these differences as a tool to expand our understanding of the teaching and learning of proof. The chapters that follow pave the way for a dialogue on how to develop students' command of proof at all stages of their education.

Notes

1. Here we define proof as a *logical argument that one makes to justify a claim in mathematics and to convince oneself and others.* The format, language and degree of formality of this argument may vary across different contexts and communities, hence, it is reasonable to expect different types of argument to be accepted as proofs at different grade levels. The authors of each chapter in this volume define proof differently depending on the grade level they examine in their work.
2. Proof, in the sense of validation, is not unique to western mathematics. Other cultures also introduced ways to validate mathematical conjectures.

I
Theoretical Considerations on the Teaching and Learning of Proof

Section I introduces the research in this volume through theoretical exposition and personal narrative that draw our thinking to the nature of proof, what students need to understand about it, and instructional situations that can help make its implementation across the grades feasible. Hersh (Chapter 1) opens the section by drawing on his own experiences as a college mathematics instructor to reflect on the question, "What would we like to see happen in high school, and even earlier?" in the teaching and learning of proof. His answer, which points broadly to what students in school mathematics need to understand about the nature of proof as prerequisite knowledge for a more formal study of mathematics in college years, suggests that students need to see proof as an activity of "careful, critical reasoning, looking closely for gaps and exceptions" (Chapter 1, p. 19) and as "careful reasoning leading to definite, reliable conclusions" (Chapter 1, p. 19).

Historically and epistemologically, the notion of proof has evolved as the field of mathematics has matured. Perhaps more importantly, an interplay between deductive and inductive reasoning has been a part of this history (Moreno-Armella, 2007), where scholars have argued for the importance of each mode of reasoning as important tools of mathematical thinking (Moreno-Armella, 2007; see also Courant & Robbins, 1996; Lakatos, 1976). Appreciating that the nature of this historical growth is not unlike the cognitive needs of students, Hersh's comments suggest that proof, as a didactic tool, must be interpreted within the context of the particular community engaged in the process of proving. He points out that the argument a child makes during arithmetic instruction constitutes a "proof" in the community of learners to which the child belongs. Moreover, not only is this accommodation of the meaning of proof to young learners not at variance with the formal, rigorous treatment of arithmetic proofs a graduate student might encounter, it is an essential building block in developing the types of reasoning in children's thinking than can serve them in later years as they study more formal mathematics.

In Chapter 2, Chazan and Lueke draw distinctions between disciplinary knowledge and school mathematics to frame a discussion of how research might help us understand what is needed to implement a longitudinal approach to teaching and learning proof across the grades (NCTM, 2000). They argue that the disciplinary practice of solving problems—not just proving theorems—requires important reasoning by which mathematicians look for insightful and elegant solution methods, and that this broader interpretation of the place of reasoning in mathematical thinking, applied to school mathematics, can bring to light more opportunities for how it can occur in classrooms.

Their discussion examines instructional contexts, such as algebra classrooms in high school or arithmetic lessons in elementary grades, where teaching and learning methods for solving problems—not proof and reasoning—are the norm. While these methods, which Chazan and Lueke define as a hybrid between algorithms and heuristics, are not proofs per se, they argue that they act as "proxies" for proofs and, thus, suggest an untapped resource for where reasoning can occur in school mathematics.

Drawing from their own work in high school algebra classrooms, they lift out the tension between "instructional situations" (Herbst, 2006) that focus on teaching methods and those that might emphasize the forms of justification that are typically seen as part of proof. They suggest that, in the presence of this tension, one might question the viability of the goal—echoed in this volume—of the NCTM's *Standards* (2000) to integrate proof and reasoning throughout school mathematics. Instead of this, they ask us to consider what forms of reasoning are possible in instructional settings that focus on methods and for which "proof" is not (or has not been) apparent. Indeed, can we broaden our vision of proof and reasoning in order to build opportunity for these activities in places we might not expect to see them or for which they are not prominent? Using Lampert's (1990, 1992) classroom research in the elementary grades as an existence proof that classroom norms, tasks and instruction can be redesigned so that classroom interaction entails reasoning and proof and, ultimately, includes learning methods, they argue that we should focus on understanding how to design instruction so that it is akin to the norms of thinking within the discipline of mathematics and supports reasoning and justification in productive ways.

Taking a distributed view of mind in which classrooms are seen as "collective, complex intelligences" (Chapter 3, p. 40), Herbst and Balacheff build towards a descriptive theory of classroom mathematical activity viewed metaphorically as public performances. In delineating the different representations of proof that can be constructed through such performances, they propose that we consider several distinctions of how proof might exist in the classroom. In particular, they develop three theoretical elements of proof that entail (1) the customary ways it is used in classroom practices and discourse and the ways that knowledge is validated and organized, (2) the "regulatory structures" in place to monitor the adequacy of problem solutions, and (3) the role proving plays in building new knowledge.

Their chapter addresses an issue, raised previously by Balacheff (2002), that posits a deadlock regarding a shared understanding of the meaning of proof—beyond its formal definition in a mathematical sense—and how it is used in mathematics education research. They maintain that the unique constraints of the mathematics classroom—constraints that include that proof's existence in the discipline of mathematics entails certain consequences regarding how it exists in school mathematics, that acknowledge the idiosyncratic dynamic between teacher and student, and that recognize that being convinced by a particular argument is intrinsic to the individual—require us to think differently about the nature of proof in school mathematics than we might in other settings such as the mathematics community. They argue that proof is not a singular conception, a universal referent that fits all contexts, but a fluid commodity in which "there are specific proofs for specific assertions about specific objects" (Chapter 3, p. 62). To this end, the descriptive theory they outline in Chapter 3 on the nature of proof in classroom mathematical activity aims to move the discourse among researchers in mathematics education beyond the impasse imposed by cultural meanings of proof.

The liberal interpretation of proof advocated in this section (and throughout this volume) disperses the esoteric and singular meaning of proof as the formalized system of logic and syntax used by mathematicians. Indeed, the work in this volume bears out a theme, common to this section, that a singular interpretation of proof as what one might call "mathematical proof" can hinder the very kinds of reasoning we need students to engage in across the grades. Instead, Herbst and Balacheff (Chapter 3) note that the longitudinal approach to teaching and learning proof advocated by the *Principles and Standards in School Mathematics* (NCTM, 2000) necessarily allows for different discourses and methods of proof and proving (or, as they describe it, customs of proving) to emerge in different classrooms, at different grades, depending on students' abilities, the representational systems available to students, the instructional resources used, and so on. From their perspective, proof can be ascribed to what a class may be doing at a particular point in time, considering its particular goals and available resources. Thus, implicit in this section—and throughout the book—is the perspective that teaching and learning proof across the grades requires fluid boundaries that accommodate different learning communities—be they students in elementary grades or those in college—and the forms of proof and proving relevant to those communities.

1

What I Would Like My Students to Already Know About Proof

REUBEN HERSH

As a college math teacher, what would I like my students to "already know" about "proof"?

The notion of "proof" is not an absolute. In the educational context, it evolves and develops up to the upper-level undergraduate and graduate school levels, where it merges with the notion as understood by researchers and other professional mathematicians.

For us college math teachers, the issue is first joined in beginning calculus. We find students who have already studied calculus in high school. They have seen "proofs" in their high school calculus course. We also offer "proofs" in beginning college calculus. These may or may not be similar to what they saw in high school. But it is, unfortunately, often found by college calculus teachers that students who "had" calculus have at best a formal or algorithmic understanding. They can correctly differentiate polynomials, even ratios of polynomials, even trigonometric and exponential functions. But they may be baffled by simple questions about tangents to curves, or rates of change, which, of course, are what give meaning and interest to the differential calculus. Does this substitution of formal exercises for conceptual understanding have something to do with the presence or absence of "proof"?

By the time the student has advanced beyond calculus to linear algebra, and then to upper-division math courses, he will have realized that "proof" is not something that first comes up in 10th grade geometry, and then returns in 12th grade calculus, while hardly being mentioned in arithmetic or algebra. On the contrary, the possibility of proof is what makes mathematics what it is, what distinguishes it from other varieties of human thought. But reaching this understanding is attained in the course of relatively advanced study (junior- and senior-level undergraduate). What would we like to see happen in high school, and even earlier?

Here we encounter a distinction that is unfamiliar to college teachers of math—the distinction between "reasoning" and "proof." To help me in writing

17

this essay, I asked the upper-division undergraduate students in my present math class at the University of New Mexico for advice and suggestions. By reading their interesting and helpful contributions, I learned that in elementary and secondary grades math, one uses "reasoning," but only in the 10th grade and beyond does one do "proof." "Proof," as understood by my students, in relation to their elementary and secondary education, seems to mean systematic, explicit, step-by-step reasoning. (Certainly not formalized, in the sense of "proof" by the so-called "first-order predicate calculus" of formal logic.) Yet somehow semi-formal, perhaps more elaborate or spelled out, than just plain "reasoning."

These distinctions actually pose fascinating questions from the viewpoint of "epistemology" and "methodology." To a mathematician, a proof is nothing more or less than a conclusive argument, one that convinces anybody who understands the concepts involved, and to which no counterexample can be given. The trouble with this honest and factual explanation of proof is that it presumes the existence of a body of qualified judges, a community of mathematicians in fact, who decide what to accept, what to question, what to reject. What logical basis do they have for these decisions? This is usually only partly explicit. There is almost always an implicit background against which a proposed proof is evaluated, and this precisely is what is not really available to a young student, or even perhaps to his or her teacher in elementary or secondary school. So the meaning of proof at the professional level is simply inappropriate at the earlier levels. But we do have *reasoning*!

For example, after children have had enough practice doing sums and products of whole numbers, using decimal place notation, they are ready to explore, say, the "commutative law of multiplication." *Any* two numbers multiplied give the same answer, no matter which is the pre-factor and which is the post-factor. Why is this? Well, children, you can see that 97 rows, each containing 24 pennies, is "the same thing" as 24 rows, each containing 97 pennies. "Turning the picture around" doesn't change the number of pennies in the array. This is certainly "reasoning." Is it a "proof"? I would say, yes. Contrariwise, in a "rigorous" treatment of arithmetic at the first-year graduate level, the proof of commutativity requires mathematical induction. After all, a picture is only a picture. Pictures can be deceptive.

Jump ahead to the 10th grade. Theorem 1 in Euclid, Book 1, is the construction of an equilateral triangle on a given base, by the intersection of two circular arcs. The diagram or "picture" that goes with the "proof" is completely convincing. You can "see" that the two arcs must intersect (both above and below the given "base"). But, notoriously, this is no "proof" in the modern "rigorous" sense. That is, it does not follow from Euclid's axioms and "common notions." There is nothing in them that says these two arcs do intersect. It is just "obvious" from the diagram. Is this a proof? Must we criticize Euclid and supply some missing axioms, the way Hilbert did? Are we corrupting young minds, misleading them, to swallow a fallacious proof? After all, pictures can be misleading.

Take something even more elementary. The diagonal of the unit square is incommensurable with the sides, having length equal to the irrational number square root of two. Since we only have the rational numbers at our disposal, it would be perfectly logical and correct to say that this line segment simply does not have a numerical length. In fact, the great philosopher Bishop Berkeley, being convinced that it was inconceivable that we could understand the notion of an arbitrarily small bit of the continuum, argued that since there must be a shortest line segment, all line segments must be commensurable, and therefore the so-called diagonal of the unit square must actually be a little bit off, must miss the opposite corner by a little bit. To us, today, Berkeley's argument is weak, but, in fact, his *logic* is impeccable. He just has different axioms. The assertion that there is a number equal to the length of the diagonal of the unit square is a profound and difficult assertion, virtually leading to the construction of the real number system—completing or filling in all the holes in the rational numbers. This is rightly reserved for college study, even perhaps graduate school. Yet we skip over this gap, this difficulty, in high school. It is likely that the rare student who finds it hard to swallow the "existence" of the square root of two is a deeper thinker than the majority who swallow anything we tell them.

So calculus teachers in college complain that their students' high school calculus course was superficial, which means, among other things, that important proofs were not given, or not fully grasped. Yet the first-year college calculus course, as given virtually everywhere today, is also notoriously "unrigorous." We differentiate without worrying that the derivative doesn't always exist. We assume that the real number system is complete, without mentioning that there is a grave difficulty to knowing whether that assumption involves a contradiction, whether the existence of the complete system of real numbers is even consistent. "They can worry about that later, when they are ready for it." So, really, what is the difference between proof in high school, at least in the upper grades of high school, and proof in the first year of college?

Not much, really!

Nevertheless, I would like my students arriving in college to know that "proof" is not necessarily something like the standard two-column format that used to prevail in 10th grade geometry. (Each "statement," in the left column, had to be accompanied, on the right, with a reference to an axiom or previous result. Then came the next statement, along with an explicit justification, and so on.) That in fact, "proof" is just "reasoning," but careful, critical reasoning, looking closely for gaps and exceptions. And that in this sense you have been proving things all along, through arithmetic, algebra and geometry.

It would be too much, probably, to expect students to understand that proof is relative to, appropriate to, the context. What is accepted as rigorous and proved in one context is questioned and reconsidered in a more advanced or more sophisticated context. After elementary calculus in the first year of college, there will come advanced calculus in the third year, where we will do it all over again, "the right way this time." It might be too much to expect this sophisticated, mature understanding of proof from college freshmen. But at least this much I must expect: proof in the broadest sense, of careful reasoning leading to definite, reliable conclusions, is what mathematics is all about, from kindergarten up. All, of course, in the appropriate language and level. Proof is not just an extra nuisance, irrelevant if you can just do the problems and get the right answers. Proof is what it is all about. I wish my new-coming freshman students already understood that!

2

Exploring Relationships Between Disciplinary Knowledge and School Mathematics

Implications for Understanding the Place of Reasoning and Proof in School Mathematics

DANIEL CHAZAN AND H. MICHAEL LUEKE

The National Council of Teachers of Mathematics' (1989) *Curriculum and Evaluation Standards* had a process standard on reasoning. The *Principles and Standards* document (NCTM, 2000), in an effort to highlight the importance of deductive reasoning in its vision of school mathematics, explicitly adds the word proof to this standard. In describing this reasoning and proof standard, *Principles and Standards* portrays a typical US education in mathematics as segregating issues of reasoning and proof in a single high school geometry course and then, for those still studying mathematics, to collegiate mathematics after the calculus sequence. In its vision of reform, the document suggests, instead, that: "Reasoning and proof should be a consistent part of students' mathematical experience in prekindergarten through grade 12. Reasoning mathematically is a habit of mind, and like all habits, it must be developed through consistent use in many contexts" (2000, p. 56). In discussing high school courses, the document suggests that reasoning and proof "should be a natural, ongoing part of classroom discussions, no matter what topic is being studied. In mathematically productive classroom environments, students should expect to explain and justify their conclusions" (p. 342).

As one reads through both the Reasoning Standard (1989) and the Reasoning and Proof Standard (2000), many goals are evident. There are performance goals: students should "develop and evaluate mathematical arguments and proofs and select and use various types of reasoning and methods of proof" (p. 341). There are philosophical goals: students should also "recognize reasoning and proof as fundamental aspects of mathematics" (p. 341) and come to appreciate how deductive argumentation is different than other modes of argumentation

21

common outside mathematics. Both sets of instructional goals are closely related to classrooms in which having students learn to prove is focal. Without suggesting that such goals be abandoned, research indicates that both of these goals are challenging (see, e.g., Senk, 1989, as well as research cited later in this chapter).

Yet, there is another sort of mathematical reasoning described in these Standards, reasoning that is not typical in many mathematics classrooms and that requires a large shift in the roles of students and teachers (from those described by Gregg, 1995). This kind of reasoning is envisioned for all classrooms, not just one in which the teaching of deductive proof is a goal. In the Standards vision, students are to develop their own mathematical ideas and to justify them: they are to make conjectures and learn to support them with evidence and argumentation.

For those who wish to strengthen connections between school subjects and their parent disciplines in academia (like Schwab, 1978), the vision portrayed in *Principles and Standards* in all of its complexity is appealing. In school, students should learn about how one comes to know in different disciplines; students should learn to act like mathematicians, historians, and scientists. Yet, while this vision has been articulated, it is rarely carried out; few students are asked to act in this way, and, when they are, it seems that many have difficulty doing so and appreciating why the sort of reasoning they are being asked to do is useful (Cuban, 1993).

Beyond the enunciation of a vision related to reasoning and proof in school mathematics, how might research help mathematics educators come to understand what is required to make this vision more common in schools? In this chapter, with this question in mind, we examine relationships between disciplinary practice with regard to proof and school practices of mathematical justification in three ways. All three of these ways share a focus on how students and teachers, consciously or unconsciously, frame their understandings of what happens in classrooms.

First, in an effort to understand why students seem to have difficulty with deductive proof, we will examine results suggesting that students do not naturally take on views of proof and proving that match disciplinary ideas. These results suggest that it is important to attend to the frames that mathematical tasks call up in students.

Second, we explore typical practice in algebra classrooms, particularly the solving of linear equations and systems of linear equations. Key to this exploration is the use of the construct instructional situation (Herbst, 2006), with its focus on the roles and responsibilities of teachers and students in relationship to the mathematics at hand. This construct helps us focus on how teacher and student classroom activity is organized and framed by participants.

We explore typical practice in order to understand how to integrate greater attention to reasoning and proof in US algebra classrooms; we argue that it is important to understand, for example, why attention to proof in the secondary curriculum has been segregated in the geometry course and why justification and proof are not currently a natural component of instruction in algebra. With such an understanding, we may be able to understand how to devise alternative classroom arrangements in which the making of conjectures and their justification is a central component of classroom interaction. Otherwise, we may fall prey to the mistaken notion that reasoning and proof can simply be added to the school curriculum without changing anything else about classroom interactions.

One side effect of this exploration is a greater appreciation of the mathematical reasoning practices that are already well established in the school curriculum. As we argue later, although portraits of mathematics often focus on deductive proof, reasoning mathematically is not limited to justifying propositional statements. Mathematicians not only prove theorems, they solve problems; in this context, they often search among alternatives for efficient, cost-effective, insightful, and elegant solution methods. A vision of connecting the reasoning in school mathematics with reasoning in the discipline could be even broader than the outline conveyed in the Standards documents. By broadening our vision, we may become aware of ways in which there is more mathematical reasoning in classrooms than is often appreciated.

Finally, we examine one design of instruction meant to change classroom norms to match disciplinary norms more closely. From examining the nature of this design, in the context of elementary classrooms, we can see how a teacher developed an alternative instructional situation in order to make room for mathematical reasoning in the classroom.

From the Literature on the Teaching and Learning of Proof

The question of the relationship between the discipline of mathematics and mathematics the school subject seems quite central to research on the teaching and learning of proof. One of the key justifications for attention to the teaching of proof in school is that proof is a key feature of mathematics. Yet, the research literature suggests that students (and some teachers) often hold views of proof and proving that are in conflict with those of mathematicians.

One key result in the literature on proof is that students, and some teachers, often give credence to arguments that are not deemed by mathematicians as valid mathematical arguments. In particular, elementary preservice teachers and high school students, when asked to evaluate arguments for a conclusion that holds for an infinite set of particular instances, seem to prefer arguments based on examination of a small number of examples, rather than a mathematical argument that covers all cases. This result is a robust one that is widely found in the literature (e.g., Balacheff, 1988; Martin & Harel, 1989; Williams, 1980; as well as numerous others).

In one study carried out in France, Balacheff (1988) distinguishes between two large categories of "proofs" that secondary students produced—pragmatic proofs and conceptual proofs. "Pragmatic proofs are those having recourse to actual action or showings, and by contrast, conceptual proofs are those which do not involve action and rest on formulations of the properties in question and relations between them" (p. 217). Similarly, Williams (1980) finds that 11th grade Canadian students in the top mathematics stream in their schools, who had "studied a modern high school curriculum designed to prepare them for study of post-secondary mathematics" (p. 10) also put great stock in empirical arguments over deductive reasoning. Martin and Harel (1989) a similar phenomenon among 101 American preservice elementary school teachers.

There is, as well, a distinct, but related result in the literature. Students who have been introduced to deductive proofs seem to underestimate the power of such arguments. In Chazan (1993), some high school students studying Euclidean geometry viewed deductive proofs in geometry as proofs for a single case, the case that was pictured in the associated diagram. Similarly, Williams (1980) found that 20% of sampled students did not realize that the given deductive proof proved a relationship for all triangles, while 31% seemed to appreciate the generality of the argument (p. 79). Fischbein (1982) describes a similar, albeit possibly not identical, set of beliefs. He found that "only 24.5% of the entire population [he refers here to the sample of students in his study] accepted the correctness of the proof and at the same time answered that additional checks are not necessary" (p. 16).

These results have potentially important ramifications for classes in which deductive arguments are used to establish results that will subsequently be used in justifying other propositions. In this vein, Schoenfeld (1989) reports on a series of studies in which students were asked to prove results deductively and then to make a construction: "[They] first produced correct proofs of the deductive results and then conjectured a solution to the construction problem that flatly violated the results they had just proven" (p. 340). To Schoenfeld: "Students give clear evidence of knowing certain mathematics, but then proceed to act as if they are completely ignorant of it" (p. 340).

What is one to make of such results and others like them? One potential set of explanations for the conflict between students' views and the desired views focuses on linguistic conventions. In the English language, the word "proof" often is used colloquially as a synonym for "evidence" in an empirical sense. Perhaps, students whose arguments proceed by naive empiricism, confuse proof—in the sense of evidence—for an assertion, with a proof of that same assertion—in the deductive sense of the word. And, similarly, perhaps students who do not appreciate the "generic" aspect (Balacheff, 1988) of the diagrams in geometric proofs are simply unaware that a generic example proof (e.g., a geometrical proof which is accompanied by a diagram) makes "explicit the reasons for the truth of an assertion by means of operations or transformations on an object that is not there in its own right, but as a characteristic representative of its class" (p. 219). They may simply be unfamiliar with conventions inherent in the two-column proof format (there is some evidence for this in Chazan, 1993). In Williams' (1980) terms, students who act in this way do not understand the generalization principle for deductive proofs; they do not understand that the validity of the conclusion is meant to be generalizable to all figures that satisfy the givens. Such interpretations are unsatisfying however because they suggest that these are simple misunderstandings that can easily be overcome. While this may be the case for some students, the robustness of these findings in a variety of contexts seems to rule out such an interpretation as the sole explanation for this phenomenon.

Another interpretation is to attribute to students a set of frames about proof and proving that operate in different contexts. From this perspective, students who make naively empirical arguments are bringing an empirical frame that is relevant in other contexts to bear on a mathematical context where such a frame is inappropriate (Harel & Sowder, 1998). In other words, they are not accustomed to using the process of proving in school as a way of coming to know or understand.

The force of the frames interpretation is to suggest that to change students' perspectives one must regularly put them in circumstances that support the desired frames. Such contexts are where deductive arguments are made and then used, curricular contexts where as Schoenfeld (1991) suggests formal and informal mathematics are not divorced: where, for example, deduction is not cut off from construction. Thus, there is a task for didactical engineering (in the French sense of the term), for the calling up of one set of frames and not another. The task is to overcome students' familiarity with measurement, the ease of collecting measurement data, the seeming incontrovertibility of those data, and students' corresponding lack of experience with formal deductive argumentation and perhaps as well with the mathematical objects and ideas which form the context in which they must argue (see Chazan, 1993 for students articulating these reasons for preferring empirical arguments over deductive ones). Of course, the invocation of such frames must take into account not only the mathematical tasks and contexts of the classroom, but also the institutional roles and responsibilities of teachers and students. In the next section, we imagine dialogue in an algebra classroom that brings this to the fore.

Exploring Typical Algebra Instruction

We now turn to exploring why attention to proof in the secondary curriculum has been segregated in the geometry course and why justification and proof are not currently a natural component of instruction in algebra in the US (for a supporting point with regard to the US TIMMS video corpus, see Manaster, 1998). To do so, we begin with a depiction of a classroom interaction that will help us examine the roles of teachers and students and how those roles and responsibilities shape the nature of the reasoning that might go on in classrooms.

A Classroom Interlude: Justification of Steps in a Method

Imagine a mathematics classroom where algebra is being taught. Students have been taught to solve linear equations in one variable. They have just been taught to solve systems of equations by adding or subtracting linear combinations of the equations in order to eliminate a variable. After students work on the problem, the teacher calls on one student to come to the board and solve the problem. The student—Blue—puts his solution on the board.[1] He has solved the problem correctly and indicates that he added the two equations in order to eliminate a variable.

Now, we take a flight of fancy. What if the teacher, influenced by *Principles and Standards* (NCTM, 2000) asks: "Why are you allowed to add the two equations?" What might students respond? And, how might they integrate their responses to this question with their work on the solving of equations in one variable? Figures 2.1 and 2.2 illustrate a conversation that attempts to imagine what might occur.

Figure 2.1 Why are you allowed to add the two equations?

Figure 2.2 Why can we add the two equations?

An Initial Interpretation of this Thought Experiment

Our first comment on this thought experiment is to suggest that the teacher's question about why one can add the two equations is not a question that one might expect in a typical mathematics classroom studying algebra. This expectation of ours is captured in the dialogue by the student's first response to the question and the teacher's clarification. After all, the teacher has taught the students to add the two equations in order to eliminate a variable. One could well expect Blue, once the question is understood, to feel that the teacher's question is unfair and indicate this by saying: "Because when you taught us how to solve systems by the method of elimination, that is what you told us to do." Moreover, given that the teacher taught this method, the teacher's inquiry about the single equation $2x = 26$, which only further serves to deepen the question, also seems out of the ordinary.

When teaching students to solve systems of linear equations, what it means to add two equations, how adding two equations is different from adding the same quantity to both sides of a single equation, and why it is legitimate mathematically to do so are all questions that do not regularly come up (for other sorts of mathematical questions of justification not typically addressed in algebra classrooms, see Chazan & Yerushalmy, 2003). And, even if a teacher did ask such a question, it is unclear what the nature of a reasonable student response might be to such a query.

With this example, we seek to illustrate that reasonable questions about the legitimacy of mathematical moves done while solving school algebra problems, questions of justification and proof, questions that our valuing of mathematical proof would embrace, are far less common in classrooms than might be expected from a disciplinary perspective. While descriptions of how mathematics is done give an important and central role in mathematics to proving, algebra classrooms as loci of mathematical activity do not seem to accord the same role to proof (see Herbst & Balacheff, Chapter 3, this volume, for a similar, but more general, point). As is suggested by Blue's initial response, in many school mathematics classrooms, students often are encouraged to use results and methods based on the authority of the teacher, rather than on a mathematical argument that the method produces the desired result. In the elementary grades, Lampert (1990) draws attention to this phenomenon.

Later we will seek to understand speculatively why it is that in classrooms this teacher's question is such an unlikely one and why mathematics classrooms as arenas for knowledge making proceed so differently from disciplinary models. To do so, we will explore the notion of instructional situations (Herbst, 2006) and introduce a model of classroom responsibilities that pertain when there are families of problems that must be solved using an algorithmic method.

Instructional Situations: A Useful Construct

Following Herbst (2006), an instructional situation is an interaction in which a symbolic exchange takes place. In classrooms, that symbolic market takes on an objective character; the teacher is expected to have knowledge that the students are not expected to have and for which the students can work and seek to claim as their own. This work comes in many forms: reading, answering questions, completing worksheets, solving problems, and performing other various academic tasks. These knowledge objects and student work both form the basis of a sort of economy, under which the teacher and students come to expect that certain work done by students can be traded for the teacher's recognition that some knowledge has been gained. Although there is certainly much more to classroom interaction than a simple exchange of knowledge for work, it is nonetheless a description of a teacher–student interaction on which the nature of schooling is based; students and teachers alike act as though knowledge can be thought of as objects that are real, and so they are, at least within the confines of the classroom (the teacher's "conferral" of the status of holding knowledge objects may or may not be acceptable in other contexts).

Herbst (2002a) illustrates that in the early 20th century, writing two-column proofs evolved over decades into a surrogate for learning Euclidean geometry. He argues that one result of this shift was the phenomenon we cited earlier, namely that students misinterpret the implications of such proofs. In addition to this evolution of trade, the process differs from discipline to discipline and even between subcategories within disciplines. The construct of situation then, is a framework that permits one to see the character of particular activities and exchanges in particular contexts. Consider a group of friends sitting down to a meal. If one of the people begins by picking up a piece of food with his fingers, this action will be seen very differently if the group of friends is participating in a picnic than if they are eating a formal meal. We would say that this action is visibly different because the actors are set within a different *situation* and it is precisely the manner in which we use the term here.

Similarly, doing proofs and solving equations are different situations that arise in secondary mathematics classrooms. In the context of doing proofs in Euclidean geometry, to contrast with the example from algebra that we just gave, each step in a two-column proof typically comes with a written justification for why this step is a legal move; why one is justified in making the claim being proposed. Although the very same students and teacher may have studied how to solve systems of two equations one year earlier in an Algebra I class, in the context of doing proofs in geometry, the behavior of the same actors is different.

There are many contextual details that may give rise to a situation, and it is unclear how finely to demarcate one situation from another. For example, should solving systems of equations be considered a different situation from solving a single equation? However, in the context of understanding proof and justification in mathematics classrooms, we are interested in those contextual details that are mathematical in nature. That is, we focus on circumstances particular to mathematics classrooms that influence and reproduce the symbolic exchange visible in school mathematics of trading student work for teachers' recognition of knowledge gained. We turn now to modeling the roles and responsibilities of teachers and students in a particular situation common to algebra classrooms, that of solving equations.

A Model of Who Does What and When in Solving Equations

In the instructional situation of solving equations, how might we understand and describe what is being traded? At a very broad level, the teacher poses a problem;

the student is expected to do some work in order to solve the equation; and then the teacher is expected to evaluate the student's solution and eventually make a decision about whether the student has "learned" how to solve equations.

Let us explore what it means to decide that a student has learned to solve equations. Whether or not one knows how to sew a button is evidenced by whether or not one is successful in the attempt. Likewise, whether or not a student knows how to solve an equation is observable in the solution(s) the student generates. But, in either case, what is the *evidence* for learning? Typically, one does not evaluate a single case. The fact that a student produces a solution does not give us enough evidence to believe the student could do it a second or third time; it is the consistency with which a student can properly solve equations which is tradable for recognition that she knows how to do so. Were it convenient, one might ask the student to provide solutions to a very large number of equations in a given timeframe and measure the accuracy with which the student does so. It would be hard to argue that a student who arrives at most or all of the correct solutions—regardless of how she does it—does not know how to solve equations.

But, this is not how conventional practice is organized; algebra teachers do not typically ask their students to take multiple-choice, timed tests on which they have to choose the answers for 100 linear equations without showing any of their work. It is the students' *work* on fewer problems, rather than their capacity to choose the correct answers, that typically serves as teachers' guide for whether or not the students know how to solve equations. And, mostly correct work, with arithmetic errors for example, is often recognized as acceptable evidence of knowing how to solve an equation. Typical practice involves having students show their work to satisfy the teachers' judgment that a student has learned to solve equations. This is one object of trade that students and teachers use as proxies for student provision of answers to equations.

What does it mean for students to "show their work?" What is it that students are supposed to show teachers in order to demonstrate that they have learned to solve equations? This is the purpose of method. We argue that methods in much of school mathematics serve as proxies for formal justification or proof. As we describe later, although methods may obfuscate the need for deductive reasoning under particular circumstances, they also open up possibilities for other kinds of reasoning. First, we seek to clarify what we mean in using "method."

Mathematically, one might suggest that an important role of method is to provide standard, well-understood solutions to problems that come up routinely in mathematical pursuits. As mathematics develops, classes of problems for which there were no known solution methods become classes of problems for which solutions have been found. When such problems come up in the context of further explorations, it is often useful to have codified ways of solving them that do not require explanation.

For example, there are many different equations and the ability to solve even a small fraction of them is enhanced when one can view them as being grouped into families each of which can be solved using certain techniques. Otherwise, one must see each equation as unrelated to the last and reason anew with each one individually. (This phenomenon was behind the mathematical duels in which the solution of cubic equations proved useful to Tartaglia and his contemporaries. See Gindikin, 1988, for one telling of this story.) Historically, this was a powerful motivation for the development of early algebraic methods. As such algebraic methods became widely known and useful in exploring questions in calculus, for example, these algebraic methods became central to school algebra.

We offer the term "method" here not as a substitute for "algorithm" or "heuristic." Rather, we use "method" to denote a process that occupies a middle ground between the two. Consider the following set of steps for solving linear equations generally (see Figure 2.3).

A method is not an algorithm because there are some small level decisions to make about how to carry it out. In the example just given, the steps do not make explicit how to simplify, or which side to collect terms with variables and which side to collect constant terms. Unlike an algorithm, students are expected to make a decision about whether to divide or to do something else (step 3) in order to simplify. What makes a method different from a heuristic is that there is a likely order and a structure in how one proceeds. Linear equation problems are meant to be solved in a particular order of steps; the process as given in Figure 2.3 is a common one. One should be able to do roughly the same steps in the same order most of the time (equations with fraction coefficients present a potential exception), and not tailor one's steps based on the particular problem at hand. Rather than describe these as vaguely defined algorithms or structured, ordered heuristics, we choose the term "method." We believe that this term can be applied to many contexts in school mathematics. In a later section, we link elementary school arithmetic with the notion of method.

Methods are important not only for their mathematical function and more than simply for their value in organizing students' learning. With the perspective of situations, the presence of methods for solving algebra problems sets up important relationships between teachers and students that demarcate who is responsible for particular classroom work and how they are supposed to go about completing it. Given the existence of these methods, in school, teachers are

---

1 Simplify (distribute if necessary)

2 Put like terms with like terms (Put all the terms containing the unknown on one side of the equals sign with the constants on the other side)

3 Divide or do what you must to simplify

4 Check to make sure your answer makes sense

Figure 2.3 A typical description of the method for solving linear equations in one unknown

Source: http://mathforum.org/library/drmath/view/57616.html

supposed to teach this valuable knowledge to their students. Students then per-form work that is intended to demonstrate their knowledge (e.g., of solving equations). The methods that are taught help teachers regulate and evaluate this work in order to make judgments about whether or not the students have in fact learned (e.g., how to solve equations).

So, when they show their work, students are expected to show evidence of having used a method, whether it is symbolic or graphical in nature, as opposed to simply showing their answers, or justifying why these steps solve the problem. In typical algebra instruction, this work is expected to take particular forms; students are supposed to write a string of equivalent equations that terminates in an equation of the form $x =$ a number. Note, also, that in typical instruction this series of equivalent equations is presented without justification (although during the new math era, some books treated these steps as a proof that the derived answer is a solution to the problem).

Similarly, one might consider a graphical method for solving equations, graphing both sides of an equation, like $3x + 4 = -x - 22$, as functions of the shared variable x and determining the x coordinate of the point of intersection. Here, again, in carrying out the method in a particular case, other procedures may need to be called on. For example, if neither function is displayed in the window on which one is graphing, then one must rescale the window. Again, from our perspective, a method is a process that is unspecified enough that it cannot be carried out without any reasoning.

We have not addressed all the questions one might have about method. For example, how does one draw boundaries between one method and another? Note that the set of steps in Figure 2.3 proceeds in a particular order. Is carrying out these steps in a different order following a different method? Consider a linear equation with fraction coefficients, such as $\frac{1}{2} x + \frac{3}{4} = x + \frac{7}{3}$.

While it may be useful strategically speaking to clear the denominators first, does this variation in the procedure qualify as a new method or not? Or, how about solving the equation $4(x - 3) - 17 = 3(x - 3) + 4$, where one considers $x - 3$ as the expression to be isolated initially? By comparison, reflect on the solving of systems of equations, where three different methods are often taught. How should one describe the relationship between the method of elimination and Figure 2.3? Is the method of elimination somehow a generalization of solving one equation in one variable to a different circumstance, or are these methods better thought of as completely different?

We set aside such ambiguities of method to think more broadly about class-room interaction: how does the existence of a method structure the nature of the trade between teacher and student? Most obviously, method regulates which solutions are deemed appropriate. Appropriate solutions give indications of using taught methods. Moreover, inefficient, or confusing, uses of a method may be seen as defeating the purpose of having a method in the first place and thus may be dismissed, even when they are mathematically correct.[2]

Second, the format of equivalent equations, for example, allows a teacher to scan student solutions quickly and evaluate them easily. If students wrote their ideas about how to solve an equation in paragraphs, such reasoning would be much harder to scan and evaluate. Having and teaching methods for classes of problems makes the teacher's task of assessing student learning less difficult.

Finally, showing work allows students to demonstrate that they have learned the method, not just guessed the answer to this particular question. Following the method allows a student to argue that they have learned to solve equations even when they do not produce the correct answer. Consider the teacher who awards credit to a student who has used a method correctly but has made a calculation error in the last step and has thus answered the problem incorrectly. A teacher is likely to recognize this work, even when a small calculation error has been made and the generated answer is incorrect, as a proxy for the desired learning.

Thus, methods regularize the relationship between teacher and student. They allow each to know the roles and responsibilities of the other. In particular, when a method is in play, the teacher is tasked with teaching the method to her students. In turn, students are expected to apply this method to problems posed by the teacher and for which the method is applicable, including challenging problems for which it is not immediately obvious that the method is applicable. The teacher then assesses students' knowledge by evaluating the strategy employed by students and the solution that their approach generates (see Figure 2.4).

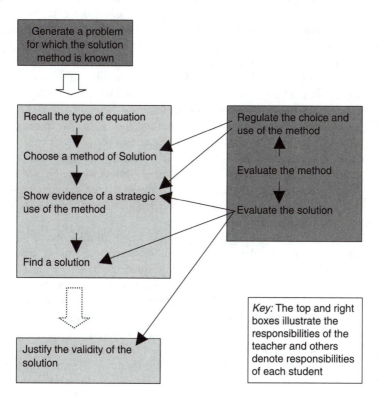

Figure 2.4 Instructional situation of solving equations

Reasoning in Algebra Classrooms: A Return to the Dialogue

If the model presented here does describe algebra classrooms when solving equations, then what are the ramifications of such descriptions for the nature of reasoning in school algebra? If teachers are tasked with introducing methods and students with using them and showing their work to demonstrate that they have applied the methods appropriately, then what sorts of reasoning might we expect to be found typically in algebra classrooms and what sorts of reasoning might be expected to be less common?

First, while *Principles and Standards* (NCTM, 2000) imagines classrooms in which "students should expect to explain and justify their conclusions," given the importance of method in organizing the situation of solving equations, one might expect a different sort of reasoning in this situation. If knowing how to solve equations means carrying out a taught method competently and efficiently, then the sorts of reasoning that one might expect on the part of students would be reasoning that contends that they are carrying out the taught method competently. Thus, reasoning about how a particular action in the context of a particular problem is an example of the method (like Blue's articulation of his purpose being the elimination of a variable) seems to be a likely candidate for reasoning with regard to solving equations. Students are supposed to learn that when solving an equation their choice of operations on the equation should isolate the unknown in efficient and expedient ways. Thus, one might expect reasoning that articulates how a step efficiently gets one toward the desired goal. Similarly, one might expect that students might be moved to explain and justify their strategic reasons for adding and subtracting particular terms, for example, "I'm trying to get all of the xs on the left side," or "I've multiplied in order to clear the denominators so I don't have to subtract fractions."

While such reasoning in the context of method seems familiar in the context of classrooms, in our dialogue with the teacher's question and Blue's response, we have tried to suggest a difference between such strategic reasoning and the sorts of explanations and justifications brought to mind for us by the NCTM reform vision. Our model suggests to us that justifying the legality of steps in a method is less likely to occur in a situation in which teachers and students have the roles and responsibilities we have laid out.

Similarly, if the teacher is in fact responsible for teaching a method and the teacher teaches correct mathematics, then it follows that a teacher would not teach a method that does not work in solving equations. The very presence of a method indicates that it is useful: it serves to solve particular problems. This does not mean that students are incapable of misapplying a method, or that a method may only have a limited domain of applicability, but simply that a method (properly) taught by a teacher cannot be mathematically wrong.

With this in mind, let us return to the question that the teacher asks about *why* Blue is allowed to add the two equations in that manner. The question that the teacher asks ("Why are you allowed to add the two equations?") seems like a genuine call for mathematical reasoning that supports the choice to add such equations, a call for reasoning about mathematical legality, rather than mathematical strategy. Yet, it seems to us that in the situation of solving equations, the question implicitly casts doubt on the method's correctness. Furthermore, discussing such issues moves teacher and students away from efficient application of a method into the development and justification of methods, tasks that are not components of students' responsibilities in our descriptive model of modal algebra teaching. Thus, we suggest that although it may be an important and meaningful path to follow, it rarely has a place in school algebra as typically conceived. Seeking a formal justification for the *method* that the teacher has already taught runs counter to the way teachers and students typically organize their work. We have integrated notion into our vignette about Blue and his solution.

We also note that in the context of solving systems of equations or the context of single equations, students can check that a method worked simply by evaluating the equation at particular values. This may obviate the need for a formal justification of why the method works. For many students, methods for solving equations may be like electronic devices; what is important is not how they work but *that* they work. In the context of the classroom, one's role as a student is to demonstrate to the teacher that one has mastered the method.

Thus, our argument is that the centrality of method in the situation of solving of equations creates opportunities for reasoning about the strategic utility of particular steps, but suppresses opportunities for justification that particular steps are indeed mathematically legal. These characteristics, at least on the surface, seem quite different from the situation of writing proofs, where there is a mechanism for supporting the articulation of legal reasons to justify assertions, but less support for the articulation of the strategic importance of particular steps along the way in creating the argument that will justify a conclusion.

Further support for the notion that in this situation the focus on method suppresses opportunities for reasoning about the legality of steps can be identified by examining how issues of verification of answers play out in the context of solving of equations. Note that in Figure 2.3, the last step of the interaction is to justify why the solution is a valid one. This can be accomplished by substituting the produced value back into the original equation. This approach to the final step is one that teachers typically emphasize. A student should always be able to check and verify that they have solved an equation correctly. However, typically students, even though one might expect they should care about the correctness of their answers, do not bother with this final step unless the teacher insists. Students seem to act as if that they know their solution is correct because they applied the method, correctly or not. While on the surface this behavior can seem perplexing, perhaps the perspective of situations helps it become sensible. If the way to demonstrate learning is to show one's steps to the answer, then independent verification that the answer is correct, while useful, is not essential to the establishment of the trade of work for the teacher's evaluation that solving equations has been learned. In this situation, students are not asked to solve problems, but rather are asked to obtain such solutions using a taught method.

A Design Experiment in US Elementary School Instruction (Didactical Engineering by Another Name?)

We have argued in the first section of this chapter that, perhaps as a function of the nature of instruction, students do not appreciate as much is desired the power of mathematical proof in coming to know and understand mathematics. And, in the second section, we have explored typical secondary school mathematics teaching with regard to solving equations and suggested that there is a tension between the learning of method and the sort of justifying that is often considered a hallmark of mathematical thought. Even though we argued that the solving of equations creates opportunities to discuss the strategic quality of students' steps, the force of the first two sections is to cast doubt on the feasibility of NCTM's vision of integrating reasoning and proof throughout the curriculum. While that is one interpretation of what we have presented, we do not share this pessimism. Rather, we suggest that reasoning and proof should be thought of more broadly in school mathematics.

We thus shift our attention to research that is not narrowly a part of research on the teaching of proof, but rather on the design of mathematics instruction that more closely approximates disciplinary norms. In this section, we apply the construct of instructional situations to analyze the possibility of designing new instructional situations within classrooms, ones that creates more space for the vision of mathematical reasoning promoted by NCTM.

In describing the design of her instruction, Lampert (1993) notes that school mathematics is characterized by cultural assumptions that:

> [A]re shaped by school experience, in which *doing* mathematics means following the rules set down by the teacher; *knowing* mathematics means remembering and applying the correct rule when the teacher asks a question; and mathematical *truth is determined* when the answer is ratified by the teacher. (p. 32, emphasis in original)

This statement has echoes of our description of the situation of solving equations we have just described.[3] We have argued that in the situation of solving equations, method regulates the roles of teacher and students in particular ways. How are those expectations to be overcome without simply jettisoning the notion that students will come to learn the methods they are currently being taught? Is there a way to teach methods, but also to create opportunity for reasoning and proof?

In a seminal article, Lampert (1990) suggests that this is possible if "The problem is not the question, and the solution is not the answer," or in our terms: if solving the problem by writing down the solution method is not how students show the teacher that they have learned. To make this more concrete, we will focus on one particular example, an example taken from a description of her instruction.[4] Lampert describes a teaching episode that exemplifies her attempt to alter the nature of the work done in elementary school mathematics classrooms. Ultimately, a goal of her instruction was that students learn to operate with exponents. But, rather than start by focusing on methods for working with exponents, she began her instruction by asking students to find patterns in the final digits in a table of squared numbers. This task was designed to foster a discussion about a mathematical justification for rules about exponents instead of focusing instruction strictly on learning to "divide or multiply by subtracting or adding exponents" (p. 44).

One way to describe the change that Lampert has wrought in her teaching in this example is that she has shifted instruction from the study of the methods of arithmetic to the study of number theory—as students engage in examining and making conjectures about the behavior of numbers under the standard operations.[5] In other words, her instructional design is not simply about insisting on particular student or teacher behaviors, but on a renegotiation of the roles of teachers and students around *content*. We note that Lampert describes this work as creating a "new" form of classroom interaction; she says as much: "[M]y role in this project has been to develop and implement *new* forms of teacher–student interaction" (p. 33, emphasis added). We think this fact is important for two reasons. First, it illustrates that typical instructional situations employ a different way of working. Second, the creation of a new instructional situation did not materialize as a simple consequence of her presence as teacher: she designed it.

Another important feature of this example is that the teacher, in this case an experienced teacher and knowledgeable researcher, is not the only component of the changes that take place. Although some may argue that teachers need only develop stronger mathematical skills for teaching in order to initiate this change, we note that Lampert fundamentally altered the tasks in which the children engaged and the accountability structure under which the children's learning would be represented (e.g., community mathematical argumentation). Thus was changed the very nature of the mathematics being done. This matches our understanding of how instructional situations determine the work seen in classrooms and vice versa: The nature of the task and its mathematical context shape the interactions that take place.

Conclusion

As we examine the relationship between mathematical activity in classrooms and in the discipline, the force of all three parts of this chapter is the importance of understanding how classroom interactions are framed by the participants and by the mathematics that is studied. Our focus on framing has led us to appreciate anew some of the existing results from the literature on the teaching of proof in schools. At the same time, it challenges researchers interested in relationships between mathematical activity in schools and in the discipline to understand better why certain kinds of mathematical activity rarely find their way into the institutional setting of school. Developing a better understanding of the dynamics of schooling that are accountable for this may stimulate new rounds of instructional design and didactical engineering that will ultimately lead to the incorporation of more activity in classrooms that bears a greater resemblance to activity in the discipline (even though that may not be the reason for promoting such activity, as articulated in Chazan, 1990, p. 22). The research on proof has much to gain from utilizing the broad range of research in mathematics education to find examples of the creation of instructional situations that enable justification and proving. Finally, this focus has also led us to appreciate opportunities for mathematical reasoning that already exist in the typical curriculum. It leads us to seek a broader picture of the nature of mathematics in the discipline: one, for example, that would incorporate the doing of applied mathematics. As a result, researchers interested in relationships between the discipline and school subjects may examine other mathematical practices, beyond justification of propositions.

For us, the construct of instructional situations is a useful tool; it brings into view structures of classrooms that are not immediately visible on inspection. In describing a situation of solving equations, we hope we have illuminated ways in which teaching and learning algebra is constrained (but not immutably) by the roles and responsibilities particular to the classroom situations that have evolved around the mathematical methods central to the subject. These obligations are distinct from other disciplines, such as history, and from other school subjects in mathematics, say geometry, or even other contexts regularly part of instruction in algebra, like doing word problems. The notion of method is a particularly powerful component of the situation of solving equations, regularizing the process of trading work for knowledge in a way that is efficient and easy to reproduce year after year.

Thus, while bringing greater attention to deductive argument in school mathematics may be a realistic goal, it will require an understanding of the constraints placed on classroom actors by situations around particular mathematical topics, like the situation of solving equations, and engineering of situations conducive to such reasoning. It may, in fact, mean that, in order to create more room for mathematical justification, classroom actors may need to reduce the amount of time devoted to work in the context of the particular situation of solving equations; the vision suggested by the Standards indicates that this skill may be overshadowed by the need for making connections between various representations of equations and other important quantitative capabilities. Developing situations based on activities that are peripheral in typical mathematics teaching (such as working and reasoning graphically, making conjectures, or *developing* mathematical methods) and making these situations more central may be conducive to a greater concentration on deductive reasoning and proof.

Of course, such efforts are likely to meet resistance; Lampert (1990) notes a variety of different challenges in transforming her own instruction. Although recent research in mathematics education highlights the gap between school mathematics and the discipline, we note that there is not consensus among mathematicians about what skills are most valuable for students to understand. Though NCTM clearly endorses the use of problem solving skills and deductive argumentation, certain branches of mathematics do not involve such work, and instead rely on the careful application of known techniques and methods. Thus the disappearance of instructional situations involving method may not be desirable from this perspective. Indeed, the skills involved in solving equations do prove valuable in many other areas in which mathematics is useful (particularly mathematical modeling). Moreover, instructional situations involving method open up many possibilities for discussing the importance of making strategic choices, itself a valuable component of thinking mathematically. We believe that regardless of one's philosophy for teaching mathematics generating new perspectives on proof and justification in school mathematics may be more realistic if one accounts for the power of instructional situations and the potential obstacles in creating new ones.

Notes

1. Although this may not be the only method the students know, the class appears to be familiar with the one that Blue uses.
2. Consider a student who adds five to both sides of the equation $3x + 2 = 14$. Although not *wrong*, we believe that the instructional situation of solving equations makes this an approach that teachers would discourage.
3. Although many of the "methods" of arithmetic, for example, feel quite algorithmic, there is often a small level of ambiguity that makes us refer to these as methods. Consider the example of teaching addition and subtraction of two fractions. While simply using a common denominator formed by the product of the two denominators is a critical part of an algorithm for this operation, part of learning how to add and subtract fractions involves learning how to find a *least* common denominator. Making such choices represents an attention to strategy that we highlighted earlier as a key component of method.
4. Other examples in the literature could replace this particular example: see Ball, 1993; Heaton, 1992; and the work of researchers who work closely with teachers, including Yackel & Cobb, 1996, and others. The key issue is that the creation of this sort of interaction requires renegotiation of the instructional situations and norms that operate in classrooms. Such a renegotiation involves both tasks and norms, as well as roles and responsibilities.
5. We would make a similar argument about Deborah Ball's instruction in examples like the well-known case of the Sean numbers (1993).

3

Proving and Knowing in Public
The Nature of Proof in a Classroom

PATRICIO HERBST[1] AND NICOLAS BALACHEFF

This chapter contributes to a descriptive theory of the mathematical performances of classrooms by proposing three senses in which classroom mathematical practices might include a notion of *proof*. The need for such a theory is based on the conjecture that public activity in mathematics classrooms embodies an epistemology, a particular way of knowing, and that it is crucial to understand this epistemology in order to inform the development of curriculum, pedagogy, and assessment. The goal appears ambitious, but is also modest in a key aspect: in considering our charge to devise a descriptive theory of classroom activities as public mathematical performances, we bracket the many important questions that one could pose about the individuals who perform. We do not address here questions of individual learning or personal meaning for the participants of classroom work; rather, we develop means to describe what they do in public. We do that not as a statement of values, but as part of a strategic plan toward eventually explaining classrooms as collective, complex intelligences where individuals contribute to a distributed cognition that might embody mathematical ideas as public performances.

The chapter considers proof as a public performance. We propose three distinct notions of proof to designate the various performances that a classroom observer might label as "proof." One of these notions is tied to the habitual uses of words such as *prove* and *proof* in a class's discourse and to the customary ways in which knowledge is validated and organized in a class. A second notion of proof, which we label *C-proof*, is tied to specific ways in which classes are able to control the fit or the adaptation between problems and their solutions within particular conceptions. A third notion of proof, which we label *K-proof*, addresses the possible role of proof in coming to know new things: it describes how the metaphorical mapping of a conception already known by the class onto a new conception can shape and warrant plausible knowledge about the new conception.

Classroom Mathematical Performances

School systems and parents take classroom activity and teaching as instruments to achieve learning goals of individual students. For other stakeholders, however, the aims of classroom activity can be different. Professional communities (e.g., mathematicians or engineers) are likely to be more concerned with the diffusion of knowledge and the percolation of knowledge experiences on collectives (e.g., a generation). Likewise the classroom teacher is a stakeholder whose first concern is necessarily collective. Unlike a tutor, a teacher rarely teaches individual students—a teacher teaches a class. It is true that the collective nature of a class is not quite like that of a soccer team or a ballet company; a class has an institutionally avowed aim to produce individual learning. Yet, that avowed aim is operationally mediated by the public opportunities to learn developed in the class. That is, if individual students learn in school, they do so not just from their individual study and not only when they have the individualized attention of their classroom teacher. They learn from being a part of a system of people, constraints, and resources—the class—whose total, systemic activity reframes individual actions as contributions to collective work, one of whose products is opportunity to learn. Students' opportunity to learn depends on classroom performance much in the same way that each of the musicians of an orchestra has the opportunity to claim to have performed a classical piece because of the performance of the complete orchestra, in which each musician may have played only a (perhaps minute) part of the entire orchestral arrangement.

Classroom performances represent knowledge as public work. They must be compelling and immersive for students to be taken as opportunities to learn; but they also must be mathematically compelling and important to be culturally valuable. The supervision on that double quality of the opportunity to learn rests on teaching (the system of human and material resources constructed with the intention to teach). The question to be considered in this chapter is what kinds of representation of *proof* can be constructed in and through a class's performances and how an observer capturing the perspective of the teacher could describe those representations of proof.

We focus here on developing theoretical means to take the teacher's perspective as observer of the performance of the class as a collective cognitive organization. And we describe proof as a public performance from that perspective. Specifically, we focus on the teacher as observer of the opportunity to learn created by classroom activity over three key timescales. This perspective is particularly important to inform teacher assessment, teacher education, and curriculum research in which a key activity is to appraise the mathematical qualities of an actual classroom performance against the backdrop of a number of other possible performances that might have happened instead or that might happen in a collection of classrooms. A description of the mathematical qualities of classroom performances from this perspective is figuratively speaking the dependent variable whose changes might be caused by the teacher's deliberate activity.

Balacheff (2002) has noted that in spite of the large corpus of research on proof in mathematics education, the field is in a deadlock due to the lack of a clear demarcation of our object of study: Is the same notion of proof useful when one wants to know why an individual believes in the truth of a statement, when a curriculum designer expects students to justify their answer to a problem, when a reader judges the merit of a mathematics manuscript, when a historian accounts for changes in mathematical ideas, or when a philosopher compares mathematics with natural science? We contend that the lack of agreement on what is meant by "proof" in mathematics education research is partially due to lack of clarity in articulating the perspective from which each researcher accounts for its object of study. To appraise classroom mathematical performances, we suggest that the extent to which classroom performances can arguably "count as proof" requires an observer to be apportioned of a special set of meanings for the word *proof*.

We use the word *performance* in alluding to classroom activity in order to stress the point that classroom activity is neither merely a set of juxtaposed individual expressions nor just an emerging social compact. Classroom activity is also a systemic response to reified cultural constructions and an enactment of institutional roles. The word *performance* attempts to profit from the metaphor of the classroom as an orchestra, where the score is the way the mathematical community views the mathematical objects at play but where the meaning lies in the experience of the educated listener—which we allude to as we speak from the perspective of the teacher as an observer. Classroom activities are not mathematical performances solely because the classroom is a mathematics classroom, and not only when their performance is faithful to a mathematically vetted score; the teacher-as-listener needs means to support the claim that a chunk of classroom activity is a mathematical performance even when they may not have used an accepted definition, a conventional symbol, or a syntactically valid proof.

Four Constraints that Shape Mathematical Activity in School Classrooms

The complexity of classroom activity and the perspective of the teacher as appraiser of this activity require attention to the singularity of classrooms as knowing organizations. These knowing organizations are singular inasmuch as they exist in response to four conditions, which we will refer to as I^4. Each of them imposes particular demands on the existence of proof in classrooms.[2]

Institutional

The school institution presumes that activity in mathematics classrooms has a subject matter, which is issued from a discipline. This discipline exists outside of school both as an academic discipline and as a discipline of live scholarly work and it relies on school to be diffused to the public. Thus the discipline of mathematics has a cultural prerogative that school citizens must acknowledge even if to no avail. The discipline of mathematics also has a cultural obligation of making itself available to oversee its representation in instruction—which is often done through the involvement of mathematicians in public commentary and textbook review. With regard to proof this means that since a notion of proof exists in the discipline of mathematics, it might be entitled to exist in classroom activity. And if it were to exist, it would be expected to exist in a form that is accountable to, if not compatible with, how it exists in the discipline.

Instructional

The encounter between the student, the teacher, and the subject matter is not a free-formed, voluntary get together. What brings them together at a certain time and keeps them together for some time also keeps them in specific places vis-à-vis one another. The student and the teacher are in an asymmetric relationship vis-à-vis the subject of studies: one of them is supposed to know the subject already; the others are not expected to know it yet, but are expected to come to know it later. There is an unwritten contract or agreement to be in that relationship for some period of time. Since breaches of this relationship are institutionally resented, the relationship has to adapt in order to continue existing for that period. In particular, assuming that proof is part of the subject matter that binds teacher and student together, proof must be "taught," "learned," "known," and "assessed" and the relevant meaning of those verbs is that ascribed to them by the school institution. Furthermore, the mathematical knowledge of students and their teacher are supposed to be different: students' knowledge is supposed to increase, adding new knowledge assiduously, while their teacher's knowledge is supposed to be relatively stable, and, in particular, the teacher is expected to know already what the students have yet to know. The teacher is expected to know ahead of the students what statements might be relevant to consider, and whether those statements will turn out to be true, even before the act of proving them turns them into theorems that students can then use.

Individual

Particularly after the 20th century, it is accepted that in spite of their shared role as students, individuals have the right to be different. It is institutionally accepted that individuals differ in how they think, what they like, the rate at which they develop, their talents, and their shortcomings. The issues in which they differ can increase in many directions without depriving individuals of their entitlement to be students. With regard to proof this means that conceivably some students may not be convinced of the truth of a statement by (what a mathematician might deem) a mathematically valid proof. Some might be convinced by compelling arguments that, however, nobody would seriously call mathematical; others may not enjoy talking or arguing, or may have difficulties reasoning logically; yet all of them will have the right to be that way in school. In particular, the timing and the timescales by which the many attributes of individuals change[3] are not necessarily expected to follow suit the timing and timescales of instruction. Hence the mathematically most valued proof of a theorem could quite plausibly not be understood by some or not be convincing to others and these could be deemed acceptable reasons for doing something else.[4]

Interpersonal

These individuals, students and their teacher, are together under relatively lax external surveillance and over long stretches of time. They don't just enact roles or express selves, they don't just apply decontextualized strategies or else improvise what they do; they actually adapt to each other by developing public identities or personae over time and they manage that interaction tactically (Erickson, 2004). With respect to proof this means that the same (communicative) acts by way of which they relate to proof are also acts by way of which they relate to each other. Thus a demand for proof, for example, is never just a request to establish truth; it can also be or be taken as an expression of distrust of someone else's word or an admission of weakness of one's understanding or a bid for a rapprochement to someone in power. The giving of a proof can also be or be taken as a gesture of arrogance or pedantry, an assumption of audience incredulousness, or a lack of respect for the audience's time. And, because interpersonal relationships continue over time, those relational bids may be consequential; they may affect relationships for good or ill.

A setting framed by those I^4 constraints is notably different from other settings in which individuals encounter proof, such as when a mathematician reviews a manuscript submitted for publication or a person tries to persuade a friend about the truth of a statement. To appraise the role of proof in classroom activity an observer needs to pay attention to several layers of activity. Within each of those layers a way to decide whether something "counts as proof" may or may not be found.

It is particularly important for mathematics education researchers to realize that our intellectual tools to describe proof need to go beyond the "institutional" constraint that classrooms are subject to. According to that institutional constraint the discipline of mathematics has the right and the duty to say what a proof is. Institutional researchers (e.g., those who might need to validate mathematically a decision to choose a particular textbook) may not have any choice other than to take the word of the mathematician at face value. But researchers in mathematics education who seek a basic understanding of classroom mathematical work can develop a more complex perspective by taking advantage of what Imre Lakatos taught to philosophers of science. In the preface of *Proofs and Refutations*, Lakatos (1976) snickers at the consequence of the formalist self-presentation of 20th-century mathematics, according to which there would have been no real mathematics before Boole stated the laws of propositional logic. When a mathematics education researcher inquires on the place of proof in classroom work they confront the same unbearable paradox that a philosopher of science confronts when looking at the role of proof in the historical record of mathematics: the methods by which classroom participants over a year's time (as much as mathematicians over historic time) produce and organize knowledge are as much the product of development over time as the knowledge itself is. One possible way of facing that paradox is to exclude any inquiry on the role of proof from any mathematics known by means other than those sanctioned by the discipline at its current state of sophistication. The alternative is to own up to the informal, heuristic nature of the proofs that serve to shape mathematical ideas, snickering also at the presumption that all the mathematics students have learned before being inducted into those practices that the discipline might recognize as proofs should be rejected as not being real mathematics. Lakatos (1978) suggests that to understand the methodology of mathematics, the latter is the best—"even the poverty of historicism is better than the complete absence of it" (p. 61).

But if we will rather own up to the informal ways in which mathematical ideas are developed earlier in students' life, how can we gauge the extent to which the performances that they form a part of are *mathematical*? Do those school performances sustain an image of mathematics compatible with Benjamin Peirce's characterization of it as "the science that draws necessary conclusions"? (Peterson, 1955). One would like to be able to distinguish between justifications that are mere experiential confirmations of an outcome from justifications that argue for the necessity of a result. And one would like to do that without making the distinction rely exclusively on the need to require clear definitions and axioms or symbolic formalism. This is what we propose to do here.

Three Timescales

There are three particular timescales (Lemke, 2000) that are important to attend to in describing the nature of proof in the mathematical performances of a class. For each of them, particular descriptive tools may be needed. Yet a comprehensive account of the role of proof in mathematics classroom work in general needs an integration of the analysis across the three timescales. The first timescale is that of the custom of a class; the basic question at this scale is what the place of proof is in the mathematical work customarily done in a class, over stretches of time of the order of the year. We will not presume, however, that a characterization of the customary place of proof in a class will provide us with a general notion of proof that individual proofs are particular cases of. After (and with) Lakatos, we are highly suspicious of taking the relationship between *proof* and *a proof* as a case of the relationship between category and member. But we acknowledge that communities, be those classes or working mathematicians, use words like proof and proving *as if* they were general categories. We propose that describing that general, customary usage is one important piece in trying to understand the nature and role of proof in a mathematics class. A class has a custom (Balacheff, 1999), a set of habitual ways of relating to the enterprise of doing mathematics. That custom may have a place for proof, which could be reflected in language uses (e.g., words like *prove*) and in habits of relating to knowledge (e.g., the expectation that after stating a claim one will provide support for its truth). The need to describe the role of proof at different timescales derives from the critique of a general notion of proof just sketched.

Lakatos's critique of formalist mathematics entails that the nature of proof in mathematical work is not formal but substantial. Proofs give meaning to (and help refine) concepts and propositions. Consequently, there is more to the nature of proof to be found in the examination of the doing of particular proofs than captured by any general idea of proof. Particular proofs of Euler's conjecture, whether Cauchy's combinatorial proof (Lakatos, 1976, p. 8) or Poincaré's vector algebra proof (Lakatos, 1976, p. 116), are different not just in what they actually prove, but also on what meaning they give to the conjecture, what steps they take as obvious, and what ideas they dare to leave unanalyzed. The nature of a proof, for instance (e.g., whether the proof is closer to a calculation than to an argument), depends on the particular system of representation available to address the ideas involved in the proof. A particular timescale at which to inspect the nature of proof and design ways of describing this role is that of particular mathematical conceptions; stable practices being done at particular moments in time and where specific means for the control of how problems are solved may be called proofs. We propose the expression C-proofs to designate these.

A third timescale is finally suggested by the classroom counterpoint with which Lakatos narrates the historical development of Euler's conjecture. In that narrative, various (sometimes competing) notions of polyhedron, versions of Euler's conjecture, their proofs, and their refutations are produced as results along the way, as the dynamics of proofs and refutations seeks to prove "the" theorem about polyhedra. That is, in addition to the particular proofs that work to verify results within specific conceptions and in addition to the custom of proving that characterizes the mathematical way of relating to knowledge, there is an operational way of coming to know in which every new proof also helps improve the state of knowledge. Proving as a tool for knowing or coming to know, whereby some changes in conceptions or incorporation of new conceptions may be made by appeal to practices that might be called "proving" whereas other changes may not. The purpose of this chapter is to articulate a descriptive protocol for the place and the role of proof at those three timescales, the customary or habitual, the local tied to specific stable practices, and the operational that helps transition to a new state of knowing.

The Possible Place of Proof in the Custom of a Class

A class may or may not trade on the words *proof* and *proving*. That is, the words *proof* and *proving* may or may not be tokens of interaction invested of official meaning. The description of proof as an element of the custom of a class needs to include accounting for the customary uses of those words, possibly comparing those uses with the uses that the words would have in mathematical discourse. But this description must also be attuned to other expressions (e.g., justify, explain) that might play similar customary role. Some basic definitions of terms from the observer's perspective may be useful to identify the register in which one could expect a class to craft its notions of proof and proving. Some 20 years ago Balacheff (1987) wrote:

> We call *explanation* a certain type of discourse that attempts to make understandable the truth character of a proposition or result acquired by the speaker. The reasons that he or she provides can be discussed, refuted, or accepted. We call *proof* (Fr. preuve) an explanation accepted by a given community at a given moment of time. The decision to accept it can be the object of a debate whose principal objective is to determine a common system of validation for the speakers. Within the mathematical community,[5] only those explanations that adopt a particular form can be accepted as proof. They are sequences of statements organized according to determined rules: A statement is either known to be true or deduced from those that precede it using a rule of deduction from a set of well-defined rules. We call these sorts of proofs *mathematical proofs* (Fr. demonstration). (p. 147–148, our translation)

Thus, we are interested in describing the place of proof (and related words) insofar as they designate discourses applied to statements, (i.e., the contents of sentences that propose an assertion as true) and that are aimed at establishing the truth of such statements. To describe the customary place of proof in a class means to describe the relative place of those discourses in relation to statements and compare that place with the relative place they have in customary mathematical work.

The extensive corpus of reflective writings by mathematicians as well as the products of their work could easily be surveyed to complement the notion of mathematical proof given earlier and its general, customary characteristics within the mathematical community. Much of this work has been surveyed, organized, and elaborated by Gila Hanna (see for example Hanna, 1983; Hanna & Barbeau, 2008; see also De Villiers, 1990). Proofs are given of general statements, often after the statements are made, to establish the truth of a statement, but at times also before the statements are made (as a way to produce the statement). Yet not all true statements are actually given a proof of their own; the proofs of some statements often yield derived corollaries and consequences whose proofs are omitted. Mathematicians' writing is full of "ideas" and "sketches" of particular proofs—thus playing out a custom where showing that a proof exists is as acceptable (and perhaps less pedantic) than providing a complete proof. And yet having an expert check that the idea of the proof can be successfully deployed has been a key gatekeeper for mathematical publication. Thus, contemporary mathematics as much as classic mathematics leaves records that might appropriately be called "descriptions of proof" (Livingston, 1999), since the actual proof is done anew every time a human uses those records to run through the argument, finding and filling the gaps. Mathematicians' reflections often address the role and characteristics of proof in their discipline in those general terms (Davis, 1972; Haimo, 1995; Thurston, 1994).

In attempting to find out whether there is a customary notion of proof in a particular classroom, the descriptive focus has to be on whether a class has the habit and resources (semiotic or interactional) for demanding and establishing the truth of statements to which participants attribute general characteristics, even if these are not the same kind of general statements that mathematicians would care to prove true. Evidently, it is possible that a class might not recognize its enterprise as one of producing and accumulating general statements which are true and whose truth character the class as community is expected to understand. It is also possible that whereas a class might be involved in acquiring true statements, its means to establish each of them as true were perceived as completely ad hoc, hence not enabling any general discourse as to the nature of its proofs. And it is also possible that while a class might be involved in the acquisition of true statements, that it had a discourse of method that addressed how truth has to be established, and even possibly made use of words like *proof* and *proving* in that context, a comparison with the elements of the customary notion of mathematical proof would reveal it as different. To describe the custom of proof of a class thus one has to track the use of key words such as *prove, proof,* and *proving* (as well as others such as justify, justification, explain, explanations, show) and observe whether they are applied to designate discourses that make understandable the truth character of statements.

In the notion of proof from Balacheff (1987) we note that in spite of the important role attributed to community and to time, and in spite of not requiring that a strict form be applied to every proof, this definition does not leave the question of what counts as proof as one of voluntary agreement within a freeform group. Communities are never freeform groups; hence consensus is never just voluntary agreement. A mathematics classroom is a special community, one that exists tied to the I^4 constraints. The institutional constraint, for example, requires classroom members to act as if the proposition whose truth character is being explained (what is at stake in the doing of a proof) exists as true beyond the communicative acts in which someone is convinced of its truth. The importance of time in that definition cannot be overemphasized: time is of the essence of what counts as proof because a community that at one moment in time might be interested in knowing that something is true is also one that at a later moment in time might build on their knowledge that such a thing is true. Balacheff's definition thus points to a necessarily circular relationship between "a proof" and "Proof" in any consideration of what is the custom of proof in a class. Historically developed customs for stating general propositions and accepting and recording proofs exist at any one time and are applied in the establishment of new propositions and their proofs. At the same time those customs are being shaped in and through the activity of proving a new proposition, for example, by the development and implementation of new heuristics or by the possibility to leave unsaid some things that can be taken as obvious. Because of that circularity, the habitual notion of proof may be useful for a community to request or expect that a proof be given to an assertion yet not sufficient to judge whether something given as proof counts as proof at a given moment in time.

In American classrooms, the word "proof" has usually been associated to the high school geometry class and this class has provided a setting for studying customs of proving. Herbst (2002a) tracked the historical evolution of a custom of proving in high school geometry—documenting what led to the establishment of the two-column form as the standard for proof writing in the early 20th century. At the beginning of the 21st century NCTM's (2000) *Principles and Standards for School Mathematics* included "Reasoning and Proof" as a Standard for all grades and across all subjects. *Principles and Standards* granted that reasoning and proof might look different across the grades—that many customs of proving might develop depending possibly on factors that might include students' developmental level or available resources (material, symbolic, etc.). Research describing the many customs of proving that could unfold in actual classrooms as curricula, districts, and teachers try to implement this Standard needs to be done (and contributions to this volume show that it is being done). But research on proof in classrooms needs to go beyond the description of the customary notion of proof embedded in a class's mathematical work.

Proof as a Regulatory Structure for a Conception

In the preceding section, we noted the need to examine the customs of classroom work for indications of whether there are recurrent concerns with the formulation and establishment of general statements and the explanation of their truth character—which could possibly be satisfied with the imperative to prove and the provision of proofs. We raised awareness that a particular class might not have such a custom—they may have no recurrent ways of deliberately engaging practices of that kind, let alone a discourse that establishes expectations on method. Yet mathematics classrooms are always involved in some work, and within particular pockets of activity some particular ways of acting may deserve to be considered proofs. We now switch to a different timescale and register.

We propose a definition of a different notion of proof, ascribable to what a class may be doing at a given moment considering what it has in hand as goal and as resources. In the following we use the expression *conception* following its usage in Balacheff and Gaudin (in press; see also Balacheff, 1995) to refer to stable ways in which students in a mathematics class relate to particular organizations of their mathematical milieu around a particular moment in time. We realize that the expression *conception* is commonly used in relation to individuals rather than to classes, and that to make it ascribable to a class makes sense only when the observer takes on the perspective of the teacher for whom the class can be represented as an abstract or *modal* student (Herbst, 1998). Balacheff and Gaudin (in press) argue that *conception* can be used to refer to stable practices (in contrast with its canonical usage naming mental structures).

Building on Bourdieu's (1990, p. 83) notion of *sphere of practice*, we have identified units of mathematical performances and called them *conceptions* (Balacheff, 1995; Herbst, 2005). Balacheff (1995) proposes that what the cognitive-based literature had alternatively called "misconceptions," "bugs," or "errors" are all comparable cases of adaptation between a cognizing agent (a student, a class) and a milieu, under proscriptive conditions of practical viability. The relative independence of different conceptions (which an observer could relate to the same concept), that the cognitive literature used to denounce as the coexistence of contradictory conceptions in the same person, is sustained by different practical conditions and by different placements in time (including changes in time across settings such as in and out of school, or within a setting such as "my class this year" vs. "my class last year"). This practice-based notion of conception is addressed in the CK¢ model, where a conception is modeled as a quadruplet $C = (P_C, R_C, L_C, \Sigma_C)$:

P_C is a set of problems or tasks undertaken successfully by an agent around a given moment of time

R_C is a set of operations that the agent could use to solve problems in that set

L_C is a system of representation or semiotic register within which those problems are posed and their solution expressed

Σ_C is a regulatory structure used to control the adequacy of the solution to the problem and warrant its solution

We contend that this model of conceptions can, in particular, be used by an observer to describe molar units of public mathematical performance in the mathematics classroom, on account of the metaphor that the class is, at least for a teacher, like a single student (the modal student). We argue that some of the conceptions of a class may be ultimately regulated, or that the adequacy of solution to problem may be ultimately controlled, by what could be called proof.

A canonical example is what could be called the "correspondence" conception of congruence in the high school geometry class. In high school geometry students are often posed problems that require them to decide whether, or to show why, two segments or two angles are congruent. Other problems that mobilize the same conception can conceivably require establishing a ratio between the measures of two segments. Unlike in earlier grades, high school geometry students are not expected to answer those questions by measuring and comparing numbers. Instead, they are expected to visually inspect the diagram to find two triangles about which the measures of some of their elements are known and that can be mapped to each other, mapping also to each other the elements to be compared. If the corresponding triangles can be claimed congruent, a comparison can be derived about the desired elements from the definition of congruence (which says that corresponding parts of corresponding triangles are congruent). A particular kind of proof, built on the two-column form (Herbst, 2002a), is often used to control the congruence of the triangles. This control strategy takes for granted the existence of the triangles to be compared—that is, triangle congruence proofs rarely include any warrant for the drawing of auxiliary lines. The proof format, however, requires the agent to list in separate lines each pair of elements of the two triangles that are known to be congruent, justifying each in reference to "given." Further, the two-column format requires the agent to put those pairs of congruent elements in relative order (e.g., a pair of corresponding congruent sides, then a pair of corresponding angles adjacent to those sides) and to compare the order in which they are listed to list of sanctioned congruence criteria (e.g., SAS—side–angle–side). More often than not, to perform these congruence comparisons constitutes the bulk of what is done on account of learning proofs in high school geometry. We have described it in this level of detail to impress upon the reader the point that, while it is undeniable that the way in which claims to congruence are controlled deserves the name "proof," the particular nature of what is expected to be involved and what goes without saying make these "proofs" particular to a conception of congruence: that which reduces congruence to triangulation and uses triangulation to organize an object-to-object correspondence. We use these as a first example of what we would like to call *C-proofs*, to refer to a particular kind of regulatory strategies of conceptions. We want to sensitize the reader to the possibility that particular ways of ascertaining (Harel & Sowder, 1998) that the solution to a problem is correct attest to a kind of reasoning that, albeit particular to a specific conception, are also singularly mathematical. In the case of triangle congruence, the particularly mathematical feature of that regulatory structure is the preference to reduce the question asked to one of a set of known cases where the information given is sufficient to assert congruence without actually confirming that congruence directly. In other words, that the solution to the problem is necessitated by the conditions of the problem and the affordances and constraints of the representation system, as opposed to contingent on who did it or when it was done.

C-proof

Let us explore the notion of C-proof further in a more elementary case, one in which the word "proof" might not even be used. In early elementary school, students learn to solve multi-digit addition problems by putting numbers in columns, taking the columns in order from right to left, adding the digits in each column, putting down the last digit of the sum for each column and carrying the other digits to the top of the next column. Place value, the notion that a number like 176 is "one hundred plus seven tens plus six units," is part of the representation system (whose basic elements are the digits 0 to 9 and the powers of 10 but also, crucially, the expression "7 tens" and its surface similarity with "7 apples"). If "46" is claimed as the solution to the addition problem "17 + 39 = ?," the adequacy of this solution could be checked by "doing it again" (as many students actually do) and either getting confirming or disconfirming results. For example a student could again get 46, 416, or 56. "Doing it again" is a regulatory structure. In some classes "doing stuff again" may be the *only* customary way to control results. We are not willing to call "doing it again" for the case of multi-digit addition a C-proof: doing it again hinges on repeating the experience of solving the problem and taking the aggregate of experiences as evidence for the correctness of a decision. It seems as though this is a clear case of empirical induction: if the majority of the times one solves the problem it happens to yield a certain result, then that result has to be the solution.

Note, however, that under similar circumstances the solution can be shown as necessitated by the problem and the properties embedded in the system of representation. The calculation could be checked by translating "17 + 39" into "16 + 40," which has to be 56 or at least much bigger than 46. This latter strategy uses the characteristics of the representational system (place value) to avoid repeating the experience of solving the same problem again. Instead of "doing it again," it changes the problem into another, simpler problem whose solution can arguably be claimed to be prior knowledge. The having of that prior knowledge entails the correctness or, as in this case, the incorrectness of the proposed solution. We consider this kind of regulatory structure for multi-digit addition to be analogous to triangle congruence proofs in the correspondence conception of congruence, despite the lack of many surface similarities. We are willing to propose both of them as C-proofs in spite of the fact that the "translation to a simpler problem" might rely on reasons which are still implicit (e.g., taking 1 from 17 and adding it to 39 to make 40 relies on the associative property, but a young child would unlikely say so).

In terms of describing the nature of proof at the timescale of the conception, the first point to make is that one needs to start from modeling the conception itself. Locating the actual and possible regulatory structures will provide means to argue that a regulatory strategy that could be considered a C-proof is within reach of the class at a given moment. For the same reason, there are evident ties between those C-proofs and the conceptions they help regulate. Congruence claims can be proved in the customary way because the correspondence conception of congruence does not require one to specify a global mapping of the plane but only a local mapping object to object. In the early arithmetic case, "transforming the problem into a simpler or known problem" can control the result of a calculation because a large amount of information is packaged within the structure of the base 10 numeration system. The observer can ascribe a conception to a class, say that at a given moment in time the class "holds" (i.e., the class works as if its members held) a certain conception (e.g., multiplication of polynomials as multiplication of numbers), or that a certain conception is present in the work that a class is doing at a given moment in time. That descriptive claim does not mean that any one student is in possession of that conception, but rather that the modal student (the normative counterpart of the teacher in classroom interaction) is held accountable for performing in accordance with that conception in the sense that CK¢ gives to conception.

A first place where an observer may recognize something that deserves to be called "proof" is thus within a conception. By definition, the control structures of a conception are such that they inject a truth value on the solution of a problem. If the injection of truth is obtained as a necessity of the formulation of the problem, that is, if the control operates by way of demonstrating that given the objects one had in the definition of the problem, and the means of representing them and handling those representations, the solution found could be reconstructed without reliving the experience of finding it, we say this control structure is a C-proof. Evidently the participants may call what they do to justify the solution to a problem a number of different things, including "proof" but not necessarily so. The process by which a real solution to a problem is rationally reconstructed beyond that which one happened to find and into that which one should have found is what we want to call a C-proof.

In particular, for a conception to have a regulatory structure that deserves the name of C-proof, the system of representation must be capable of representing not just the referents in the problem but also the means of operating on those objects (so that the experience of operating cannot add actions on which the result depends that the system of representation cannot handle) and what the collective holds itself accountable for knowing about those objects. These proofs are conception dependent and not all conceptions would have control structures that satisfy the conditions for being C-proofs in this sense. That is, in some conceptions the validation of solutions to problems is empirical, contingent on experience, in the sense that it requires recurrent engagement in the solution process (and confirmation by reproduction of the same result). Furthermore, in some classes the custom of mathematical work might include a discourse on method that legitimizes determining empirically the truth of general mathematical statements.

Thus, the notion of C-proof just defined and the notion of proof proposed by Balacheff (1987) and appropriated here to find and describe the custom of proof in a class are not easily related in terms of object and category. C-proofs are not only conception specific but also, and for that very reason, they may hardly adhere to a common form. Furthermore, classroom participants might not even call them "proofs." To sustain a custom of proof, some form is needed (and by form we are being inclusive of logical form as well as graphical and linguistic form).

Any conception that the class is held accountable for has some means to regulate solutions to problems (at the very least by having the teacher check the student's answer against a key). It seems appropriate however to keep calling the totality of those possibilities "regulatory structures" or "control structures" and to reserve the word *C-proof* for more special ones. Even a class whose custom of proving avowedly defers to mathematical proof, such as that of the high school geometry class, might potentially regulate some of its conceptions in ways that are contingent on experience.

In general, some conceptions are sufficiently formalized (that is, have a system of representation sufficiently rich and precise) to dedicate control structures that operate on semiotic translations of the experience of solving a problem, translations related to the model or "theory" of the objects and operations involved in the problems, a theory expressed in terms of the system of representation. That is to say, a technical "text" or "discourse" replaces the action of solving a problem by expressing why the result *should* be what it is. Note that this notion of C-proof does not a priori distinguish symbols like $\bigcirc\bigcirc$ or 17 from symbols like x or pq. They all are signs, even though some may be more amenable to an operational grammar than others. Each of them may or may not be part of a system of representation that supports C-proofs for a certain conception in a certain class, but the extent to which they do depends not on what they look like but on how their grammatical/rhetorical behavior within the system of representation allows them to contain the experience of solving a problem and in such a way obliterate the need to run that experience again to confirm a result.

Proof as a Tool to Know with

The previous consideration of the differences between a custom of proof and the specific notion of C-proof shows that those two notions together do not fully encompass the whole of the work of proving. A custom of proof closer to the custom of mathematical proof would be epitomized not in the cursory C-proofs of stable conceptions but in the breakthroughs of particular ideas that happen at particular moments. The having of great ideas, such as to draw an auxiliary line in a diagram in order to be able to consider two triangles that were not previously considered, is often outside the usual triangle congruence proofs in high school geometry. Yet, those great ideas are what often gets recorded in the "description of proofs" (Livingston, 1999) that mathematicians write. Lakatos (1976) showed that the notion of informal proof is one of the key methodological tools for the mathematician to exercise and control their plausible reasoning. Picking up on Pólya's (1954a) ideas, Lakatos described mathematical reasoning as not deductive, inductive or intuitive, but *heuristic*. How can we describe this notion of heuristic, informal proof in ways that we can use it for classroom observation? If the dialectic of proofs and refutations was to exist in a class as the means by which new knowledge is generated, how could we be attuned to see it?

The prior discussion of conception, which led to defining C-proofs, provides tools for ascribing conceptions to students on the basis of their capability to do some things, not necessarily of being aware of having any knowledge that warranted such actions. But the practices that an observer models as a conception could also be an object of public reflection for the participants. This is the case in classrooms as a consequence of the instructional constraint noted above: students are in classrooms not just to do things but also to learn. Thus, they need to acknowledge some transitions in their responsibilities, from a state in which they were entitled not to know something to a state in which they are accountable for knowing that thing. Thus, some public discourse about public action needs to be done by participants (usually the teacher) to install some practices as objects of knowing and to claim that the class now knows them. Hence, from an observer's perspective, participants' work does not just attest to the existence of a conception but might also sometimes attest to the class's reflective awareness of *having a conception*. The particular form taken by the knowledge that the class is held accountable for having can, of course, vary: they might just acknowledge knowing how to do "this kind" of problem, or perhaps knowing how to perform a general process instantiated in "those and other" problems, or even knowing that "such and such" thing is the case. On that accountability for knowing one can build the basis for a third sense of proof, anchored in a timeline where transitions are located from one to another state of public knowing. To develop a third way to answer the question of what counts as proof we will be especially attentive to reflective work that involves using a known conception to create or understand a new one. We locate a third place for proof in those moments of reflection when practices that ordinarily exist separately are juxtaposed, related, and compared. To illustrate this point we refer to Herbst's (2005) work on conceptions of equal area.

Within the CK¢ model, the expression "knowing of μ" or Kμ is used to denote a set of conceptions that an observer (provided with a mathematical idea μ) can ascribe to a person or a collective. For example, an observer could note within an interval of time in which a class works on area problems, different conceptions of equal area, the whole set of which constitutes by definition the class's *knowing of equal area*. In Herbst (2005), students' work on the "triangle problem" (to find a point in a triangle that will split it into three triangles of equal area) was used to show four different conceptions of equal area: empirical, congruence, equal dimensions, and quantitative. The "empirical" conception (EMP) indicates that two triangles are of equal area if the area formula applied to the measures of their bases and heights yields equal areas, whereas the "congruence" conception (EAC) indicates that two triangles are of equal area if they are congruent. The "equal dimensions" conception (EAED) indicates that two triangles are of equal area if their bases and their corresponding heights are congruent (no matter what they measure), and the "quantitative" conception (QEA) indicates that two triangles are of equal area if they can be shown to be produced by scalar operations (adding or subtracting shapes, and multiplying shapes by a common number) on triangles known to be of equal area (see also Herbst, 2003, 2006). An observer could formally establish a set of relationships among the conceptions in that set by defining translation functions among those conceptions. For example, the "quantitative" conception subsumes logically the "equal dimensions" and the "congruence" conceptions as particular cases. Yet the QEA conception emerged in classroom work after the EAED conception, and this one after the EAC conception. Students initially approached the triangle problem with the EMP or the EAC conceptions—in one case by accepting the problem as one about any triangle and in the other case by choosing a triangle where they could construct a solution that could be proved correct. This work on equal area is relevant to our present purpose because it afforded the following observation: the transition from EAC to EAED and then to QEA was made possible by explanations that, while responding to the demand to prove that was customary of claims in those geometry classes, could not be considered examples of the C-proofs of triangle congruence. We describe those explanations insofar as their resemblance with and difference from the C-proofs of triangle congruence next. We then proceed from this example to propose a third sense of proof, which we call *K-proof.*

The EAED arguments used by students to control why they claimed two triangles were of equal area were rarely written, let alone written in two columns. They involved little precision of notation (e.g., no semiotic distinction between "equal" and "congruent" was made) and incomplete justification (e.g., not all statements were justified). Some things were taken for granted, such as the validity of the area formula, which was neither a theorem nor a postulate at that time, but just prior knowledge. Rarely would those arguments have been recognized as proofs that met the usual standard of a two-column congruence proof. Yet, as we noted in the reports of that study, students responded to the request to prove that non-congruent triangles could be equal in area. They did so by way of using bits and pieces of the control structures normally used to prove triangles congruent. The emergence of the EAED conception was based on resources from both the empirical (EMP) and the congruence (EAC) conceptions. The empirical conception contributed the expression of the area formula (which is an operator in EMP) as an element of the semiotic register of EAED. In EAED the area of a triangle T could be *represented* as $A_T = \frac{1}{2}b_T \cdot h_{b_T}$ (as opposed to merely *calculated* by executing that formula, as in EMP).

The congruence conception contributed a way of tracking corresponding congruent elements across two triangles to be proved of equal area: students looked for sides, one in each triangle, that were known to be congruent and could be taken as bases, then looked for the corresponding altitudes (which some times had not been drawn), and checked whether they were known or could be argued to be congruent; they marked these congruencies in the diagram with hash marks, and equated (in speech or using inscriptions) literal expressions of the areas of the two triangles being compared. In view of this, we submit that the C-proofs of triangle congruence were being used as a metaphor to create a viable way of controlling equal area claims for the case when the triangles compared were not congruent. The metaphor mapped *congruence* into *equal area*, elements (sides, angles) into measures of dimensions (measure of bases, measure of altitudes), and the practice of marking congruent parts with hash marks mapped into marking congruent dimensions with hash marks. Furthermore, the metaphor mapped *congruence criteria* (ASA, SAS) into the *area formula*, to produce the sense that to assert that areas were equal it sufficed to show bases and heights are congruent without actually calculating either one of the areas. Thus, in spite of the facts that the explanations of why areas were equal were not in general written in the two column form characteristic of triangle congruence proofs, that the area formula conveniently hid the problems associated to geometric quantity and its measure, and that the area formula for triangles had not yet been proved (and actually no axiomatic definition of area had yet been provided), those explanations were accepted as appropriate responses to the request to prove that triangles were equal in area. The productiveness of considering these kinds of metaphorical mapping between control structures becomes more apparent when we consider the control of claims under the quantitative conception.

One problem where we observed what we called the quantitative conception of equal area was that of picking a point *inside* a triangle so that when this point was connected with the three vertices two of the three triangles created would be equal in area. Students picked a point on a median (O on BM, in Figure 3.1) and claimed that two of those triangles (△AOB and △BOC in Figure 3.1) were equal in area. To prove that claim, the proposed argument used the observations that,

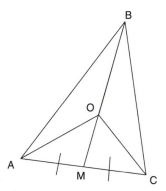

Figure 3.1 Two triangles of equal area

on the one hand $\triangle$ABM and $\triangle$CBM and, on the other hand, $\triangle$AOM and $\triangle$COM were equal in area (which students warranted using equal dimensions). They then argued that one could subtract the second pair of equal area triangles from the first pair of equal area triangles to get the targeted pair of triangles and that since they were subtracting equals from equals, the results would be equal. Once again, we submit that the C-proofs of triangle congruence were used as a metaphor for a proof of area equality. In particular, the imperative that statements be warranted by reasons made possible to identify a property of area—the property that asserts that equals subtracted from equals yield equals. We consider the correspondence to be metaphorical (as opposed to considering this case of proof to be just an adaptation to area of the kind of C-proofs used in triangle congruence) because at this particular moment, the reason given was *not* a known postulate, theorem, or definition, but rather a *new piece of knowledge* about equal area (see Herbst, 2005, 2006).

The two examples discussing the way equal area claims were explained in the EAED and QEA conceptions of equal area demonstrate that the emergence of awareness of new conceptions of equal area was made possible by controlling claims of equal area in those new conceptions using metaphorically the C-proofs of triangle congruence. We have used the notion of metaphor (i.e., a correspondence that maps systems of meanings; e.g., congruence and equal area; Black, 1962) to argue for the "likeness" of control structures for different conceptions. We now use that metaphorical mapping between control structures to identify a third notion of proof.

Take the particular correspondence between, on the one hand, observing three pairs of congruent parts and deriving that two triangles are congruent, and, on the other hand, observing two pairs of equal area triangles and deriving that a third pair of triangles (made out from combining those) are of equal area. The presumption that such a correspondence should exist is manifest in the making of one and the other procedure be responses to the same habitual word—*prove*. The taking for granted of that presumption warrants, in this case, the introduction of a new piece of knowledge—that equals subtracted to equals are equals.[6] It becomes reasonable that such property of areas would be valuable to have insofar as it helps bolster claims about equal areas. We would like to call *K-proofs* those metaphorical mappings between the C-proofs of established, known conceptions and the possible C-proofs of new conceptions. We submit that the existence of those metaphorical mappings, the presumption that one can operate on a new conception *as if* it were a known conception, and, in particular, the presumption that one should be able to control the truth of claims in the new conception as if they were claims in the old conception, witnesses of a practice that one would want to associate with proving. Indeed, the work of such metaphors can both produce new knowledge and establish such knowledge as plausible, reasonable. Unlike C-proofs that control the truths of claims within a conception, K-proofs control the provisional mapping between conceptions and in particular help develop new conceptions. Similarly to C-proofs where the mechanism is one of reducing an unknown problem to a known or simpler problem, the mechanism of K-proofs enlarges a state of knowing into a larger state of knowing by projecting metaphorically the control structures of old, known conceptions into controlling claims about (new) solutions to (new) problems that could not previously be handled. K-proofs are key contributors to reflections on known conception when these are brought over to control metaphorically claims that pertain to the new conception.

Thus we contend that such segments of time when a class engages in nontrivial reflections about conceptions (by nontrivial we mean reflections that actually involve at least two spheres of practice) are places where a third sense of proof can be found. Specifically, we propose that those reflections are candidate moments to observe mathematical performances that might count as proof and whose purpose is for participants to publicly warrant the bringing in of new ideas, to organize how newly acquired elements of public knowledge integrate with the old, and to argue for the plausibility of work they anticipate doing in the future. Evidently not all nontrivial reflections will necessarily include performances that deserve the label "proof": these reflections might organize spheres of practice in many different ways, for example as juxtaposition of cases organized in a taxonomy. More important than elaborating on what strategies they might use to structure their old and new public knowledge is to specify what proof might look like when doing this job.

Earlier we introduced the notion of C-proofs to refer to conception-specific control structures characterized by a translation of the experience of solving a problem in terms of the system of representation used to solve the problem and express its solution. That is, a C-proof waives the need to rerun the experience of solving the problem to control correctness. To define K-proofs we will use a similar strategy where what we handle are now spheres of practice rather than problems.

We will postulate that for reflections that involve different spheres of practice (different systems of meanings) to include K-proofs, the observer must see the class engaged in mapping one system into the other in such a way that the C-proofs of one of those spheres of practice are applied to control the solutions of problems in the other. This kind of mapping we consider to have the same formal structure as what Max Black (1962) called a *metaphor*, whereby parts of two systems of meanings are associated by a mapping that permits to project the structure of one system onto the other one and, possibly, use the projected structure to "create" elements in the least developed system. In language use, metaphorical projections start from the proposition of a metaphor such as "the poor are the Blacks of Europe" (Black, 1962) which suggests a projection of the known system of social relationships articulated by race in the US onto a system of social relationships articulated by class in Europe. By the use of the known system ("race in the US"), metaphorical projection warrants the posing of questions such as "which policies sustain the segregation of classes in Europe?" or "do the Europeans have anything like affirmative action?" A similar metaphorical mapping between number and operations, on the one hand, and polynomial algebra, on the other, is often used in school algebra to warrant the algebra of polynomials: the control structures of number calculations control the good definition of operations among polynomials—at a time in algebra studies when, since no formal definition of polynomial has been provided, the operations among them could hardly be defined other than by metaphorical projection. A metaphorical projection postulates a metaphor between systems of meanings and affords a heuristic to find out things about one of the systems. Participants may engage in mapping different spheres of practice and develop public knowledge of any one of those spheres of practice.

Public knowledge of new mathematical objects can develop as a result of engagement in proving that hinges on the "making believe" (or "metaphor") that some new, unknown objects are like old, known objects and can be controlled with similar control structures. In the case of polynomials and numbers this means that the practices associated with polynomial algebra may appear to be rational in spite of the vagueness with which the indeterminate is usually introduced: in spite of the fact that rarely people know what the X means in the polynomial $X^2 - 1$, they can accept that $X^2 - 1 = (X + 1) \cdot (X - 1)$ by virtue of the distributive property, which they expect to hold in this domain as a result of the metaphorical projection between numbers and polynomials. Students can in that way organize a "theory" or proof-organized knowledge of polynomials even when the basic objects and their properties have not been formally constructed.

The example discussed at length on the metaphorical mapping between congruence and equal area demonstrates how it is possible for a class to develop proof-organized knowledge of equal area even when the notion of area itself is informally defined, tied to the application of a formula that does not yet rest on an axiomatic definition. Herbst (2005) makes the point that characteristics of two-column proofs of congruent triangles (a C-proof) were projected metaphorically into those new conceptions as a result of the expectation that areas of non-congruent triangles be shown equal. The projection was metaphorical because the two-column proof format was applied on a set of problems where it usually did not apply and where not all conditions for application were available. In making this point we must underscore that the capacity to develop proof-organized knowledge rests, on the one hand, on the existence of a habit or custom of demanding and providing proofs to decide on truth and falsity, and, on the other hand, on the existence of conception-specific means for control of solutions to problems that replace the experience of solving them with a surveyable record of the solving inscribed in the semiotic register of the conception. K-proofs are therefore informal (metaphorical) instruments of knowing that respond to the presumption that a general sense of proof exists. But K-proofs operate on the presumption that one particular proof of a particular conception will afford and warrant the new ideas that will flesh that proof out.

What can an Observer do with the Theoretical Elements Provided?

The previous account provides ways of thinking about the places that proof might occupy in the development, management, and organization of public knowledge in a class. To summarize, one can expect proof to exist at three distinct levels in a class.

First, proof can be a general element of the custom (the *habitus*; Bourdieu, 1990) of a mathematics class, as evidenced by a number of general expectations about official knowledge and uses of language at work: if propositions are made that can be true or false, if speakers can be held accountable for explaining the truth of what they propose, if the request to prove, explain, or justify can be met with more than blank stares, we can say that a class has a customary notion of proof and we might be able to describe it in detail. Second, proof (as C-proof) can be a specific regulatory structure for some conceptions: the solutions of some problems may be controlled by procedures on the representation of those problems and the objects involved. Specifically, a C-proof is what shows that a proposed solution to a problem is necessitated by the givens of a problem and the structure of the system of representation. Third, proof (as K-proof) can be a way of relating two or more conceptions, providing ways of operating on one conception "as if it was like" operating in the other. Thus (1) helping impose a(n) (informal) reasonableness onto other new conceptions whose vagueness might compromise their capacity to have native C-proofs as control structures, and (2) helping create the impression of rational organization in the aggregated public knowledge of a class.

The question of what counts as proof has three possible answers in any given class; those three answers are meaningful at least from the perspective of the teacher. The observer can use the foregoing account to locate those levels of activity and describe how the performances of a particular class relate to the performances that the observer judges possible given the conceptions and knowings that the observer can ascribe to the class. When this observer is the teacher, or somebody else, attempting to describe the extent to which the work of teaching in that class has made room for proof, these three notions of proof can help operationalize that description.

The observer can tell whether or not the work of teaching in the observed class has fostered a custom of making general statements and requesting proofs of their truth. The observer can bring to the analysis of those practices the set of values that constitute the mathematical habitus: are memories of past knowledge stored in general statements (statements about general objects) that could be true or false? Are these statements confronted in regard to their truth or falsity? Are expectations of what can be done next expressed in general statements that could be true or false? Is truth or falsity ever addressed as if it could be decided a priori, before experience?

Our observer could also look at the work done to control problem solving within specific conceptions. In each of the *conceptions* held by a class over a certain span of time, proof may or may not exist as the (ultimate) control structure for judging solutions to problems in specific spheres of practice. In examining the performances of a mathematics class at various points in time, the observer can determine whether the students or the teacher could have validated a solution to a problem by showing that the solution found experientially could have been anticipated rationally. If the solution could have been anticipated from transformations of the problem using the symbols and grammar of the system of representation with which they encode the problem and its solution, our observer could say that such conceptions are controlled by C-proofs.

The two strata noted earlier, that of the habitus and that of conceptions, are opposed insofar as they possibly allocate a role for proof. The former is general and concerns the values that characterize a mass of knowledge over a long stretch of time, the latter is specific and concerns the validation of specific knowledge claims at given moments in time. In the production of mathematical knowledge by mathematicians, custom and conceptions support each other: General, shared ideas of what a proof is and what sorts of things are interesting to prove guide the proposition of specific problems and their solutions where original proofs can be found that solve those problems and that at some points inform the customary notion of proof as well. The notion that to incorporate these specific solutions to problems to what the community knows, they need to be represented in ways that make its organization memorable and rational (for example, by defining explicitly new concepts and making explicit which pieces of prior knowledge a proof relies on) could also be inherited from that custom. But a class is a very different organization than the mathematical community, particularly as it has to do with the I^4 constraints noted at the beginning that make the developmental paradox cited at the beginning much more of an issue than in the case of history of mathematics. The decision to teach X at a certain point in students' develop-ment often entails the need to compromise the discipline's views on X for what can be done with and about X in schools (pace Bruner, 1960). Thus the question—can the public knowledge of a class be proof-generated knowledge in any sensible use of the word proof?

The previous section, in which we developed the notion of K-proofs as meta-phorical projections of the C-proofs that control selected conceptions, shows how one could observe whether and how proof plays a role in the development of new knowledge. Specifically, an observer who has used the notion of *conception* to parse the stream of classroom events into spheres of practice that he or she can associate to conceptions, would also, as a result of that parsing reveal moments in classroom discourse where some reflective discourse is engaged. In particular, the observer could identify discourse that elaborates on those spheres of practice by relating some of them to others. Whenever the C-proofs that operate to control problem solving in one of those conceptions are used metaphorically to produce further reflections on another sphere of practice (e.g., to improve a system of representation, to control the solution to a problem, etc.) that work could be called a K-proof. In these circumstances, the work of doing a K-proof can serve the development of new, public, knowledge even when the formalization of such K-proof (e.g., into a C-proof) might have required that new knowledge to be already acquired.

A customary, general, and vague notion of Proof; a local, specific notion of C-proofs; and a reflective, metaphorical notion of K-proofs are the three senses in which proof can have a place in a class. An observer can use each of these notions to inspect the performances of a class over time, at specific times, and over segments of medium duration. The observer can use these notions to express "what counts as proof" in a class and to contrast those descriptive observations with what one might be able to observe in other classroom contexts that work under similar constraints.

If the field is in a deadlock as regards to what we mean by "proof," we contend this is so partly because of the insistence on a comprehensive notion of proof that can serve as referent for every use of the word. We suggest that we need to pay particular attention to the arena of mathematical activity where one wants to observe the work of proof. We have indicated four constraints that make class-room work singular insofar as a public mathematical performance. Its critique by an observer requires tools different than a critique of mathematical performances in other arenas, such as the interaction of an expert mathematician and the research paper she or he has to review. We have argued that to make it operational for understanding and appraising the mathematics of classrooms we need at least three meanings for the word. Proof as a general thing is just an expression of the general values that might be the basis of a class's mathematical performances; as those performances are looked individually, proof dissolves into many proofs, there is no such thing as a unique and universal characterization of proof; there are specific proofs for specific assertions about specific objects. In some of these practices, objects are sufficiently known so that the particular proofs actualize the general values noted earlier. In others, the dependence on vague terms or on isolated pieces of empirical or perceptual knowledge might defy any attempt to decide once and for all whether they are mathematical proofs. The notion of K-proof, establishing a possible lineage for those practices by way of meta-phorical projections, may help operationalize how the public knowledge in some classes appears to be more rational than that of other classes even if their public knowledge is not more formalized than the public knowledge in others.

Notes

1. The writing of this chapter has been supported in part by a grant from the National Science Foundation REC-0133619 to the first author. Any opinions, findings, and conclusions or

recommendations expressed in this material are those of the authors and do not necessarily reflect the views of the National Science Foundation. The authors acknowledge valuable comments by Michael Weiss and Takeshi Miyakawa.

2. Note that in stating these constraints we are not *prescribing* or endorsing them as obligations that if fulfilled might create any desirable state of affairs. We are merely *describing* them as they appear to play out in school classrooms.

3. Biological growth, identity, learning style, disability status, race, etc.

4. Obviously, that differs greatly from how a proof is accepted as proof in mathematical publication.

5. *The* mathematical community is assumed to be the contemporary community of mathematical researchers that proves theorems and publishes these proofs as scholarly work.

6. This analysis is simplified. To complete it one would have to consider also a metaphorical mapping between semiotic registers, for example between a segment and a triangle, between the measure of a segment and the area of a triangle, and between the properties of the morphism that maps segment union into number addition on the one hand and the properties of the morphism that maps plane figure union into area addition.

II
Teaching and Learning of Proof in the Elementary Grades

Proving is a human activity that extends beyond formal mathematics. While there are rules of logic that govern a rigorous notion of mathematical proof, the idea of proof has its genesis in human activity, in the need to establish certainty. Even young children, without formal training in logic and deduction, exhibit an innate curiosity to know why things work as they do, to understand why things are as they are, to be convinced. Indeed, as toddlers begin to acquire language skills, one of their most frequently asked questions is most likely "Why?"

Regarding natural tendencies in young children's thinking, Mason (2008) argues that children's capacity to generalize, to identify structure and relationships within the details of the particular, is present from birth and that it is the obligation of formal schooling to cultivate these nascent abilities for mathematical purposes. Applied to teaching and learning proof in elementary grades mathematics, this suggests that young children bring to formal schooling a natural inclination to prove, to convince or be convinced (see also Maher, Chapter 7, this volume), that can be cultivated in their mathematical thinking.

From this premise, Section II explores young children's intuitive, informal notions of proof and how their natural tendencies toward proof can be nurtured in ways that build a foundation for a more advanced study of mathematical proof in later grades. But, by "proof in the elementary grades," we are not referring to how a mathematician might view proof. Instead, this section adopts a broader interpretation of proof as a socially constructed process whose norms for what constitutes "acceptable" arguments have fluid boundaries that are, to an extent, unique to the community it serves. What it takes to convince a child is different than what it takes to convince a mathematician. As Schifter (Chapter 4) illustrates, understanding what it means to prove includes understanding not only who needs to be convinced and what they need to be convinced of, but also appreciating how the nature of the argument might change relative to the audience. It is over time as children experience building arguments that they can come to appreciate the notion of an acceptable argument in a mathematics classroom (NCTM, 2000).

The *Principles and Standards for School Mathematics* (NCTM, 2000) maintains that "reasoning and proof should be a consistent part of students' mathematical experience in prekindergarten through grade 12" (p. 56). This is especially significant given the abundance of research, noted by several chapters in this section, that even college students still use empirical arguments. Such research underscores the need for school experiences that challenge students' thinking *as early as the elementary grades* regarding what constitutes a logically valid argument. This section points to what these experiences might look like for young children and provides evidence that children can learn to develop informal deductive arguments in preparation for a more rigorous study of proofs and proving in later grades.

Developing Children's Understanding of Proof

While each chapter in this section presents a different aspect of the story of proof, there are common principles that can be lifted out to help us understand teaching and learning proof in the elementary grades and how it can prepare children's thinking for its more formal study in later grades. In Schifter (Chapter 4), proof originates as an issue of certainty when children explore questions such as "How can we know this is true for all cases?" Schifter explores how arguments built from children's representations of story contexts are an important form of proving in the elementary grades because they contain an element of generality, essential in more rigorous proofs, which empirical arguments do not have. Through "representation-based proofs" as she terms it, number becomes a placeholder for "any quantity" when children build arguments in which they reason from the structure of numerically quantified representations rather than with specific numbers. Thus, while children's arguments might reference particular numbers, their use of number reflects a generality that removes the argument from a particular case or example. Through this, Schifter describes how representation-based proofs serve as a tool by which children can confront the limitations of arguments that look only at particular examples or that simply appeal to an authority source. As such, she maintains that: "[A]rgument from representation . . . is an effective route to establishing general claims in the elementary classroom" (Chapter 4, p. 76) and, thus, to building a foundation for proof thinking in later grades.

Morris (Chapter 5) continues this emphasis on children's use of representations as a tool for developing their deductive reasoning skills. However, while Schifter looks at children's deductive reasoning in a mathematical context that originates with explorations in number, Morris focuses on a mathematical context that is based on reasoning with quantities that are intentionally not measured numerically. In particular, she examines how children reason with representations that embed their actions and operations on physical quantities within the representation itself, where the act of representing involves capturing actions on quantities that are not numerically quantified. Morris argues that representing and reasoning with arbitrary, non-numerical quantities (e.g., "Quantity A") builds the idea in children's thinking that an action is applicable to an infinite set of objects rather than a single case and, thus, helps children build general arguments that apply to a global domain. Morris concludes that the development of children's capacity to deduce relationships about arbitrary quantities, quantities that are expressed symbolically and are not ascribed a numerical value, is compatible with the nature of mathematical proof and, as such, is an important way to prepare children for a more formal study of proof in later grades.

Fosnot and Jacob (Chapter 6) also focus on children's deductive reasoning in a study that examines how a mathematician's process of reflecting on arguments, reorganizing ideas, and building justifications is replicated in young children's attempts to convince their peers of mathematical claims. They maintain that the act of refining connections between mathematical statements is an important aspect in the notion of mathematical proof because it facilitates the development of proofs that are simple, elegant, and without redundancy. They describe how open number lines can support children's understanding of justification and proof as children express and compare relationships between quantities. In their study, they found that the open number line served as a tool to build rules of deduction in children's thinking about equality and inequality. Perhaps more importantly, they observed children using these rules in a "chain of deductive moves" (Chapter 6, p. 110) to prove conjectures about equivalent relationships. However, the activity of eliminating redundant information to build simple, elegant arguments about equivalent relationships was not a simple process for children. Moreover, because of its complexity, Fosnot and Jacob argue that children need multiple experiences in re-examining and simplifying their ideas in order to make concise arguments if they are to build a foundation for understanding a mathematician's view of proof.

Maher (Chapter 7) draws on her longitudinal work on children's reasoning to focus our attention on children's natural inclination towards the idea of proof. Through a combinatorial task to find the number of different stacks of colored cubes, Maher found that students were able to reason in ways that included forming conjectures, explaining findings, discussing ideas, and inventing ways of convincing themselves and others about the validity of their ideas. In doing so, children used methods of exhaustion, contradiction, recursion, and cases to develop justifications about the numbers of stacks of cubes with certain color combinations.

In addition to the story it tells regarding children's discovery of the idea of proof, this chapter underscores an important point, implicit in other chapters in this section, regarding curriculum in the elementary grades and children's natural tendency to prove. In particular, issues of proving can be lifted out of rich problem solving tasks elementary teachers *already have on hand* to support children's proof thinking. Teachers do not need a specific "proof curriculum." Part of the task's appeal is its accessibility to teachers and students; it simply asks children to find the different possible combinations of stacks of four cubes using cubes of two colors. It is arithmetic, and children can model their thinking in concrete ways by stacking cubes or drawing stacks of cubes. However, like many of the tasks described in this section, in the hands of a skilled teacher (researcher) this task became a significant means by which principles of proving, such as formulating conjectures, establishing an argument's validity, or building arguments through techniques such as exhaustion and contradiction, could surface. Thus, not only is the task easily placed in an elementary grades curriculum, it can be extended in ways that bring out children's natural tendencies towards proof and reasoning.

Teaching Practice that Supports Proving

The bigger challenge, perhaps, is to understand how to develop elementary teachers' intuitions about proving so that their instruction can support these tendencies in children's thinking. The work by Reid and Zack (Chapter 8) foregrounds teacher practice by examining aspects of instruction that support proving in the elementary grades. The premise of their work is that proving requires—among other things—a culture of problem solving, communicating, and conjecturing. As their work suggests, and as we can infer from other studies described in this section, teaching practices in which children are encouraged to question ideas, to understand one's own and others' thinking, to develop conjectures and build arguments for these conjectures, or to write or talk about one's reasoning, are all forms of practice that are foundational for building a habit of mind toward proving.

Schifter (Chapter 4) notes that a teaching practice that values asking and answering "Why?" helps children develop a habit of mind for establishing certainty. This "impulse to prove" (Chapter 4, p. 79) can then be honed over time to precise forms of conjectures that need rigorous arguments. But it requires a community of discourse—an initiation into a mathematical community that is guided by standards of communication that govern the process through which children learn to build and appreciate the structure of convincing arguments (Fosnot & Jacob, Chapter 6; Reid & Zack, Chapter 8). Maher (Chapter 7) adds that classroom environments which provide interesting investigations for students to pursue and which make available appropriate materials and sufficient time to think deeply about problems support a culture of proving that is leveraged by children's natural inclination towards sense making of mathematical ideas. In contrast, a traditional classroom culture which views mathematics as arithmetic skills and procedures to be mastered and which privileges timed processes to measure children's success are antithetical to the goals of teaching and learning proof in the elementary grades.

It is to our advantage that practices of teaching that support a culture of proving are consistent with practices advocated by reforms in mathematics education. This economy of ideas strengthens the argument for developing children's capacity for proving in the elementary grades. Teachers who understand, for example, good practices of communication can use these skills in all aspects of their mathematics instruction, including proving. The challenge is, perhaps, the development of elementary teachers' mathematical knowledge. The particular kinds of mathematical knowledge elementary teachers need in proving have not historically been a strong component of their preparation and continuing education. For example, they need experiences with building mathematical arguments, an understanding of the levels of arguments children might use and how to scaffold children's ability to build more sophisticated arguments, an understanding of an appropriate notational system for expressing arguments and how to help children develop this, and the flexibility for finding opportunities in their curriculum to integrate the activity of proving into their daily instruction. Embedding this knowledge into practices that support a culture of proving can transform children's mathematical thinking.

Common Themes and Points for Discussion

Although the studies described in this section reflect different approaches to research on children's understanding of proof—some based on reasoning with number (Schifter, Fosnot, & Jacob) and some based on reasoning with non-numerical quantities (Morris), some rooted in children's natural language (Schifter, Maher) and some involving symbolic notation (Morris, Fosnot, & Jacob)—they share the important goal of helping children build general arguments, based on reasoning with arbitrary quantities, that support claims about classes of objects, not single instances. In this sense, they challenge children's notion that empirical arguments are logically valid. They provide evidence from a variety of contexts that children *can* learn to reason deductively—an essential prerequisite for formal studies of proof. And, in the dialectic of signs serving as both tool and result of mediated thinking (Vygotsky, 1962), they explore how representations either help children develop proofs or reflect children's mediated thinking about proofs. Finally, they are embedded in a culture of practice that emphasizes problem solving, reasoning, communicating, conjecturing, and justifying ideas, and they point to curricula that involve the use of rich problem solving tasks that children have ample opportunity to explore.

These common themes help define some of the types of foundational thinking and practice in the elementary grades that can support children's capacity for proof in later years. But the studies described here also raise other points for us to consider. First, the chapters depict an underlying mathematical context rooted in the development of children's algebraic reasoning. For example, Schifter (Chapter 4) describes children's arguments about sums of arbitrary evens and odds, Morris (Chapter 5) explains how children reason about relationships in quantities that are symbolically notated, Fosnot and Jacob (Chapter 6) look at how children reason algebraically about relationships between equal and unequal quantities, and Reid and Zack (Chapter 8) think about teaching practices in which children analyze patterns and functional relationships.

Because algebraic reasoning has the inherent goal of developing and justifying mathematical generalizations (see Kaput, Carraher, & Blanton, 2008), it can motivate the study of proof in the elementary grades. Thus, advocating the development of children's algebraic reasoning as a point of reform (NCTM, 2000) mutually supports the development of children's understanding of proof. This reciprocity, in turn, strengthens the argument for the study of both algebraic reasoning and proof in the elementary grades.

The contrast in approaches described by Schifter (Chapter 4) and Morris (Chapter 5) foregrounds another point regarding the language or notational system children use to build their arguments. For example, are children constrained by the lack of a symbolic notational system for expressing general arguments or does their natural language and familiarity with numbers provide a more meaningful channel for expressing generality? In this sense, how does the language system children use, whether symbolic or natural, shape their capacity for developing proofs? In Schifter's study (Chapter 4), children reason with number in their own natural language system as a way to build and express arguments. As a result, they use number as a type of pseudo-variable. Morris (Chapter 5), by way of contrast, illustrates how children might reason in a symbolic notational system without the constraints of number. But what are the constraints and affordances of using a notational system that co-evolves with the child's understanding of deductive reasoning? Perhaps Fosnot and Jacob (Chapter 6) offer a solution to this question with their use of representational tools such as the open number line, where symbols and variation can be represented in meaningful ways to bridge numeric and symbolic reasoning.

The introduction of proof and proving in the elementary grades in ways outlined in this section is a relatively new idea. The studies described here open our thinking to the possibilities of what children can do in a classroom culture that nurtures their natural tendencies towards proving. Moreover, they point to how these experiences can prepare children for more formal studies of proof in higher grades.

Research is needed, however, to connect children's experiences in elementary grades to their understanding of proof in more advanced studies of mathematics. How, in particular, does instruction that develops children's deductive reasoning and their ability to build general arguments and express these arguments in symbolic or pseudo-symbolic forms impact their notions of proof in secondary grades and beyond? If children enter secondary grades with an emergent understanding of what counts as a logically valid argument and how to develop one, in what ways can this extend their mathematical understanding in more advanced studies? Moreover, what are the implications for curriculum and instruction in secondary grades (and beyond) when children bring these understandings to the classroom? The subsequent sections in this book, which address the teaching and learning of proof in later grades, offer a starting point for making the connections between children's experiences in elementary grades and their study of proof in later years.

4
Representation-based Proof in the Elementary Grades

DEBORAH SCHIFTER[1,2]

The mathematics classroom is the preeminent context in which students can learn to rely on their own powers of reasoning. Given a classroom environment in which they are encouraged to share their mathematical ideas, even students in the primary grades can and do engage in impressive mathematical reasoning (Ball & Bass, 2003; Cobb et al., 1992; Lampert, 1990). Investigators have found that young children are able to explore and discuss the regularities they observe in the number system, articulate the generalizations they see, and deal successfully with such questions as, "How can we know this is true for all cases?" Thus, at an early age, these students are already engaged in the process of proving claims of generality.

This chapter investigates the possibility and actuality of proof in the elementary grades. How do young children consider mathematical claims that apply to a general class? Which criteria for proof are appropriate to elementary students, yet would support distinctions essential in the later grades? And how does such work on proof support the work that is already at the heart of the elementary mathematics program?

The Context of Our Work

Since 1993 the author, with colleagues Susan Jo Russell and Virginia Bastable, has worked in collaboration with groups of teachers to investigate students' mathematical thinking in classroom contexts (Schifter et al., 1999). More recently, we have focused on early algebra (Bastable & Schifter, 2007; Schifter, 1999, Schifter et al., 2007).[3] From 2001 to 2004 monthly meetings of project staff with collaborating teachers were organized around a set of mathematical tasks designed to help teachers explore arithmetic generalizations that might arise in elementary classrooms. In addition, teachers read and discussed cases—reflective and usually detailed descriptions of classroom episodes—produced in earlier projects.

As teachers engaged together on the mathematics and on the cases, they refined their understanding of mathematical generalization, analyzed children's thinking, and considered which classroom activities would best support children across the grades as they take up these ideas. The teachers introduced these or related mathematics activities to their students, then wrote their own cases —meant to be shared with the group—documenting the resulting classroom process. They reported on:

1. their students' thinking as this was reflected in classroom conversation
2. the ways in which representations were used by their students
3. the questions this episode brought up for them about their own teaching practice.

Staff members read and responded to each case, highlighting in particular the mathematical–conceptual issues in play.

These cases, together with videotapes from the classrooms of a small subset of teachers in the group, provided data we used to refine our own thinking about the key algebraic ideas to be addressed in the curriculum, the development of those ideas across the grades, and the classroom tasks that could draw them out. The data also quite quickly led us to the question of what counts as justification of a mathematical generalization in the elementary grades.

The work of generalizing and justifying emerges quite naturally from ordinary classroom activities in which children engage in their study of arithmetic. Once students have noticed and described a regularity in the number system, the questions *why does this pattern hold?* or *will this always work?* lead to the search for justification. As we studied the nature of arguments presented by grades K–5 children, we asked: what forms of explanation and justification do children offer as they begin to realize the problematic nature of making claims about infinitely many numbers? Through what means do they address the problem of making such arguments? What aspects of students' arguments align with ideas about proof they will encounter in later grades?

Students' Responses to the Challenge of All

Consider the set of exchanges in Box 4.1, a composite drawn from several 3rd grade classrooms.

Box 4.1 Responding to Challenge

Students have been studying odd and even numbers and have established that one way to ascertain that a number is even is if that number of objects can be arranged in pairs with nothing left over. A number is odd if that number of objects, when arranged in pairs, has one left over without a partner. Today, in the course of a discussion, the class conjectured that the

---

sum of two even numbers is even. When the teacher asked students to explain how they knew this will *always* be the case, no matter what two even numbers are added, they offered the following responses:

Paul: I know the sum is even because my older sister told me it always happens that way.

Zoe: I know it will add to an even number because 4 + 4 = 8 and 8 + 8 = 16.

Juan: Also, 6 + 12 = 18 and 32 + 20 = 52.

Eva: We really can't know! Because we might not know about an even number and if we add it with 2 it might equal an odd number!

James: We can never know for sure because the numbers don't stop.

Claudia: We don't know because numbers don't end. One million plus one hundred. You can always add another hundred.

Melody: Your answer will be even because you are using even numbers.

As Melody spoke, she pointed to arrangements of cubes in front of her.

Then she continued:

Melody: This number is in pairs (pointing to the light colored cubes), and this number is in pairs (pointing to the dark colored cubes), and when you put them together, it's still in pairs.

The teacher, Ms. Emerson, has challenged her students to address a problem central to the doing of mathematics: how can you defend a claim that applies to an infinite number of instances? Her students' responses represent four categories that might be expected of a group of 3rd graders:

appeal to authority
inference from instances
assertion that claims about an infinite class cannot be proven
reasoning from representation or story context.

The first category is illustrated by Paul who is ready to accept the claim by appeal to an authority, in his case, authority ceded to his older sister. Students might accept a claim because a teacher told them, because it is "in the book," or because the "smart" kids in the class say it is true. As one moves through school and is socialized into our culture, there are many things, including mathematical definitions and conventions, that students must accept as given. However, the mathematics classroom provides a context for students to learn to rely on their own powers of reasoning, and the challenge offered by Ms. Emerson is an instance of this.

The second category of response is reliance on particular instances to support a general claim. Zoe attempts to demonstrate the claim that the sum of two even numbers is even by adding 4 + 4 and 8 + 8. Students are often quite satisfied with this means of justification: "It works in the cases I tried, so it must be true." In fact, research reports indicate that even many college students are satisfied to accept a general claim on the evidence of a few examples (Harel & Sowder, 1998, 2007; Knuth et al., 2002; Martin & Harel, 1989).

Juan may have recognized that, in choosing equal addends (doubles), Zoe has picked a special class of example, and the claim might be true only for that class. (In fact, by an alternative definition of even, the double of any whole number is even.) So Juan uses unequal even addends to test the claim.

In fact, Juan's strategy illustrates a useful habit when testing a general claim. By choosing specific instances of different types, one might succeed in refuting a claim. Finding examples in support of the claim points in the direction of its truth and may yield greater insight into the claim, itself. But ultimately an accumulation of instances is not adequate mathematical justification.

A third category of response is found among those children, illustrated here by Eva, James, and Claudia, who recognize the inadequacy of relying on particular instances to justify a general claim and thus conclude that certainty is impossible. As students become aware that numbers are infinite, they also come to see that a conjecture cannot be proved by testing examples, since one counterexample is enough to refute the claim. Since "numbers go on forever," it is impossible to check every instance. Eva points out there *might* be an even number out there, which they don't know about, that when 2 is added, yields an odd number. She reasons, since they can't test all numbers, they'll never know.

Melody takes a different approach, representing the fourth category of response: using a representation (in this case, a collection of cubes) to make an argument based on reasoning. Melody refers back to the definition of even number and shows that one number represented by pairs of cubes, joined with another number of paired cubes, yields a larger set of pairs.

However, although Melody doesn't speak about particular numbers, her representation is necessarily of specific numbers of cubes. How do we know that she is, in fact, making an argument about the sum of any two even numbers, rather than the particular numbers 10 and 16? She offers further evidence when the teacher questions her, and Melody again explains her reasoning:

Melody: Because 2s don't get to odds. If they're two even numbers, they're both counted by 2s, and if you put them together, you keep counting by 2s, and that always equals an even number.
Eva: So she's saying she already knows it that it always equals even.

Thus, even though Melody points to particular numbers of cubes, she says she is thinking about *any* pair of even numbers. The same argument would work, even if she had an arrangement of six cubes and four cubes, or eight cubes and 22 cubes. That is, the conclusion follows from the structure of her representation, rather than the specifics of the instance she happens to have chosen.

What Constitutes Proof at the Elementary Level?

Consider a slight variation of Melody's argument as compared to one that might be offered by a mathematician (see Figure 4.1).

Each argument states a premise (two even numbers are added) and argues to a conclusion (the sum is even). The mathematician shows how each step, from premise to conclusion, is justified by a definition, fact, or principle already established, relying on the laws of arithmetic, definitions that apply to the domain of integers, and algebraic notation to communicate the argument. Whereas Melody employs a spatial representation that embodies a definition of even number applicable to the domain of counting numbers and demonstrates the action of addition as the joining of two sets to show how the conclusion follows from the premise.

The variation of Melody's proof in Figure 4.1 uses dots to indicate any number of pairs of cubes, representing *any* two even numbers. However, the dots were inserted by the author of this paper to communicate how the spatial representation can accommodate the infinite class even numbers. This device increases the complexity of the argument without necessarily increasing its persuasiveness to an elementary classroom (Monk, 2008, p. 175).

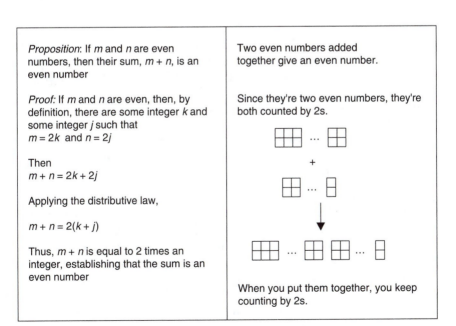

Figure 4.1 Mathematician's proof versus a variation of Melody's proof

Clearly, the tools of the mathematician's proof are not generally available to children in the elementary grades. At this level, most students are still coming to understand the kinds of actions that are modeled by the four basic operations. The laws of arithmetic cannot be the basis of their justifications when these laws are still in question for them. Indeed, what mathematicians call the commutative, associative, and distributive laws, for example, might arise in classrooms as conjectures to be proved. Neither is algebraic notation typically available as means for expressing generality.

However, young children *are* capable of justifying claims of generality. Reasoning from visual representations to justify general claims, as Melody does, appears to be accessible, powerful, and generative for students. Our work has convinced us that argument from representation (physical objects, pictures, diagrams, or story contexts) is an effective route to establishing general claims in the elementary classroom.

The next question then: what criteria for representations can be offered for such arguments? Through studying examples of such arguments, we have identified three:

1. The meaning of the operation(s) involved is represented in diagrams, manipulatives, or story contexts.
2. The representation can accommodate a class of instances (for example, all whole numbers).
3. The conclusion of the claim follows from the structure of the representation.

We hypothesize that if this form of reasoning were to be explicitly encouraged in the elementary grades, it would enhance students' work with proof in later grades. Students would have experience with the ideas that (1) relying on specific instances to prove a general claim is logically flawed and (2) proving the truth of a claim that applies to an infinite class is nonetheless possible. In order to highlight these aspects of reasoning about generalizations, as opposed to reasoning about specific numbers, we refer to such arguments as proofs—specifically, "representation-based proofs."[4]

Studying Computation and Representation-based Proof

One common objection to working on the ideas of proof in the elementary grades is that the curriculum is already very full; it is unreasonable to assume that more and more content should be moved into the elementary grades because older students do poorly with that content in the later grades. Our research demonstrates that working on representation-based proof actually can enhance the study of the content everyone agrees is at the heart of the elementary curriculum.

Two vignettes have been selected to illustrate this point. The first is taken from Margie Riddle's 4th grade classroom. When a computational question arose among her students, Riddle provided time for her students to explore the issues that underlay it. In that context, they devised a proof to resolve for themselves a mathematical relationship they initially found surprising.

The second vignette comes from Karen Schweitzer's 1st and 2nd grade combination class. In this case, Schweitzer asked her students to come up with a proof for a proposition that everyone in the class already accepted. Once their proof was before them, they realized it supported a more general claim, one that applied to more complex computational contexts.

Proof as an Aid to Understanding: An Example from a Grade 4 Classroom

As part of her morning routine, Margie Riddle had given her class some subtraction problems. Since her students were in the middle of a science project in which they weighed apples as they gradually dried out, she set the problems in that context. Included among them were 145 − 100 and 145 − 98. As the children began to consider the latter problem, Riddle realized here was an opening that held much potential for learning, and so she deferred further discussion until the math lesson later in the day.

What Riddle had seen was this: many of the children realized there was a connection between the two problems. However, after calculating 145 − 100 = 45 and before actually solving 145 − 98, they weren't sure if the answer to the second problem would turn out to be 2 more or 2 less than 45. Once they did solve it and they knew the answer was 47, they wondered, why was it 2 *more*? After all, when they changed 100 to 98, they *subtracted* 2, so why *add* 2 to get the right result? Their puzzlement, Riddle felt, could lead her students to a deeper appreciation of the meaning of the operation of subtraction.

When the class returned to the problem later that day, several children offered their ways of finding the answer to the problem: 145 − 98 = 47. Riddle describes what happened next:

> Brian was waving his hand in the air, insisting on explaining his thinking, too. He struggled to find the words. "It goes with the problem before," he declared. "It's like you've got this big thing to take away and then you have a littler thing to take away so you have more. Can I draw a picture?"
>
> I nodded, and he came up to the blackboard, thought for a while, and then drew a big blob like this [Figure 4.2].
>
> "See, this is the apple at first," he explained. "And you take some away and have some left. Then you take away 98 grams instead, so it's over here." It appeared to me that Brian had a very clear mental image that was helping him think his way through the problem, but that he was having a hard time communicating it to us.
>
> However, his classmates were watching and listening fairly intently. Suddenly, inspired by his presentation, Rebecca said excitedly, "Yeah, it's like you have this big hunk of bread and you can take a tiny bite or a bigger bite. If you take away smaller, you end up with bigger."

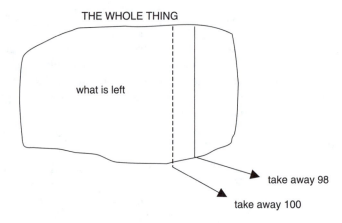

Figure 4.2 Brian's thoughts on subtraction

"Do you think this will always be true?" I asked.

"I think so," she answered.

During the discussion up to this point Max had been quiet. Now, inspired by Rebecca's explanation and Brian's picture, he continued further with the thinking that was unfolding. He raised his hand and said, "Yeah, the less you subtract, the more you end up with. AND . . ." he continued with great emphasis, "in fact the thing you end up with is exactly as much larger as the amount less that you subtracted."

These 4th graders' exploration of a generalization began with a simple arithmetic problem: 145 − 98. Since that problem was juxtaposed with a similar problem, 145 − 100, students were immediately drawn to compare them. And although their initial question was formulated in terms of specific numbers—since 145 − 100 = 45, and since 98 is 2 less than 100, shouldn't 145 − 98 be 2 less than 45?— a more general question lay just underneath—if you decrease a number in a subtraction problem by a certain amount, shouldn't your answer decrease by the same amount?

The class discussion about this question began with students' strategies for calculating 145 − 98. For example, Jillian explained that since 47 + 98 = 145, then 145 − 98 must equal 47. Although Jillian's (and other students') explanation verified that the answer to 145 − 98 is 47, not 43, it did not satisfy the class's desire to understand *why* the numbers worked out that way.

But Brian's "blob" took the class into different territory. His representation illustrating the action of subtraction—removing a part from a whole—showed not only *that* the result of 145 − 98 (the amount to the left of the solid line) is greater than the result of 145 − 100 (the amount to the left of the dotted line), but *why* it has to work out that way.

Although Brian's representation was labeled with specific numbers (98 and 100), his classmates interpreted it in general terms. Indeed, if we ignore the labels

"98" and "100," Brian's representation satisfies our three criteria for proof: (1) Subtraction is represented by removal of a part from a whole. (2) The whole can be any positive value, the subtracted amounts any positive values less than the whole. (3) The premise that two different amounts are subtracted from the same whole is shown, and the representation demonstrates that the conclusion must follow—when the lesser amount is subtracted, the amount remaining is larger.

Rebecca read the representation in these general terms and declared, "If you take away smaller, you end up with bigger." Max offered a more precise statement of the generalization: "The less you subtract, the more you end up with, and in fact the thing you end up with is exactly as much larger as the amount less that you subtracted."

However, Max's elegant statement was not the end of the lesson. After all, there were other children in the classroom with thoughts and questions of their own. Riddle writes:

> I asked if anyone else wanted to comment, and Riley raised his hand. Having experienced several moves in the past 2 years of school, he is a student who is caught between traditional procedures that he has trouble remembering and doesn't understand, and trying to catch up to his class-mates who are more accustomed to figuring out strategies for themselves. He often feels lost during math discussions, but today he seemed eager to get involved. "I used to think it was 43," he said, "but now since I saw it on the calculator and heard everyone talk, it's 47, but I don't get why."
>
> Often Riley was completely lost during these discussions, but this time I suspected he was actually at the brink of understanding. "Let's think about a different context," I suggested. "Pretend . . . you have 145 pennies, and, the first time I take 100 from you. Now go back to you have 145 pennies again, but this time I take 98."
>
> A huge smile broke across Riley's face. "Oh, now I see," he said happily, confirming my expectations. He added, "It's like you replay it in your mind, and now it makes sense." Like Brian, he now seemed to have a mental image that helped him solve the problem.

In this classroom episode, the impulse to prove grew naturally from the mathematical question these 4th graders posed to themselves. This episode illustrates that the work of proof need not be an add-on to an already full curriculum. Rather, it can support students' understanding of calculation and how the number system behaves under the four basic operations.

Furthermore, Riddle's example shows that the work of proof should not be reserved for those students who tend to excel in mathematics. In this case, the proof developed from whole class discussion, engaging students who represented the entire spectrum, from those who excel in mathematics to those who often struggle. Brian (who, Riddle later explained, more often was quiet during mathematics discussions) provided an image, but struggled with the language to articulate the idea it represented. Rebecca offered a straightforward verbalization of the generalization along with another image from students' everyday lives. Max articulated the generalization more precisely. These three students, and, presumably, other members of the class whose ideas were not recorded in the case, provided an argument in support of a claim that applies to an infinite class, i.e., they were engaged in the process of proof.

Riley, a struggling student, couldn't quite follow the discussion as it played out, but did have a sense that he could understand. After he asked for help, Riddle talked him through the ideas, now with a different context, which satisfied Riley. Although he may not have followed the ideas through to the generalization, he could make sense of why 145 − 98 must result in an answer 2 more than 145 − 100. That is, since Riley was still thinking in terms of specific numbers, he was not involved in proving the generalization. Yet, the whole class discussion was nonetheless productive for him, as well.

Proof as a Route to Conviction: An Example from a Grade 1–2 Classroom[5]

In solving particular subtraction problems, Margie Riddle's 4th graders had come across a question that confused them all—if you *decrease* by a certain amount a number you're subtracting, do you *add* or *subtract* that amount to/from the difference? Their proof not only showed them *that* you add that amount, but it also explained *why*.

But a proof can serve other functions, as well. For example, a proof might be used to convince someone else of a claim, even if you already believe its truth.

First and 2nd grade teacher, Karen Schweitzer, had a series of discussions with her students about what happens to the sum when the order of two addends changes (what they will later learn to call the commutative property of addition), and by now they were no longer counting to test it out; they were all convinced that the sum remains the same. But beyond that, when she asked her students to explain their thinking or to say why they felt so sure, they didn't have much to say. Schweitzer had a hunch there was more to be mined from this question. And so one day she returned to the idea that the order of the two addends doesn't matter, the total stays the same, and told her students they would spend time showing how they knew this was true.

But what could motivate this exercise? Unlike Riddle's 4th graders who had formulated their own puzzlement, Schweitzer's students were already convinced of their generalization; they felt no need to explore it further. And so Schweitzer called on another function of proof: to convince others of the truth of a claim. Schweitzer explained to her students:

[T]o prove it meant to convince someone. For example, if the principal came in and they had to convince him that what they said about the order was true. I told them they could use any tools they needed to help them explain and I listed cubes, diagrams and number lines as possible options. With that, I sent them off in pairs to work on their proofs.

We came back together as a group after about 15–20 minutes of the children working in pairs. Some children brought with them cubes or base 10 blocks, and some brought written work. I wanted to make sure they really had the idea of needing to convince someone. I thought this might help us get past the notion that we all know this so what else is there to say. I began the conversation with this reminder. "I want you to make sure that when you are doing your explaining, I want you to pretend that Mr. Valen is standing here and he doesn't believe you so you've got to be really careful to convince him. Don't assume anything. Say all the things you need to say and really convince him, like prove it."

I chose Kathleen to start us off because she is someone who often fits into the category of saying something is true but not being able to say why, but this time she had found a way to express why. She brought a clear diagram with her, in which she had drawn 10 cubes and 20 cubes, showing that they could be added in either order and the total remains the same [Figure 4.3].

Kathleen: If you take a 10 and a 20 and then you switched them around, it will just equal the same thing. You put the 10 in one place and then you put the 20, and you put the 20 where the 10 was.

Schweitzer posed several questions to Kathleen, asking her to point to her picture as she spoke, connecting the visual image to her words. Although the discussion had been particular to the image of 10 cubes and 20 cubes, soon Kathleen spoke in terms of a generalization: "Because it's like any way you switch it, it would just be the same thing. Like, any number you could probably do it with. You could probably do it with any number."

Although Kathleen left room for doubt ("You could *probably* do it with any number"), she now extended her thinking beyond 10 and 20 to consider *any* two numbers. Schweitzer emphasized this move by posing a question to the class [Figure 4.3]:

Teacher: So are you saying that you think it would work with any numbers? Like with any two numbers you could switch them and it won't change the answer?
Andrew: Any, any number.

Once the cubes in Kathleen's image moved from representing the specific numbers 10 and 20 to representing *any* two addends, it now satisfied the criteria for proof:

1. Addition is represented by the joining of two sets.
2. The representation can accommodate any two whole numbers.
3. The representation shows that switching the placement of the two quantities does not change the total.

When the question of the order of addends arises in other classrooms, students often produce an image similar to Kathleen's. As they exchange the position of

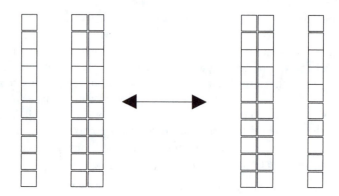

Figure 4.3 Kathleen's thinking on addends

the two sets, students explain, "You don't add any more on and you don't take any away, so the total stays the same."

Now, with this image solidly in mind, the students in Schweitzer's class were ready to extend their generalization:

Teacher: You are agreeing with any, any number you can do that?
Kirsten: Like 17 and 33.
Nathan: And you don't just have to use two numbers.
All of a sudden three or four children were talking at once. I tried to get all of the ideas heard.
Teacher: So Kirsten said you're just switching to a different spot. Molly, what did you say?
Molly: I said you didn't have to use only two numbers.
Teacher: You didn't have to use only two numbers.
Kirsten: I know you could use 10 and 5 and 3.
Teacher: Wow, you could use three numbers. Like 10 and 5 and 3?
Molly: You could use eight numbers.
Kirsten: It doesn't matter.
    At this point there were children joining in and talking all at once saying that you could use more than two numbers and in fact that it didn't matter how many numbers you used. Even a billion numbers!

The lesson had begun with a discussion of what happens when the order of two addends is reversed. Although the students in this class had already been convinced that the total remains constant, their teacher had insisted that they develop a proof—an argument that would convince the school principal had he been standing at the door and been doubtful. Whereas the students created demonstrations that necessarily represented specific numbers—10 and 20 cubes, tally marks, circles, etc.—these could stand for *any* two numbers: when the quantities are switched around, the total is unchanged. With that image in mind, the students extended the generalization even further: not only does each stick of cubes stand for any number, but the argument still holds no matter how many sticks you have. You can start with any number of addends—"Even a billion numbers!"—and changing the order will not change the sum.

Although several members of the class had enthusiastically reached this conclusion, Schweitzer felt there was more to explore. By continuing the discussion with different representations offered by the students, more children could be brought into the argument, and those who were already convinced could consider applications of this idea. As the class discussion continued, students looked specifically at the sum of 4 + 6 + 2—an easy problem for students who were fluent with combinations that made 10 and could add any one-digit number to 10. But now they understood that if they came across those same addends in a different order—4 + 2 + 6 or 6 + 2 + 4—they already knew the total without have to tackle a more challenging calculation.

Another group showed their representation of 140 with base 10 blocks—one flat of 100 and four sticks of 10:

Corey: You're just switching them around [switching the position of the blocks on the carpet] and not putting any more on or taking any away. You're not adding some or you're not taking any away. You're just switching them around and putting them in different spots.

Teacher: Okay.

Marissa: And you also can break the numbers up only if you don't take any or add any more on.

Teacher: Marissa, can you say more about that? You can break the numbers up as long as you don't . . .

Scott: Yeah.

Marissa: . . . take any away or add any more on.

At this juncture, the students were reinforcing the core idea behind their proof: when addition is represented as the joining of any number of sets, changing the placement of those sets—as long as nothing is removed and nothing more is added—does not change the total. Then Corey went on to show that the blocks could be arranged as 10 + 10 + 100 + 10 + 10, or as 10 + 10 + 10 + 10 + 100.

These students agreed that you can change the order of *any number* of addends of *any* size without changing the total, but now the base 10 representation moved them into decomposing multi-digit numbers and reordering the components. Joining the idea of reordering addends with place-value decomposition will be an essential skill in developing computational fluency with multi-digit numbers.

In this lesson, the teacher had challenged the class to *prove* a conjecture, a claim about addition, which students had already agreed on. By asking that they devise an argument to convince someone outside their community that changing the order of two addends does not change the total, the students created images of addition that not only accommodated addends of any size, but they saw that the same argument held for *any number of addends*—a new idea that delighted them. Then, with this new, extended generalization in mind, students and teacher considered applications to the work of their curriculum.

Some readers may question the notion that students were *proving* the generalization that two addends can be exchanged without changing the sum. After all, the commutative property of addition is an axiom, accepted without proof. However, these 1st and 2nd graders are not yet operating within this axiomatic system. As they are coming to understand what the operations are, these properties may still be in question. In a later grade, students working on multiplication may notice that 3×4 and 4×3 give the same result, as do 2×5 and 5×2, and wonder whether "switch-arounds" (as many children call them) work here, too. Again, in the context of the knowledge and experience of these elementary grade students, such a conjecture—when the order of two factors is changed, the product remains the same—asks for proof (cf. Brown, 1996).

Furthermore, once Schweitzer's students presented their argument for two addends, they realized that the same representation—enhanced by their imaginations that could see their cube towers representing addends of any number *as well as any number of addends*—offered proof of a more general claim. Indeed, their second claim—that the order of any number of addends can be changed without changing the sum—is one that mathematicians would prove.

Conclusion

When elementary classrooms are structured around student thinking and students are invited to share their ideas, they frequently offer observations of regularities they notice in the number system. When these observations are generalizations that apply to an infinite class, teachers (and students) may ask: are we sure this is *always* true? How do we know? In this way, students are challenged to consider how to argue for the truth of a claim about an infinite class. At a young age, they already have the opportunity to learn that although it is insufficient to test specific instances (even lots of them!), a claim about an infinite class nevertheless *can* be proven. We have found that representation-based arguments provide a mechanism that is accessible and powerful for young students to engage in the process of proof.

The episodes chosen for this chapter illustrate what this process can look like in the elementary classroom. The claims proved in these examples are:

The sum of two even numbers is even.
The less you subtract, the larger the difference.
The order of addends does not affect the sum.

The cases written by teachers in our project show students proving a wider set of claims. For example:

The sum of two odd numbers is even; the sum of an odd and an even is odd.
The order of factors does not affect the product.

In an addition problem, if you subtract a certain amount from one addend and add it to the other, the sum remains constant.

In a subtraction problem, if you subtract (add) the same amount from (to) both numbers, the difference remains constant.

If you double one factor and halve the other, the product remains constant.

The factor of a number is also a factor of that number's multiples.

In a multiplication problem, you can decompose one factor, multiply the parts by the other factor, and add the subproducts.

The goal of having students prove such claims is not merely one more piece of content for teachers to squeeze into an already full agenda. Rather, as the examples presented in this chapter illustrate, the challenge to prove generalizations about numbers and operations engages students in many facets of mathematical proficiency (Kilpatrick et al., 2001). Through these discussions, students develop richer understandings of the meaning of the operations and their relationships, which in turn support greater flexibility with computational procedures.

Neither is the challenge to prove an enrichment activity reserved for a single category of student. As one teacher wrote: "When I began to work on generalizations with my students, I noticed a shift in my less capable learners. Things seemed more accessible to them." When generalizations are made explicit—through language and through visual representations used to justify them—they become available to more students and provide foundation for greater computational fluency. Furthermore, the habit of creating a representation when a mathematical question arises supports students in reasoning through their confusions. They come to see mathematics as sensible and develop confidence in their own efficacy.

At the same time, students who generally outperform their peers in mathematics find this content challenging and stimulating. The study of number and operations extends beyond efficient computation to the excitement of making and proving conjectures about mathematical relationships that apply to an infinite class of numbers. As one teacher explained, "Students develop a habit of mind of looking beyond the activity to search for something more, some broader mathematical context to fit the experience into."

Finally, at the same time that such an approach to proof deepens students' understanding of the number system, it engages them in a process central to mathematics. As they experience a variety of modes of coming to believe the truth of a claim, reasoning becomes the standard for its acceptance. They learn to privilege this kind of reasoning over appeals to authority or testing instances.

Notes

1.   The ideas expressed in this chapter have been developed through collaborations with Virginia Bastable, Susan Jo Russell, Steve Monk, and groups of teachers. Acknowledgment is also owed to the Professional Development Study Group, the DMI Advisory Board, the *Investigations* Advisory Board, and Alan Schiffmann for their helpful feedback and suggestions.

2.   This work was supported by the National Science Foundation under Grant Nos. ESI-9254393 and ESI-0242609 awarded to Deborah Schifter at the Education Development Center and

Grant Nos.ESI-0095450 and ESI-0550176 awarded to Susan Jo Russell at TERC. Any opinions, findings, conclusions, or recommendations expressed in this chapter are those of the authors and do not necessarily reflect the views of the National Science Foundation.

3.   This work informed the revision of the K–5 curriculum, *Investigations in Number, Data, and Space*, as well as the development of two modules of the *Developing Mathematical Ideas* professional development series.

4.   We do not claim that representation-based proof is the only form of proof that appears in the elementary grades. For example, students may argue from generalizations that have already been established by the classroom community.

5.   A version of this classroom episode appears in Schweitzer, 2006.

5

Representations that Enable Children to Engage in Deductive Argument

ANNE K. MORRIS[1]

How can we develop children's ability to engage in deductive argument at the elementary school level? This chapter focuses on the critical role of representations in developing children's ability to engage in deductive argument. It describes one type of representation that may make deductive argumentation more accessible to elementary school students. These representations are schematics, or diagrams, that show mathematical actions and relationships.

A Dilemma

"Teaching proof" at the elementary school level is especially challenging. It presents us with a dilemma in that children's ideas about mathematics are frequently rooted in their actions on the physical world. For example, children develop ideas about part–whole relationships by taking objects apart and putting them together again. Children develop ideas about the operation of addition by joining physical amounts. Elementary school teachers build on these informal conceptions by using concrete objects to represent numbers and quantities. Actions are carried out on the concrete objects to develop mathematical ideas and the actions are represented symbolically. For example, teachers and students represent the addition of five blocks and three blocks as $5 + 3$, or the addition of one-third of a circle and one-eighth of the circle as $1/3 + 1/8$. These representations refer to a particular case—to a particular action on particular objects.

In contrast, proving frequently requires representing, not a particular case or a particular action, but a generalization that applies to all objects of a given kind. What kinds of representations (e.g., diagrams, symbolic expressions) would help children to think about, make, and prove generalizations—in particular, a generalization that a particular type of action can be performed on all objects of a given kind, or that a relationship holds for all numbers or quantities of a given kind? How can we help children move from representations of actions on particular objects to representations of generality? How can teachers help children "see" and represent the general relationships and types of actions that are often at the root of deductive argument and that allow them to reason deductively about a class of objects?

Representing Generality

The challenge of developing representations that support children's reasoning about general classes of objects has been addressed explicitly in an elementary school mathematics curriculum developed by Davydov and his colleagues in Russia (e.g., Davydov, Gorbov, Mikulina, & Savel'eva, 1994). (This curriculum will be referred to as the "Russian curriculum.") Because this curriculum, and the conceptual arguments that lie behind it, focus intentionally on facilitating children's deductive reasoning and because they have not been disseminated widely for American audiences, it is worth describing briefly the pedagogical approach taken by this curriculum along with sample representations that provide a primary tool for children's reasoning. My aim in this chapter is not to prove that this approach is superior to other approaches, although the data are promising (e.g., Morris & Sloutsky, 1998), but rather to introduce the ideas and the representations for further consideration and study in American contexts. Given the current attention to alternative curricula for school mathematics, it is useful and timely to consider an approach that is different from most current standards-based curricula in its explicit attention to deductive reasoning (e.g., Laudien, 1998).

A primary goal of the pedagogical approach developed by Davydov and colleagues is to help children to recognize, apply, and deduce generalizations. In this approach, representations are used to help children connect their ideas about actions on particular quantities to concepts about actions and relationships that apply to classes of objects. For example, children are provided with representations that help them move from, and relate, the action of taking physical objects apart and putting them together again, to the concept of a part–whole relationship between any quantities of a given kind. The representations may also help children to use the abstract relationship as the basis for a deductive argument.

In this approach, curriculum developers and teachers ask the following question: what actions on the world underlie the individual's conception about a particular mathematical relationship? A representation is then designed that shows both the action and the relationship. The representation allows the child to initially create meaning for the representation based on his/her actions on particular quantities; i.e., when the child looks at or uses the representation, it initially represents his/her actions in the world. However, when the child is ready, the representation can be given additional interpretations: it represents an action that can be performed on any objects for which this kind of action is applicable, and it also represents a relationship between quantities. Depending on one's interpretation, the representation shows an action on particular concrete objects, an action on any objects of a given kind, a relationship between particular concrete objects, and a relationship between any objects of a given kind.

Example 1: The Part–Whole Schematic

An example of this type of representation from the Russian curriculum and its potential uses in deductive argument will now be described. In grade 1 of the curriculum, children have many experiences that are designed to develop their understanding of part–whole relationships. For example, they pour water from a single container into two or more containers, and pour separate containers of water into a single container. They cut a paper rectangle into parts, or put the parts together to form a whole rectangle (e.g., Mikulina, 1991). The children represent their actions on concrete objects with a special "schematic." The quantity that is the whole is designated with a letter—for example, A. The parts are also designated with letters—for example, B and C.[2] It should be emphasized that the letters do not represent numerals. Rather the letters represent physical amounts that are not numerically quantified, such as amounts of water or sand, the area of a piece of paper, the length of a ribbon, and so on. The children represent their action of breaking the whole, A, into its parts, B and C, with an "inverted V" schematic, as shown in Figure 5.1A.

The schematic helps children recognize and focus on the nature of their actions on physical amounts in the world. If one moves along the schematic from top to bottom, the schematic itself suggests the action of separating the whole into its parts. If one moves along the schematic from bottom to top, the representation depicts the action of putting the parts B and C together to form the whole, A. Thus there is a close correspondence between the child's actions and the representation of this action. In addition, when the children initially use this representation, it is always accompanied by actions on quantities. For example, the children separate a pile of sand, A, into two amounts of sand, B and C, and they represent their actions with a schematic like that in Figure 5.1(A). The teacher sometimes presents children with this type of schematic, and children have to illustrate this action with specific concrete objects of their choosing. Both of these factors—the close correspondence between the nature of the child's

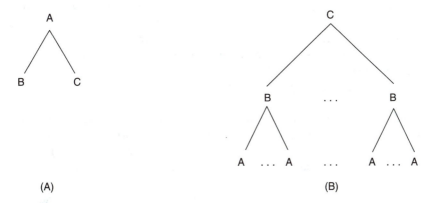

(A)

(B)

Figure 5.1 Breaking the whole, A, into its parts, B and C, with an "inverted V" schematic

actions and the nature of the schematic, and the initial use of the representation in conjunction with children's actions on specific concrete quantities—appear to make it a developmentally appropriate representation and provide multiple opportunities for the child to create meaning for the representation through actions on concrete objects.

Although the schematic is initially viewed as a representation of an action on particular physical quantities, it also focuses attention on the depicted actions themselves. The schematic shows the relationship between the actions of joining and separating, and the idea that the actions are reversible and one action can "undo" the other. The picture of these actions and the relationship between them helps children to study the characteristics of these actions, and to deduce other related facts. For example, the children use the schematic, and concrete materials and contexts, to figure out the following facts: if I know the size of the parts and I want to find the size of the whole physical quantity, I need to join the parts. If I want to find the size of a part, I need to take away the known part from the known whole to find the size of the remaining quantity.

The schematic also helps children to move to a discussion of "any quantity." The children gradually move from actions on particular quantities to treating the representation as a picture of an action that is applicable to any quantities that can be manipulated in this way. A, B, and C can be any quantities where this kind of action is applicable. The representation becomes meaningful to children because it represents their actions on objects, but after numerous experiences with concrete objects they understand the meaning of the representation and they do not need to perform actions on concrete quantities to understand its meaning. The children begin to talk about *a* quantity B or *any* quantity B, rather than *this* quantity B (Lee, 2002). The representation can potentially help children move from actions on particular quantities to thinking about the characteristics of this action for "any" quantities.

What is unique about this representation is that, although it can be interpreted as, and suggests an action, it can also be interpreted as showing the relationship between A, B, and C. Children can interpret it as an action, a relationship, or both. Thus the representation represents the action that is the source of children's conception of the relationship—it is the action of separating a quantity or joining quantities that is the source of one's conception of part–whole relationships, and the representation shows both—but it simultaneously shows the relationship between these quantities. That is, A is the whole comprised of two parts B and C. The children also go on to study the properties of part–whole relationships using this schematic along with concrete physical quantities. For example, they learn that a whole is bigger than its parts. Children also learn to use part–whole relationships as a tool for solving problems by repeatedly identifying and representing part–whole relationships in a wide range of contextualized problem situations (described later).

Representations like the part–whole schematic also provide a bridge to the world of algebra. With the aid of the schematics, children can learn to express generalities algebraically. For example, 1st grade students in the Russian curriculum use the part–whole schematic and concrete materials to figure out how to write algebraic statements that represent actions on physical amounts:

B + C = A (adding the parts makes the whole)
A – B = C (the whole minus the part equals the other part)
A – C = B (the whole minus the part equals the other part)

Again, the letters represent unmeasured physical amounts, not numbers. Children create meaning for these statements on the basis of their actions on physical quantities. For example, they take a "whole amount of water," A, and pour a portion, B, into a container; what remains? (the other part, C). Although these equations initially refer to particular physical quantities, children eventually move away from performing actions on quantities without losing the physically based meaning of these equations (Lee, 2002; Mikulina, 1991).

Using the Part–Whole Representation to Prove

How can this type of representation of actions and relationships help children to engage in proving? Consider the following classroom situation. A group of elementary school students are asked to prove the following conjecture: if $a|b$ and $b|c$, then $a|c$. Ben, a 3rd grader, lays out 120 snap cubes (Schifter, Bastable, & Russell, 2008a). He then partitions them into 15 groups of eight cubes, showing that 120 is divisible by 8. He then shows that each group of eight cubes can be partitioned into two groups of four cubes, four groups of two cubes, or eight groups of one cube. Ben uses his snap cube model to support his argument that all the factors of 8 are also factors of 120. Ben has created a clever representation, but it refers to a particular case.

The type of representation that is described in this chapter suggests *one* way to help children move from arguments involving numeral-based arguments (like that offered by Ben), to making and representing general deductive arguments. Children can be provided with instructional experiences and representations that allow them to represent the idea that an action (in this case, the action of breaking a quantity into its parts) is indeed applicable to an infinite set of similar objects, to represent a relationship among quantities as well as actions on them, and to represent a class of objects rather than a single case. The representation could help this child link his actions on particular quantities to the idea of a more general relationship that explains why the factors of the factors of any counting number are also factors of the number. The child's thinking can be represented in a more general way with part–whole diagrams (see Figure 5.1(B)).

The diagram shows the action of breaking any measured (i.e., numerically quantified) or unmeasured quantity C into its factors or parts. The child's reasoning can be represented in a more general form: if a quantity of size C can be separated into a number of equal parts of size B, and each part of size B can be separated into a number of equal parts of size A, then it must be the case that the quantity C can be separated into a number of equal parts of size A. While the schematic shows the action of breaking the quantities into parts, it also shows the relationships among them. It also moves beyond Ben's representation because A, B, and C can represent *any* quantities where a part–whole relationship is applicable. It is a step toward representing generality, and using general relationships as the basis for a deduction.

By identifying, and developing children's understanding of, the general relationships that are fundamental to mathematical argument, children's deductive argument making can be extended to several content domains, and can increase in sophistication as their mathematical knowledge develops. For example, in the Russian curriculum, students learn how to write algebraic equations involving multiplication using part–whole diagrams. The part–whole schematic in Figure 5.2(A) shows m parts of size A, and students learn to represent this action/relationship as $B = m \times A$.

With this development, students could potentially write a more sophisticated argument for the conjecture, "If $a|b$ and $b|c$, then $a|c$," with the aid of a schematic. Ben's argument can be elaborated (see Figure 5.2b):

$B = m \times A$ (m = the number of parts)
$C = n \times B$ (n = the number of parts)
Therefore $C = n \times m \times A$

The schematic can potentially help students to explicitly recognize, represent, and use part–whole relationships to make general deductive arguments.

In summary, an intriguing characteristic of this type of representation is that it is concrete and accessible to children because meaning for the schematics is developed from actions on concrete quantities. However, the representations have the potential to lead to discussions of "any." They allow a child to represent the

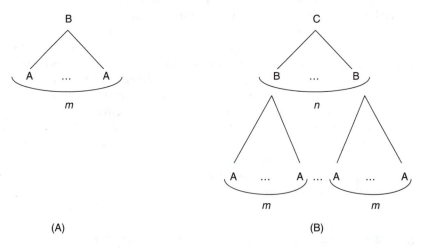

(A)   (B)

Figure 5.2 m parts of size A; students learn to represent this action/relationship as $B = m \times A$

idea that an action, a relationship, or a deduction can apply to any number or quantity, something that can be difficult to convey and represent with concrete materials and particular cases. The schematics are also sensitive to developmental considerations because they are a dual representation of an action and an abstract relationship, where the action is the one that underlies the individual's conception of the relationship. In this way, children can move from understandings about actions on particular quantities to understandings about actions and abstract relationships that are applicable for all quantities of a given kind.

What changes over time for children is the meaning of the letters—from particular concrete quantities to "a quantity" or "any quantity"—and the dual interpretation of the schematic as an action and a relationship. Students can then use the representation to make deductions that hold for all quantities of a given kind and can use it as a bridge to writing algebraic expressions and equations, and using algebra to prove.

From Quantitative Reasoning to Arithmetic Reasoning

One feature of the Russian curriculum that is associated with the schematics is unique and warrants further explanation. Understandings about the properties of mathematical actions and relationships are first developed in a non-arithmetical context. These understandings are initially developed on the basis of children's actions on physical amounts—physical amounts that are not described numerically by the teacher or the students. The children then apply these ideas in numerical contexts. The curriculum develops the idea that the relationships that are depicted by the schematics hold, not only for physical quantities, but for their numerical measurements as well (Mikulina, 1991).

In most cases, this instructional sequence is repeated for each new topic. For example, part–whole relationships are first explored in the context of unmeasured physical quantities. Letters are used to represent unmeasured physical amounts. After this initial work with physical quantities, students are asked whether numbers can be divided into two parts, and use part–whole diagrams to show, and explore, part–whole relationships in numerical contexts (e.g., 4 and 2 are the parts, and 6 is the whole). At this point in the instructional sequence, letters may also be used to represent numbers.

Part of the rationale for this instructional sequence is based on the observation that children frequently fail to develop explicit understandings about the properties of mathematical relationships and actions from their experiences with numbers and arithmetic (see, for example, Kieran, 1992). In their arithmetical work, children's attention may be directed toward computing and "finding the answer," rather than mathematical structure. The approach taken in the Russian curriculum develops these ideas from a different source (children's actions on physical quantities). The curriculum then helps children transfer these ideas to arithmetical and numerical contexts.

Moving from general ideas about mathematical relationships and actions to numbers turns the usual sequence on its head. Instead of beginning with number and helping children to generalize from specific numerical cases to classes of objects and the properties of mathematical actions and relationships, the Russian curriculum begins with the latter and then treats number as a special case. As children identify, represent, and use general relationships and the properties of mathematical actions in the context of number and arithmetic, they are expected to develop a richer understanding of the ideas. Children's explicit understandings about mathematical structure are intended to guide their observations and learning in the domain of arithmetic.

Another example of the general to specific sequence is the following. After children learn to identify part–whole relationships between physical amounts, children analyze story texts in terms of part–whole relationships. Texts such as the following may be presented for analysis: *there are Q pounds of apples in a basket. Z pounds are added, after which T pounds of apples are in the basket.* Children are asked to analyze the relationship between the quantities and to make a part–whole schematic that shows the relationship. (In this story, T is the whole, and Q and Z are the parts.) Students are also presented with schematics, and asked to make up their own story texts for the schematics.

Later in the instructional sequence, one of the quantities in a story text is made into an unknown, which converts the story text into a story problem. Children may be given a problem such as the following: m children were in the park when some more children ran into the park to play. Then there were t children at the park. How many children ran into the park? The children write the equation that describes the actions of the problem: $m + x = t$. In order to analyze the part–whole relationship in the problem, children draw a part–whole schematic. This enables them to identify the missing quantity as a part or a whole, and to determine the appropriate arithmetic action for solving the problem. Since the missing quantity is a part, they need to take away the known part from the known whole to find the size of the remaining quantity. Thus, they write the equation $x = t - m$ as the solution. Story problems with numbers are also presented at this point in the curriculum (e.g., Jim has some cards. His brother gives him six more cards. Now Jim has 11 cards. How many cards did Jim have in the beginning?) The children carry out the same type of analysis for the numerical problems (see, for example, Schmittau & Morris (2004)).

In summary, the Russian curriculum uses actions on unmeasured physical amounts as the starting point to develop children's ability to recognize, represent, and use general relationships and the properties of mathematical actions. These understandings then are extended and enriched by applying them to the specific case of number and arithmetic. (See Carpenter, Franke, & Levi, 2003, for an approach that uses children's arithmetic reasoning as the starting point to develop these competencies.)

Example 2: The Multiplicative Schematic

To elaborate the schematics in the Russian curriculum it is useful to consider a second schematic for multiplicative relationships between quantities. This schematic is first introduced in the following form (see Figure 5.3(A)) in grade one of the Russian curriculum (Davydov et al., 1994).

When the schematic is first introduced, children interpret it as meaning, "repeat or lay off the physical quantity B four times to make (or build) the quantity C." In this type of schematic, the arrow is always drawn from a smaller quantity to the bigger one. As with the part–whole schematic, this schematic initially represents children's actions on quantities: the tallies indicate the action of laying off the quantity B four times (one tally for each action). Later in the 1st grade curriculum, the concept of number is introduced. The schematic is changed to reflect this new idea (see Figure 5.3(B)): the tallies (e.g., "I I I I") above the arrow are replaced by numerals (e.g., "4").

In grade 1, this schematic is used to develop ideas about the action of constructing (or building) a physical quantity from another physical quantity, comparing quantities, measuring a quantity with another quantity, and number. In grades 2 and 3, the same schematic is used to develop ideas about multiplicative relationships between quantities. For example, Figure 5.3 shows that B < C by four times and C > B by four times: if B must be repeated or laid off four times to construct C (an action), then B is "four times smaller" than C, and C is "four times bigger than B" (a relationship).[3]

Children first carry out this type of action in a series of problems. The schematic is introduced to record their actions. After the introduction of the schematic, children solve various problems involving the schematic. Children might be presented with this schematic, where the letter "C" has been replaced by a question mark, along with a physical quantity labeled "B." B might be the length of a line segment, three circles, five squares on grid paper, two cups of water, and so on. The children have to build, or make, the unknown larger quantity by repeating or laying off B four times.

In another type of problem, children are presented with the schematic where the letter "B" has been replaced by a question mark, along with a physical quantity labeled "C." C might be a line segment that is 12 squares long on grid paper, for example. The children have to find the unknown smaller quantity: what type of action is required? Children figure out that they need to divide the physical quantity C into four equal parts in order to find the smaller quantity (since the schematic indicates that the unknown smaller quantity was repeated four times to build C). Alternatively, children might be presented with the schematic where the tallies have been replaced by a question mark, along with a physical quantity labeled C (e.g., a larger area on grid paper), and a physical quantity labeled B (a smaller area on grid paper). To solve the problem, the children might build the

Figure 5.3 Multiplicative schematic in the Russian curriculum

quantity C by laying off or repeating B; they write a tally or some other equivalent notation to record each copy of B (one tally drawn for each action of laying off B).

Again, the schematic can be interpreted as an action or a relationship: in addition to the action, it can be interpreted as showing the relationship, "B fits into C four times, or C is four times bigger than B, or B is four times smaller than C." Again, this schematic can be interpreted as referring to a specific quantity, or *any* quantity. For example, the schematic in Figure 5.3(B) can be interpreted as showing the relationship between two specific physical amounts where C is four times bigger than B, or B and C can be interpreted as any quantities where this kind of action and relationship is applicable.

Notice again that this kind of representation also depicts the relationship between types of actions. The schematic depicts the action of repeating a smaller quantity *n* times in order to build or find a larger one. However, if I begin with the larger quantity and I want to find or make the smaller one (i.e., if I want to move the other way along the schematic), what type of action should I carry out? If I have to copy or repeat the smaller quantity *n* times to make the larger one, how can I begin with the larger quantity and find the small one? How can I "undo" my action? I can break the larger quantity into *n* equal parts—that is, the number of repetitions. The size of one of the equal parts must be the quantity that I repeated. Thus, the schematic shows the relationship between the actions of repeating and partitioning (repeating a smaller quantity to build or find a larger one and partitioning the larger quantity into equal parts in order to make or find the smaller one). Understandings about actions that can be applied to classes of objects are often required in deductive argument, as well as understandings about related actions that can "undo" these actions.

Using the Multiplicative Schematic to Prove

One of the intended benefits of beginning with reasoning about general cases and then applying this form of reasoning to specific cases is to prevent the common misconceptions that frequently arise when students try to generalize from specific cases to general cases. A classic misconception that develops about the notion of "mathematical proof" is to assume that collections of instances of a phenomenon constitute a proof. Consider the findings of Lee and Wheeler (1987), who gave the following problem (the "girl problem") to 118 10th graders:

A girl multiplies a number by 5 and then adds 12. She then subtracts the original number and divides the result by 4. She notices that the answer she gets is 3 more than the number she started with. She says, "I think that would always happen, whatever number I started with." Using algebra, show that she is right.

Twenty-six of the 118 students tested a few numbers and based their conclusion on the numerical results; if the answer were 3 more than the original number in each case, they concluded the generalization was true. Sixteen students set up the expression $(5x + 12 - x)/4$, but then used the expression only as a formula for creating and testing numerical examples; they again based their conclusion on the numerical results. Thirteen set up the equation $(5x + 12 - x)/4 = x + 3$ and then substituted numerical values for x to demonstrate this equality. Only nine of the 118 students set up the expression $(5x + 12 - x)/4$ and then algebraically simplified it to $x + 3$. Four of the nine then went on to "demonstrate further" by substituting numerical values for x to see if they did get 3 more.

Several studies have yielded similar findings. The studies suggest students often interpret or use a large number of supporting numerical instances as conclusive proof, and cannot represent and use mathematical relationships and the properties of operations in order to prove a generalization (e.g., Bell, 1976; Chazan, 1993; Fischbein & Kedem, 1982; Martin & Harel, 1989; Morris, 1999, 2002; Porteous, 1986; Williams, 1980). For the girl problem, the results suggest the majority of the 10th graders did not understand that they could use the multiplicative and additive relationships involved in the problem to establish a general relationship between the original and final numbers.

One central assumption of the Russian curriculum is that reasoning about non-specific physical quantities first—by acting on quantities and recording the actions symbolically with special schematics—helps students to recognize and deduce generalizations about specific quantities represented as numbers. This sequence is intended to intentionally develop their ability to recognize, use, and deduce general relationships. A brief description of one case of using this curriculum in an American school in New York (Lee, 2002) illustrates the way in which richer forms of deductive reasoning might develop.

Lee (the teacher) worked with six elementary school students for approximately 3 years, completing the grade 1–3 curriculum.[4] Lee presented children in grade 3 with the task of creating expressions for the missing quantities in the schematic displayed in Figure 5.4.

The left-hand portion of the picture is the type of schematic just discussed: the "t" above the first arrow indicates a multiplicative relationship; the quantity in the first box (i.e., a) is t times bigger than the quantity in the middle box (i.e., ?). The "by b" above the second arrow indicates that the quantity in the right box is bigger by b than the quantity in the middle box (that is, the expression "by b" denotes an additive relationship). Lee reports that the children easily decided that the missing number in the middle box was $a \div t$, and the missing number in the right box was $a \div t + b$.

The children's ability to solve this type of problem is based on their earlier work involving actions on physical amounts—their actions of repeating and

Figure 5.4 Creating expressions for missing quantities in a schematic

partitioning physical amounts and representing these actions symbolically. For example, children can reason in the following way: if I have to repeat an unknown smaller quantity t times to build a, then I will have to divide a into t equal parts in order to find the unknown smaller quantity. Therefore the unknown quantity is $a \div t$. Notice that the schematic shows the actions involved but it also shows the relationship between the resulting quantities.

Once again, there is a gradual movement toward talking about "any." For example, Lee describes the following debate involving a problem of this type in grade 3. Lee presented the following schematic to the children (see Figure 5.5).

This schematic means that the second quantity from the left is 2 times bigger than a, and larger by 5 than the third quantity from the left. The quantity on the right (i.e., ?) is 3 times bigger than the quantity to its left. Therefore students should write the following answers in the second, third, and fourth boxes respectively (moving left to right): $a \times 2$, $a \times 2 - 5$, and $3 \times (a \times 2 - 5)$.

The children solved the problem individually and then discussed it. They decided the final answer (i.e., ?) was "$a \times 2 - 5 \times 3$." Disturbed by the children's failure to use parentheses, the teacher asked the children if they could replace the letter "a" with a number, and if their conclusion would hold in this specific case. Lee reports, "All the children answered with confidence, 'Sure.' They said that we could do so because the letter 'a' could be any number" (Lee, 2002, p. 134). The teacher then replaced "a" with "5" and asked the children for the final answer in this case. One child concluded, "It doesn't make sense. It is 10–15 [and the answer should be 15]." Another child, Chris, concluded, "It happened because you chose a wrong number. The letter 'a' must not be 5." The class rejected the explanation. One child explained, "The letter 'a' can be any number. It can be 5. It doesn't matter." Chris then withdrew his explanation. The children eventually decided that parentheses were needed. One child explained, "We have to multiply the whole thing, $a \times 2 - 5$, by 3. Without this we can multiply only 5 by 3, and it is not the way the schematic represented" (p. 135). The class subsequently changed their general answer to $(a \times 2 - 5) \times 3$.

This discussion shows the children interpreted a variable as "any number," but it also shows their ability to focus on the general relationships involved in the problem (in this case, the multiplicative and additive relationships involved in the solution of the problem). Both are important competencies involved in deductive argument.

To reiterate a central claim, by developing children's understanding of the general relationships that are fundamental to mathematical argument, children's deductive argument making can be extended to several content domains, and can increase in sophistication as their mathematical knowledge develops. Returning

Figure 5.5 Solving a problem in a schematic

to the girl problem, it appears the 3rd grade children in Lee (2002) have developed many of the component understandings that would allow them to represent and make sense of the problem.

Using the schematics for multiplicative and additive relationships, the problem can be represented as seen in Figure 5.6 below.

Because the children interpreted a letter as "any number," it is reasonable to hypothesize that they will interpret the relationship between x, and the resulting expression, $x + 3$, in the rightmost box, as holding for any particular case.

Can Young Children Engage in Deductive Argument?

This chapter has assumed that it is developmentally appropriate, and important, to begin working toward deductive argumentation in elementary school. Is it appropriate to engage children in this form of argument? Why aren't numerical inductive arguments (like those observed by Lee and Wheeler (1987) on the girl problem) sufficient at the elementary school level?

Elementary schoolchildren show some competence with respect to deductive reasoning, and many findings suggest that it is appropriate to build on, and develop, this competence during the elementary grades. Children, even at the preschool level, perform very well on simple deductive inference tasks (e.g., Moshman, 2005). When young children are presented with premises representing many logical domains (e.g., class logic, propositional logic), they commonly reach the same correct deductive conclusions reached by adults (e.g., Braine & Rumain, 1983; Hawkins, Pea, Glick, & Scribner, 1984). At around age 6, children begin to recognize that inferences are a potential source of knowledge (Pillow, 1999; Sodian & Wimmer, 1987). Children commonly begin to develop a more explicit, metacognitive awareness of their logical reasoning, and explicit understandings about the distinction between logical and nonlogical arguments, at around age 11 to 12 (Morris, 2000a; Moshman & Franks, 1986). Elementary schoolchildren make deductive arguments in academic domains other than mathematics (e.g., Anderson, Chinn, Chang, Waggoner, & Yi, 1997). With appropriate instructional support, some elementary schoolchildren can begin to use and/or understand general forms of argument in mathematics at around grades 3 or 4, and can begin to expand single-case numerical arguments into general arguments (Ball & Bass, 2000, 2003; Carpenter, Franke, & Levi, 2003; Schifter, Bastable, & Russell, 2008b). Many elementary schoolchildren can learn to use literal symbolism to represent general relationships and the properties of mathematical operations, and can interpret a variable as "any number" (Carpenter et al., 2003; Lee, 2002; Kaput, Carraher, & Blanton, 2008; Morris & Sloutsky, 1998). American elementary schoolchildren have shown an ability to use the schematics in the Russian

Figure 5.6 Using schematics for multiplicative and additive relationships

curriculum to represent general mathematical relationships, and an ability to use the schematics to represent and deduce mathematical ideas that they have not been exposed to in prior instruction (Dougherty & Slovin, 2004; Lee, 2002; Morris, 2000b).

There is evidence that elementary schoolchildren's ability to make general mathematical arguments improves when they are given appropriate instructional support (e.g., Ball & Bass, 2000; Carpenter et al., 2003; King, 1973; Maher & Martino, 1996a), and that the majority of middle and high school students do not develop an ability to form or understand deductive mathematical arguments unless they are provided with appropriate instructional experiences (Kieran, 1992). Moreover, if children's deductive reasoning competence is not developed and children are encouraged to rely on inductive mathematical arguments throughout elementary school, they may develop the belief that inductive arguments in mathematics are logically valid. Adolescents and adults commonly appear to believe that a large number of numerical instances of a generalization prove the generalization (e.g., Lee & Wheeler, 1987; Martin & Harel, 1989). Morris (2007) found this belief can interfere with adults' ability to apply their understandings about indeterminate premise–conclusion relationships. These findings suggest that it is beneficial to develop ideas about mathematical justification that are more compatible with the nature of mathematical proof, from the very beginning of instruction.

Conclusion

The type of diagram or schematic that is described in this chapter has the potential to help children *explicitly* recognize and use the general relationships that underlie many deductive arguments. Such representations also have the potential to help children understand the nature of, and the relationship between, particular types of action that can be performed on all objects of a given kind. They can help children move from particular cases to talking about "any" (Lee, 2002). They simultaneously appear to help children create meaning for algebraic representations, another tool for representing general relationships and for proving (Mikulina, 1991; Morris & Sloutsky, 1998). With these "interim" representations, students have additional tools to explicitly represent the general, and consequently, to engage in deductive argument.

Given the plausibility of the arguments that support the use of such representations, the value attached to developing students' deductive reasoning abilities, and the promising preliminary data, it is worth considering how such representations might be used in an American context. It is likely that the schematics presented in this chapter are only illustrations of a range of schematics that could be designed. Researchers, curriculum developers, and teachers could stimulate this work by thinking about powerful "interim" representations that suggest both an action on, and depict a relationship between, particular quantities or all quantities of a given kind. They could identify actions on physical quantities that help children "see" general, fundamental mathematical relationships. This could lead to designing representations that suggest both an action on physical quantities and a relationship between the physical amounts. Testing the effects of using such representations with appropriate age students could define a productive research agenda.

Notes

1. Preparation of this chapter was supported by the National Science Foundation (Grant #0083429 to the Mid-Atlantic Center for Teaching and Learning Mathematics). The opinions expressed in the chapter are those of the author and not necessarily those of the Foundation.
2. Literal symbols are introduced very carefully, in a sequence of instructional steps (see Davydov, 1975).
3. In English, the terms "bigger than" and "smaller than" should only be used for additive comparisons; it is incorrect to say "C is four times bigger than B" or "B is four times smaller than C," as I have written here. Instead I should write "C is four times as much as B" or "B is one-fourth as much as C." I have used the incorrect expressions for several reasons. First, in Russian, "bigger than" can be applied for addition and multiplication. For example, "8 на 2 больше чем 6" literally means, "8 is on 2 bigger than 6" and "8 в 4 раза больше чем 2" means, "8 is in 4 times bigger than 2." Second, the "n times smaller" expression avoids the use of fraction language prior to the introduction of fractions. Third, the similarity of the terms "n times bigger" and "n times smaller" appears to make the reversibility of the two related actions more apparent—i.e., the action of repeating a smaller quantity n times to make the larger one, and the action of partitioning the larger quantity n times to find the smaller one. The lack of similarity in the terms, "n times as much" and "$1/nth$ as much," may make this relationship less apparent.
4. The children were 7 to 8 years old when they started the 1st grade curriculum.

6

Young Mathematicians at Work

The Role of Contexts and Models in the Emergence of Proof

CATHERINE TWOMEY FOSNOT AND BILL JACOB

Professional mathematicians acquire their understanding of what proof is through participation in their community, and to them it involves rigorous reasoning, without gaps, that establishes the validity of a mathematical statement based on clearly formulated assumptions. Although courses on proof have emerged in the undergraduate curriculum in recent decades, for the most part, mathematicians have developed their conception of proof in settings where their early attempts were questioned: "How did you get from here to here?" or "I believe this, but why is your next statement true?" What then follows is an act of reflection, a reorganization of ideas, and hopefully reasons that "fill the gaps" so a proof can emerge. Mathematicians then engage in this process throughout their careers. In this chapter, we explore the emergence of a small piece of this development as 2nd graders (7 year olds) and 5th graders (10 year olds) participate in "communities of discourse" attempting to convince their classmates of the truth of particular mathematical statements, and as they solve for unknowns. We look for evidence of reflection and reorganization of ideas about the connections between mathematical statements in elementary classrooms, in contexts that raise issues of equivalence and rules for obtaining new equivalences from previously given or established equivalences.

As opposed to argumentation,[1] which relies on the interpretation or the accumulation of arguments from different points of view, we assume proof is based on establishing the validity of mathematical statements which are assembled in an order where accepted rules of inference allow one to infer each statement from previous statements or hypotheses. We do not necessarily equate proof with formal deduction; however, we see the act of re-examination and explanation of the relationship between mathematical statements as an important component of the development of an understanding of the mathematician's notion of proof. At times, such re-examination or explanation leads to a deductive rule (for example, substitution, cancellation, universal generalization, etc.) In this work we are looking for evidence that students are able to re-examine the connections between mathematical statements, and when they are, we examine the ideas students construct while thinking about a justification for a particular connection. We take the perspective (following Carpenter, Franke, & Levi, 2003), that there is a developmental link between forms of justification, generalization, and proof and we illustrate and discuss the role of context and models, in particular the open number line, in supporting this development.

The data we report in this chapter are from an ongoing developmental research project on the development of algebra—a project that is the focus of a "think tank" at Mathematics in the City at the City College of New York.[2] The early work of MitC focused primarily on the development of number and operation, PK-8. Integrating the RME approach (e.g., see Gravemeijer, 1994) with constructivism (as exemplified in projects like Summermath for Teachers; see Schifter & Fosnot, 1993), we enabled teachers to employ the didactical use of realistic contexts and models (e.g. the open number line, ratio tables, arrays), and turned classrooms into workshop environments whereby children engaged in exploration, inquiry, and math congresses where they discussed and defended strategies and ideas (see Fosnot & Dolk, 2001a, 2001b, 2002). This work is situated in this environment, the students' teachers were participants in MitC workshops, and the students were acculturated to the MitC workshop approach. This enabled us to employ the didactical use of realistic contexts and have students use models such as the open number line while we examined their role in students' emergent conceptions of proof. Students' use of the open number line is a central consideration in this chapter.

While examining students' early attempts to use deductive reasoning in the context of equivalence, we found that their construction of certain mathematical ideas was prerequisite. Of particular importance is students' ability to work with expressions as objects, instead of as describing a procedure. For example, if students are to recognize that $10 + 5 + 1 = 10 + 3 + 2 + 1$, without first determining that both sides add to 16, they need to be able to use their understanding of $5 = 3 + 2$ and work with both 5 and $3 + 2$ as objects that fit inside the original equation. This distinction becomes more problematic with the introduction of variables,[3] and we investigate the use of the number line as a tool in facilitating understanding of the object–process distinction.

Grade 2

Building an Understanding of Equivalence

We began in grade 2 by focusing on the establishment of equivalence. As many researchers have pointed out, young children often think the equal sign means "the answer is coming" (Carpenter et al., 2003). To hinder this notion from forming, we began with a coin game from *Investigations in Number, Data and Space* called *Collect 50 Cents.* Boards were made for pennies, nickels, and quarters, and children played with a partner. In time, we began to record truth statements with them. For example, if one child had 1 quarter, 2 dimes, and 8 pennies and another had 4 dimes, 1 nickel, and 8 pennies, they would write 1Q + 2D + 8P = 4D + 1N + 8P. When the boards were not equivalent they wrote ">" or "<."

Initially, most children worked procedurally, doing the arithmetic left to right to justify the truth of the statement, or physically traded coins to make the boards appear identical. To encourage alternative actions, such as canceling out the 8P on both sides of the equation and/or treating expressions as equivalent objects, we began to explicitly discuss strategies for "how you know" and is there "any other way to know"? An early discussion (in October) between Philipp and his partner, Isaac, and Trish (the teacher) follows:

Philipp: I have 41 cents. [Philip's board had p, n, d, and q (one penny, one nickel, one dime, and one quarter)].

Trish: How do you know?

Philipp: I added it up. 25 + 10 . . . 35 . . . plus 5 . . . 40, 41.

Trish: How much do you need to get to 50 cents?

Philipp: Um . . . 9. Because it's 40 and then it's 10 to 50.

Trish: What about you Isaac?

Isaac: I don't know. 3 and 5 . . . [Gets ready to give up, starts counting on his fingers.] [On his board he had 3p + n + d + q.] I think maybe 47.

Trish: How do you know?

Isaac: I'm just guessing.

Trish: Do you have more or less than Philipp?

Isaac: Oh he has lots more.

Trish: Let's look at the board. Is there another way to tell without adding it all up?

Isaac: He has more, I know. Wait . . . maybe I do. Cause we both have this [puts his hand on top of the n, d, and q]. I have 3 [referring to the pennies]!

Trish: So who has more?

Isaac: I do!

Trish: How much do you have?

Isaac: [Looking at Philipp] I guess I have 43 cents, cause I have 2 more!

Philipp and Isaac begin procedurally doing the arithmetic (adding), but in this dialogue there is evidence of Isaac at the end stepping back and comparing the relations. He examines what is identical, covers it, and compares only the pennies. He then substitutes and concludes that if Philip's expression is equivalent to 41, and his is 2 more, he must have 43. The logic behind these actions is:

The statement $3p + n + d + q = p + n + d + q$ is not true because:
  $n + d + q = n + d + q$, thus we only need to analyze the truth of $3p = p$.
Since this statement is not true, the first statement is not true
  $3p = 2p + 1p$ is true, therefore
If $p + n + d + q = 41$; then $3p + n + d + q = 43$

We are not trying to claim here that Isaac is using proof by deduction in any formal mathematical sense. First, the coins have specific values and his action and logic are directly tied to the specific case of the coins, and, of course, there is no written proof. But he re-examines his thinking, and there is an emerging logic of deductive rules to his actions that involves equivalence, elimination, and substitution, and this kind of thinking began to become more prevalent with his classmates as well. And so we asked ourselves: "What could we do to support the development of this thinking and how far could these 7 year olds go?"

We began to work on encouraging the children to treat expressions as objects to be analyzed, rather than to proceed with addition. Initially, we stayed within the context of the coins and boards, but as children became more comfortable stating that equivalent coins could be disregarded, we began to encourage this same kind of action with numbers. To begin, we considered situations where students could consider strategies for establishing equivalence without adding all. Trish showed students Isaac's board, a picture of six pennies and one dime, and asked how much money he had. After several student explanations regarding the amount Isaac had (16 cents), Trish described the trade Isaac made which resulted in p + n + d on his board. After a child noted that the amount was still 16 cents, Trish led the following discussion:

Trish: Oh! So, can I write this? [She writes the equation d + 6p = d + n + p.]
Child: Yes, because it's the same amount just different coins.
Trish: How could we write this using numbers?
Child: 10 + 6 = 10 + 5 + 1
Trish: Is this equation true?
Child: Yes, because it's the same amount, just with different combinations.
Trish: Oh . . . that's a nice shortcut. So I don't have to add it all up and I can just see that both sides have 10 and 6.

After several more equations in a similar fashion, Trish introduced a "secret coin" as a way to begin a conversation on canceling out equivalent amounts:

Trish: Another time, I saw Devin and Mia Chiara playing. Mia Chiara said, "I know I have more than Devin. I know without counting." Here's what the two boards looked like [shows 2P + 2N + D on one and P + N + 2D on the other, with letters, not coins]. How did Mia know?
Child: Because two nickels and one dime is less than two dimes and one nickel.
Child: And they both have a nickel and a dime and a penny, so forget about that part. A penny and a nickel is less than a dime.
Trish: What if I gave them each a secret coin? I won't tell you what it is. Let's call it C, for coin . . . [After signaling a few moments of pair talk, Trish initiates whole group discussion again.]
Michael: It's still not fair. If you just add the same amount to each they each have more but the difference is still the same.

Trish: Doesn't it matter what C is?
Michael: No. C could be any number. As long as you give both the kids the same it doesn't matter what the amount is.
Chynna: You could also take an amount off. As long as you do the same to both kids it stays the same. If it wasn't fair before, it still won't be fair.

In the last two comments, there is evidence that the idea of adding or subtracting C to both sides of an equation will not affect the equality or inequality of the expressions. Acceptable rules (actions) for deduction are beginning to emerge in the community.

Representation: The Open Number Line as a Tool

Once the children were comfortable examining the coins and boards and determining equivalence (or inequality) by rules of exchanges and canceling, we removed the context of the game and began to focus solely on examining strings of equations. Once again our purpose here was to encourage the children to substitute equivalent expressions as objects that could be exchanged and/or canceled—to continue to develop a system of accepted deduction *rules* within the community. But we were also interested in whether children would begin to use these emerging rules of the community to justify the connections in a chain of mathematical statements.

To provide a representational tool where conjectures, actions, and arguments could be explored, we introduced an *open number line.* The number line model allows learners to perform concrete actions that might allow for dynamic mental experiments with corresponding statements. In contrast to a traditional number line, the open number line only includes a representation of the statements children make as they justify their thinking. By using the top and the bottom of the line, emerging rules such as canceling and /or adding like amounts can be represented and explored. When both the top and bottom of the same line are used, we refer to the representation as a *double number line.* Sometimes students use two number lines in parallel or adjacent to each other and then discuss the relationships between them. We refer to this as an *emergent double number line.*

We introduce the number line using a measurement context. Students measure lengths with multilink cubes and the measurements are transferred to an adding machine tape (Figure 6.1). This representation is used to represent the iterated units of measurement (the length of each cube) as a way to mark measurements on the tape, which over time will emerge as an open number line model as the sequence progresses. (For readers unfamiliar with our work on this, or for those who wish more detail, see Dolk & Fosnot, 2004; Fosnot, 2007; Fosnot & Dolk, 2001a, 2001b; Gravemeijer, 1999).

Once the number line model was constructed, we began to use it to represent the equivalence of symbolic expressions and the deductive rules the children were developing as they worked to prove their thinking about equations. During a

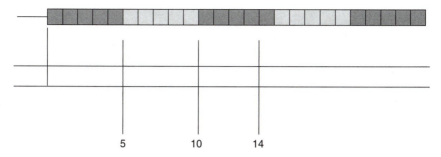

5   10   14

A string of alternating groups of five Unifix cubes (two colors) is strung above the tape to enable children to find the measured lengths

Figure 6.1 Constructing an open number line

conversation in February, Trish asked students if 8 + 6 = 5 + 9 was true. Trish recorded Lela's argument that it was true "because 9 is 1 more than 8, and 6 is 1 more than 5" on a double number line (see Figure 6.2). Another child explained, "It's like taking 1 from 6 and giving it to the 8. You get 9 + 5 = 5 + 9."

As children continued to consider more equations, we found that the double number line seemed to facilitate their noticing equivalent parts of the expressions. They invented terminology to discuss their observations, referring to characteristics of the representation rather than formal terms. In the following dialogue students use the word "mirror" for the symmetry provided by the commutative law and refer to cancellation as looking like a "banana" or reducing to chalk dust. Trish prompts the students to consider the relationships between a string of related equations as they progress. She writes 13 + 8 = 5 + 9 + 13 − 6 and asks if it's true:

Child: Equal. Because 13 is on both sides so I can cancel those out. 5 + 9 is 14 and 14 − 6 is 8. And 8 = 8.

Child: My way is different. I started with the 9 and took 6 away. That left me with 13 + 8 = 8 + 13. And I know that is equal because the numbers can be turned around.

Trish: I'll share my way, but I don't know if it will work all of the time. Tell me what you think. When I wrote the problem, I used the problem before it: 13 + 8 + 6 = 5 + 9 + 13. I took six away from both sides, and since I knew the first problem was true I thought this one had to be, too. What do you think?

Child: That works! All the time . . . because if you add any number to both sides and it's the same number it will stay the same.

Child: It's like a mirror on the number line . . . like the symmetry we did with Michael [the student teacher]. What happens on the top happens on the bottom. And like if you add a number and then take it away, it makes a jump but then you jump back to where you started. Looks like a banana!

Child: Or like chalk dust! You make a mark and then you erase it. A number minus the same number is 0.

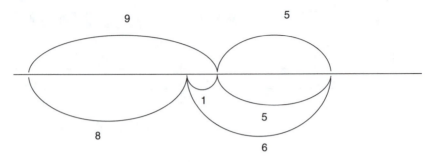

Figure 6.2 Recording equations on the double open number line

Trish next introduces an equation containing a variable to see how this would influence their concept of equivalence. She writes $8 + 6 = 5 + 9 + N$ and asks if this is true. The children seem to be ready to incorporate a variable into their context. When a child argued they were "not equal because if you add a number on only one side it won't work," Trish asked students to discuss this with partners. She asked Roxanna about her group's thinking:

Roxanna: Mostly we agree, but what if the secret number is 0? It doesn't work then [referring to the statement of the previous child].
Trish: So this is true only for when N equals 0?

Extending the Use of Variables and Solving for them

Because children were beginning to solve for the variables and explore cases when statements were true and not true, and to analyze for when $N = 0$, we decided to explore this further. Cathy (first author) pulled together a small group of eight children from the class who were very enthusiastic and seemed to be able to be challenged further with the use of variables. They worked together at a table. She asked them to analyze $C + 2N + 6P = 6P + D + C$:

Devon: I just cancelled out the Cs and the 6Ps because it is the same on both sides. That leaves $2N = D$. It's sort of like nickels and dimes but it is true for any number as long as D is twice N.
Haille: Or if $N = D − 1/2D$
Cathy: Tell us more. Can you prove that to us?
Haille: Can I show what I wrote? [Haille's work is shown in Figure 6.3.]
Haille: [She shows her work and explains.] I thought of 5 and 10 first, too. But then I tried 24 and 99. I wanted to prove it would work for any number. I cancelled out the Cs and the 6Ps, then I took away half of D from both sides.
Mia Chiara: Why did you do that?
Haille: I thought that if you needed two Ns to make a D, then half of D is an N. [This is exactly what Mia Chiara had written. See Figure 6.3.]

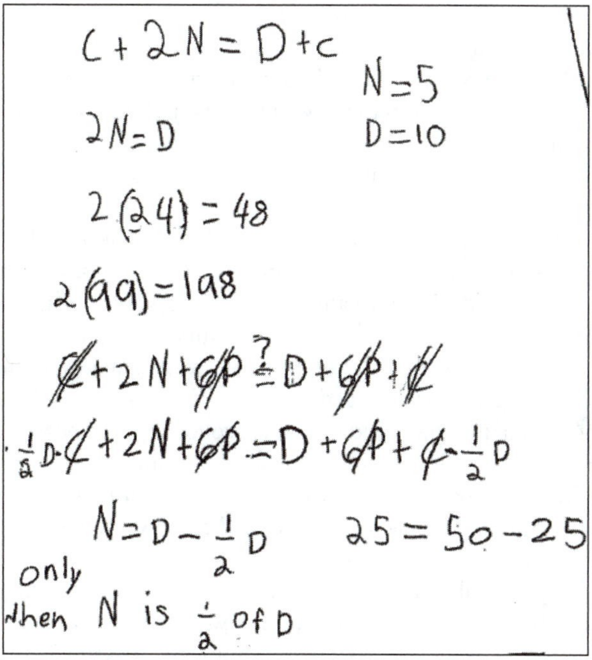

$$C + 2N = D + c$$
$$N = 5$$
$$2N = D$$
$$D = 10$$
$$2(24) = 48$$
$$2(99) = 198$$
$$\cancel{c} + 2N + \cancel{6p} \overset{?}{=} D + \cancel{6p} + \cancel{c}$$
$$\cdot \tfrac{1}{2}D \cdot \cancel{c} + 2N + \cancel{6p} = D + \cancel{6p} + \cancel{c} - \tfrac{1}{2}D$$
$$N = D - \tfrac{1}{2}D \qquad 25 = 50 - 25$$

only
when N is $\tfrac{1}{2}$ of D

Haille's Work

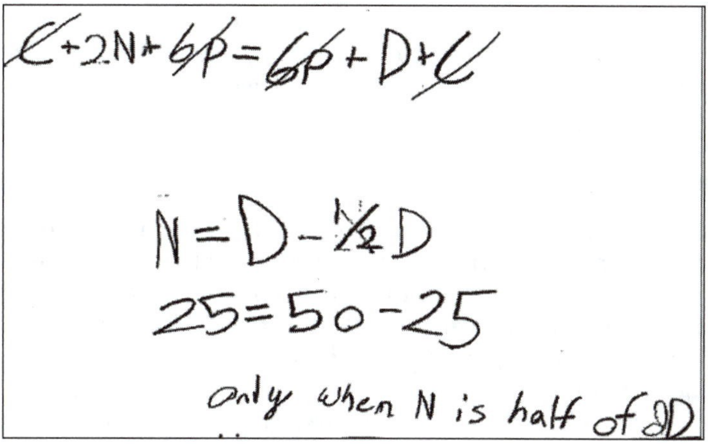

$$\cancel{C} + 2N + \cancel{6p} = \cancel{6p} + D + \cancel{C}$$

$$N = D - \tfrac{1}{2}D$$
$$25 = 50 - 25$$

only when N is half of 2D

Mia Chiara's Work

Figure 6.3 The work of Haille and Mia Chiara

Cathy: Does everyone agree? If it takes two Ns to make a D, is N equal to 1/2 of a D? [Everyone appears to agree so Haille goes on.]

Haille: So that gets me N = D − 1/2D, but this is true only when N is 1/2 of D.

Cathy: Mia Chiara, on your paper you wrote "if D = 0 then works." Could you tell us about that part?

Mia Chiara: Half of nothing is still nothing!

Around March, it seemed apparent to us that a system of shared principles and rules of deduction were emerging in this young community of discourse of 7 year olds. They understood equivalence and knew they could exchange and substitute equivalent expressions. They had constructed compensation (what is added must be removed to maintain identity), commutativity, ignoring (canceling out) like amounts, and adding and/or subtracting N to both sides to simplify for analysis. They were even realizing they needed to analyze for cases of N = 0. We began to ask ourselves, if we provided an audience, a reader, and asked them to write up their proofs would these shared principles and rules of deduction begin to be used in a chain, building on each other? Would their written arguments take on any early form of proof?

Trish wrote up several equations on a paper and we asked the children to write out their proofs to us to read. Several pieces of work are included in Figure 6.4.

What is interesting to note is how their arguments are sequential. Second, the moves they make are based on the deductive rules established by the community as accepted truths within the field of experience of the number line. One child writes, "First I knew 4 + 5 = 9 and there was a 9 on the other side, so I crossed those out. On one side was 5 − 5, so I crossed that out. Then there was a 2 on the other side so it wasn't equal." Another writes, "There are two Ns so those are out. If I switch the seven from the 17 and put in the five, then I have 15 + 7 = 15 + 7."

Discussion

Several points can be made. First, the children in this 2nd grade class clearly have constructed a system of several rules of deduction and these are held as accepted truths in their community. These rules have been explored and developed in a context (measurement of iterative units) as represented by the double number line, generalized, and they are subsequently used in a chain of deductive moves to prove equivalence (or inequality).

Several questions can also be raised, however. Are these chains of reasoning more than argumentation? Can we classify them as emergent forms of proof by deduction? If children continued and were supported in this type of work how might their work continue to develop? What might older children do with a double number line representation, and could this be used to foster a notion of cancellation in algebraic equations? Before we attempt to answer these questions we will present further data from grade 5.

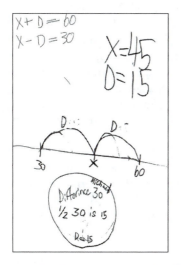

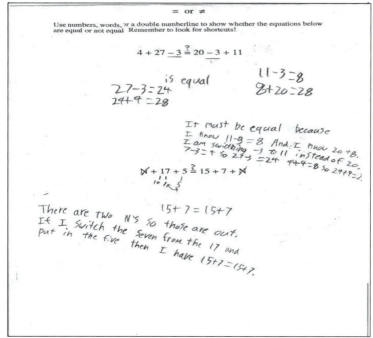

Figure 6.4 Children's work on proofs of several equations

Grade 5

The California Frog Jumping Contest: A Context to Introduce the Number Line in Grade 5

Unlike our 2nd grade discussion, which spanned 6 months' work with the class, the work reported here is from one 90-minute session. Two problems were investigated. The first problem was designed to introduce the number line representation and the second set of problems provided an investigation where we could see how and to what extent students constructed cancellation and whether they would use the representation as a tool to solve algebraic equations. The 20 students had used open number lines previously to represent computational strategies with whole numbers but to our knowledge not in the context of an algebraic problem. The introductory problem was the following:

> *Problem 1*: MT is a bullfrog. He is world famous for his long jump. Every time he jumps he travels exactly the same distance. When he takes four jumps and eight frog steps it is the same as taking 52 frog steps. How many frog steps are in two jumps and four frog steps? How many frog steps are in each of MT's jumps?

As anticipated, most of the students solved this problem in 5 to 10 minutes using relatively direct calculations such as $52 - 8 = 44$, $44 \div 4 = 11$, and $22 + 4 = 26$ frog steps. The purpose of this initial problem was to set up a situation where the teacher would represent students' strategies on an open number line. The discussion included time for students to discuss the relationship between the following two representations where the upper diagram represented four jumps and eight frog steps and the lower diagram represented two jumps and four frog steps (see Figure 6.5).

Students discussed how the upper diagram was two of the lower diagram. This conversation about the relationship of the two equations was an attempt to provide the double number line as a representation for work on the next investigation.

> *The jumping contest*: MT decides to hold a jumping contest. The three contestants are Cal, Sunny, and Legs. In this contest, all frog steps are the

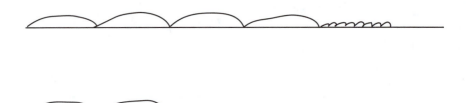

Figure 6.5 California frog jumping contest

same size. Also, when a frog jumps, he always travels the same distance. Your problem is to find out which frog has the biggest jump.

When Cal jumps three times and takes six steps *forward* he lands in the same place if he jumps four times and takes two steps *backward.*

When Sunny jumps four times and takes 11 steps *forward* he lands in the same place if he jumps five times and takes four steps *forward.*

When Legs jumps two times and takes 13 steps *forward* he lands in the same place if he jumps four times and takes five steps *backward.*

The students worked on this investigation for about 30 minutes. Fifteen students used open number line diagrams (as prompted in the earlier discussion) in their work on this investigation, and conversations with them indicate most were trying to use the diagrams as a tool in their reasoning. All were emergent double number lines, two number lines either parallel or adjacent (the double number line as used in the 2nd grade classroom had not been introduced). (Also, since frogs jump up, it may seem unnatural to indicate their hops under the line.) Four students reasoned using arithmetic, some drawing diagrams subsequent to finding jump lengths. One student found Cal's jumping distance with parallel lines and then worked numerically on the other two. Many students had difficulty comparing jumping scenarios (such as 3 jumps + 6 steps = 4 jumps − 2 steps) because they would create two open number lines and either didn't start jumps at the same points or didn't align the jumps equally. [In hindsight we wondered if it would have made a difference had we been more explicit in the use of the double number line earlier, rather than two open number lines.] After pair work, the class was pulled together for a congress to discuss their findings.

Our interest was how this classroom, as a community, might make use of the open number line, and particularly how strategies employing cancellation of equivalent amounts with variables might develop. This interest shaped our organization of the discussion. Bill (second author) led the discussion and picked three students to present a solution for each part: Maria, Yolanda, and Sam. Maria had struggled initially but seemed to be making sense of the cancellation. She drew a diagram on the board for the class (see Figure 6.6):

Maria: I drew three jumps and six small steps, and another three, I mean, four jumps take away two small steps. And I drew this a certain way showing the line because I decided to ignore these three jumps and one can see how the six steps and the jump taking away the two steps were equal. So . . .

At this point, Maria hesitates, and after a bit of silence the class is asked to put in their own words what she has done so far:

Bill: Try to put Maria's strategy in your own words. Tell us what the big bar in her drawing is.

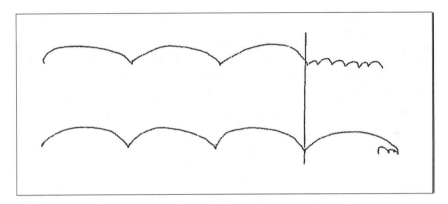

Figure 6.6 Maria's diagram for the California frog jumping contest problem

After a few minutes of pair talk, a second student explains:

Student: I think it is kind of like a separator, so because the three jumps and the three jumps on the bottom both meet up. But we want to see how the six little steps and the big jump minus two steps meet up because it is supposed to end up in the same place . . . Because it's not really important where the other three jumps are at. It's really just important to find out how they meet up.

Bill: You say this is like a separator. Right? You separated the stuff that doesn't matter so much?

Student: Well, it is not that it doesn't matter. It's just that you could tell that they automatically meet up and that you don't have to worry about that.

Bill: You could tell they meet up because . . .

Student: Because three jumps and three jumps are the same thing.

After some further discussion, Bill asks Maria to resume and she uses her hand to show on her line how the six and two are equivalent to one jump. Bill then asks Yolanda to present her solution to Sunny's jump length. She begins by showing the diagram in Figure 6.7.

Yolanda: Well, what I did is I knew that there were three equal jumps here so these two jumps and these four (pointing to four frog steps) were supposed to equal 11 jumps.

Bill: Show us the three jumps.

Yolanda: The four jumps are the same amount. So . . . one jump and four has to equal 11 frog steps. So, well, I saw four here so this and this were equal (pointing to four frog steps at the right-hand side of the diagram) so that left seven frog jumps here so that meant it has to equal one jump because there were only seven left and there were four before.

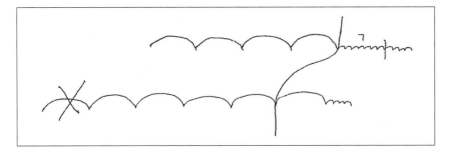

Figure 6.7 Yolanda's diagram for the California frog jumping contest problem

The class was asked to discuss in pairs how the two strategies were similar and how they were different. Two students reply:

Student: I think the strategies are pretty much the same because in each strategy she took away the jumps because she knew they didn't really help her, and also with Maria she took [away, too] because they wouldn't help her solve how many steps there are . . .

Student: They both used the separation bar. That was the biggest, I think, similarity, because it was like the root of their, um strategy. So I think that was the biggest . . .

Bill: So the separation bar enables you to take away things that are unnecessary?

Student: like a storage box, kind of . . .

The class discussed why one can take away equivalent jumps, or as some preferred, to put the jumps in a storage box. Students were also asked if it mattered that they didn't know the sizes of the jumps before they put them in a storage box, and none of the students who spoke was concerned about it. Sam was asked to present a solution to the third question, involving Legs. Sam presents what might be considered a generalization of the cancellation strategy. Figure 6.8 shows what he writes during this conversation.

Sam: Depending on the difference of the number of jumps and whether the steps are forward or backward you can always figure out how many steps are in a jump. For Legs [it is] 13 plus five divided by two is nine. That's how many it was because with two jumps and then there are 13 steps [draws] and then there are four jumps and five steps backwards and so since there are two extra jumps that's why you need to divide by two. [He writes $13 + 5 = 18 \div 2 = 9$, which after a short discussion on the use of the equal sign is rewritten as $13 + 5 = 18$ with $18 \div 2 = 9$ below it.]

Bill then asks, "Show us on the picture . . . where is the 13 + 5?"

Sam: If these two were the same jump it would just be 13 + 5 because two plus 13 equals up to here and then the 5 would go back here because they are in the

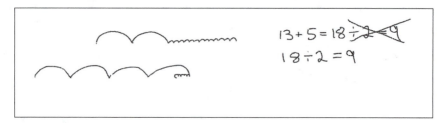

Figure 6.8 Sam's generalization of the cancellation strategy

same place. So you need to know 13 + 5 to get here. And then because this is all one jump it would just be 13 + 5. But since it is two jumps you need to divide it by two because they are separate jumps.

The class is then asked if they think the earlier rule was used in this shortcut:

Student: Yes.
Bill: And why?
Student: It was kind of like an invisible line, because the first two jumps and the first two jumps on the bottom he didn't care about because they wouldn't help him solve the problem. So, so he just paid attention to the two last jumps and the 13 little steps so technically he is using the rule but it is just like an invisible line.
Bill: Do you think you used the same strategy, Sam?
Sam: I don't know. I guess.

Discussion

In this discussion, we see a class working as a community where part of their efforts are devoted to constructing and defining an algebraic rule for which one cancels equal amounts of an unknown. The conversation revolves around certain lines drawn to locate equivalent points on two open number lines. It was problematic for some students to figure out where to draw lines on their diagrams, which the teacher unintentionally referred to as "bars,"[4] but most students eventually succeeded in using the lines as a tool in problem solving.

Also in the discussion, we witnessed an evolution of invented terminology that gives us some clues about the development of students' notion of a cancellation rule. In explaining the line, the second student describes it as a *separator*. This terminology of a separator is quite appealing, because as is clear from the discussion, many students are unwilling to remove the equal amounts from consideration. They seem to view the separator as breaking the problem into two parts, even though they recognize the equivalent amounts on the left are not necessary to solve the question. This hypothesis is supported also by the fact that one student refers to the separator as providing a "storage box." The equal amounts in the unknowns are not thrown away; they are stored, not "cancelled." The students prefer the word "unnecessary" (in terms of answering the question) rather than terminology of the teacher/researcher who used words like "ignore" or "remove."

Sam did appear to cancel equal amounts and remove them from consideration. His rule considers only the extra jumps and the forward and backward steps, and on his written paper he uses equal signs to represent operations rather than equivalence, writing $13 + 5 \div 2 = 9$, together with the check $9 \times 2 = 18 + 13 = 31$ and $4 \times 4 = 36 - 5 = 31$. In the congress he creates a diagram that includes all the jumps, indicating he understands the representation, but he does not include a separation line and his discussion only includes the right portion of the diagram. We find it significant that other students realize he is using a version of their earlier rule in the form of a "hidden line," which indicates the community is making progress towards identifying cancellation as a deductive rule. But it is not clear that Sam sees that he is using the rule. In fact, on the back of his paper he records his justification to the second problem as: "4 jumps + 11 steps is the same as 5 jumps + 4 steps, so the difference between 11 steps and 4 steps is the difference between 4 jumps and 5 jumps." Assuming he used similar reasoning in the third case, we can understand why he was not convinced he was using an invisible line. He is able to work with the remaining jumps (as if they were separated out) but the entire context remains fixed in his view while working on the question.

Implications for Proof

We have been interested in elementary students construction of deductive rules (in this case a "cancellation law") and their ability to use such a rule in subsequent problem solving as a developmental step in proof. We find that use of the double number line does appear to facilitate the construction of, or at least the awareness of, such a process and students are subsequently able to make use of the representation as a powerful tool for thinking. Moreover, the notion evolves in a community of discourse where the representation is the object of discussion. So even though some students may construct the rule without using the double line, a secondary benefit of it is that it provides a context for discussion of the rule in the community. The terminology of "separators," "storage box," and a "hidden line" emerged because of the use of the representation and we doubt such terms would arise in conversation around the usual symbolic formulation.

The conversations observed in the 5th grade class indicate that the term *cancellation law*, while suitable for a fully formed conception of the deductive rule, *given $a + b = a + c$ deduce $b = c$*, may not be a suitable term for early understandings of this rule. It appears that some students are more comfortable with the deductive rule, *given $a + b = a + c$ deduce $a = a$ and $b = c$*, and that separating off the $b = c$ in isolation may be a second hurdle. In fact, we noted resistance to such isolation of $b = c$. If so, this raises a more general question. To a mathematician, part of the elegance (and power) of proofs is their reduction of key ideas into a compact form, eliminating redundancy or irrelevant components (recognizing that the work going into creating a proof is something different). But for the learner of proofs (or in this case a deductive rule) the rules of proof may come to them "situated" (either in context or in a representation) and that the stripping off of redundant or irrelevant information is not a simple action to take.

In contrast, in the 2nd grade classroom, where we worked with students over an extended period of time, the students repeatedly used the word "cancel" with ease. We suspect this is the result of the extended time this community of students had to develop the concept. It is also the case that in the 5th grade class, the students were canceling unknowns, and to them, this may be substantially different than canceling a known amount. Further, as is evidenced by Sam, who wrote $13 + 5 = 18 \div 2 = 9$, the class may not have a fully formed notion of the = sign and this could slow their willingness to accept a cancellation law. Yet at the same time, Sam's success in using a form of cancellation shows that understanding of equivalence is not the same as understanding the equal sign. This illustrates the importance of development, and that the requisite understanding of the equality of expressions and equivalence can start in the elementary grades.

The act of creating a mathematical proof is not a simple task and usually requires re-examination of initial reasoning to simplify and align ideas into a logical order. In this sense it involves a "meta-process," much like that of an author reconsidering an initial draft and rewriting the material. We feel that this learning should be an objective for students as well. The second author teaches a problem-solving course to undergraduate mathematics majors where, inspired by work of Schoenfeld (1992), his students engage in non-routine problem solving, discuss their strategies, and explicitly consider metacognitive aspects of the process (self-regulation). He has found however, that the request "now give a nice proof of what you found" presents a new difficulty. Quite often, after a successful problem-solving experience, a proof will be a retelling of all of the steps that went into the solution process rather than a re-examination and simplification. (For example, displaying an initial collection of detailed calculations instead of going back and simplifying using basic theorems of linear algebra that all the students had previously studied.)

We are encouraged that the representations used by the classes we studied here provided a context for the students, as a community, to re-examine their reasoning. This re-examination provided better understanding of the deductive rule mathematicians refer to as cancellation. The analysis of Sam's work that Leg's jumps are nine steps led to a notion of an "invisible line," possibly a step closer to conventional cancellation than a "separator." Similarly, in the 2nd grade class, students expressed the commutative law by saying it was "like a mirror on the number line," or that cancellation "looks like a banana." The emergent understanding of these deductive rules resulted from re-examining the equation using the double number line representation. Subsequently, these 2nd grade students did incorporate forms of cancellation into their proofs. Reviewing these conversations one might be tempted to characterize this type of re-examination as "sense making about one's reasoning." The next step in a study of such students' ability to engage in proof could be, for example, to determine if students are able to use representations to reformulate initial lines of reasoning and rearrange and simplify them into more of a compact form of proof. This would represent a yet higher development of mathematical maturity.

Notes

1. In the mathematical research community, one often hears the word "argument" used synonymously with "proof." In the context of this chapter, argumentation is used differently—a good argument would be one supported by the preponderance of the evidence rather than rigorous reasoning from a set of clearly formulated hypotheses.

2. Mathematics in the City was begun in 1995, initially as a collaborative professional development project (between New York City Schools, the City College of New York, and the Freudenthal Institute). Funded initially by the National Science Foundation and the Exxon-Mobil Foundation, the center is now a national center of in-service, research, and curriculum development funded by various sources.

3. For further discussion of the two different conceptions of abstract algebraic notation, structural (as objects) and operational (as procedures) see Kieran (1992) or Sfard (1991).

4. The researcher had not intended to use this terminology. It crept out, presumably because of the analogy with music. The students did not adopt the word "bar."

7

Children's Reasoning

Discovering the Idea of Mathematical Proof

CAROLYN A. MAHER

Proof making has ordinarily been associated with mathematical reasoning of high school and college-age students. The expectation that young children can invent the idea of proof is relatively new (Maher, 2005; Maher & Martino, 1996a, 1996b; Yackel and Hanna, 2003). However, there is growing evidence that the invitation to justify and convince others of the validity of solutions results in sophisticated reasoning in young children (Fosnot & Jacob, Chapter 6, this book; Reid & Zack, Chapter 8, this book). Results from a program of research studies at Rutgers University show that when children are placed in supportive learning environments, the idea of proof can come naturally to them (Maher & Martino, 1996a; Uptegrove & Maher, 2004). This chapter extends the work by showing *in detail* the *process* by which convincing arguments are built by children as young as 8 years old.

Theoretical View

The theoretical foundation for the program[1] of research on which this chapter is based comes from recognition that individual learning can occur most effectively within a community whose members have access to and are influenced by the thinking of others. Engagement with others opens up possibilities for sharing and comparing ideas. The relationship between the individual learner and other members of the community is interconnected. A person's way of knowing may influence or be influenced by the ideas produced by others. Opportunities for sharing invite the building of new mathematical ideas. Children enjoy working on tasks that elicit their thoughtfulness and the opportunity to engage with the ideas of others. In the activity of problem solving, children bring forth, compare, and communicate with their own or another's representations. In listening to the ideas of others, learners have an opportunity to see how another's representations and reasoning compares to their own and how it relates to ideas of others. By examining a variety of representations, children can build a deeper understanding of the mathematical idea (Davis & Maher, 1997; Maher & Davis, 1990).

The interplay between concrete experience and building meaning is fundamental, especially for young learners. Concrete experience can motivate engagement in mathematical tasks and invite students to notice relationships between the physical entities being studied. Resulting images, patterns, and relationships can become mathematical objects that one sees and works with mentally in extending ideas, building meaning, and exploring abstract concepts (Gravemeijer, 1999). It is helpful for children to investigate concrete cases very deeply. The details of the concrete cases can motivate and then be compared with their abstract representations (Maher & Speiser, 1997).

Children's representations of mathematical ideas—that is, the ways in which students think about and make sense of these ideas—are fundamental for their reasoning and learning. Indeed, we believe that, in large part, growth in students' representations of a mathematical concept constitutes learning about that concept (Davis, 1972). The representations that a person builds for a mathematical idea or procedure can take different forms. As children gain more experience in a mathematical domain, their representations of the concepts within that domain become more sophisticated. Working on different tasks encourages students to elaborate and extend their representations and to build connections between representations (Davis & Maher, 1997). The connections that a person makes when analyzing and developing solutions to well defined, open-ended problems provide further opportunity for growth in knowledge. Bruner (1960) indicated that understanding the fundamental structure of a subject makes the subject more comprehensible and reduces the gap between "elementary" and "advanced" knowledge.

Proof making is a special type of mathematical activity in which children attempt to justify their claims by deductive argumentation. In our view, students' learning of proof making occurs as they develop more sophisticated representations of mathematical argumentation. Consistent with our views of how mathematical knowledge develops, as children's experience with argumentation grows, students' conceptions of argumentation will become more sophisticated and strategies for deductive argumentation will be developed.

Background

The data for this chapter come from a program of research detailing the collective building of mathematical ideas. The children's reasoning is documented by their actions—that is, what they do, say, build, and write—as they work on a task. These actions are recorded on videotape. In studying how the children make sense of complexity of the tasks on which they are working, we observe the representations they use, heuristics they invent, the modifications they make, as well as the support they give for their arguments.

The 3rd graders in this chapter come from a school district that was partnered with the Robert B. Davis Institute for Learning (RBDIL) in a cross-sectional study of the development of mathematical ideas and ways of reasoning.[2] The children initially worked together in pairs. Later, they presented their ideas to the teacher/researcher and then to the entire class. Follow-up individual interviews were conducted with several of the children. An expectation was that the children work together on their investigations[3] and that the teacher/researcher intervention be limited to asking the children for explanations, verifying the reasonableness of their arguments, and, when necessary, seeking further clarification or elaboration.

Data and Analysis

Videotapes were made of two sessions, each approximately 90 minutes in length. At each session, there were two cameras and two sound technicians. By videotaping the children as they worked together, we were able to monitor the *process* by which the children built a justification for their solutions. The video data were studied and transcribed to uncover the meanings that children gave to mathematical situations. Detailed analysis made it possible to trace the origin and evolution of the children's arguments, that is, how the children expressed their ideas through spoken and written language, the physical models they built, the drawings and diagrams they made, and the mathematical notations they invented.

The video data were described and critical events were flagged and coded according to the identification of modes of reasoning. All sessions were transcribed. The connected series of critical events that formed a trace led to the emergence of the narrative (e.g., Davis, Maher & Martino, 1992).

Results

The session began with Jackie and Meredith seated together as partners in a 3rd grade class. They were asked to find all possible four-tall towers that could be built selecting from red and yellow plastic cubes that were made available. Meredith was seated to the right and Jackie to the left. As Meredith built towers, she placed them to the right side of the row of towers; Jackie placed her towers to the left. Meredith monitored the inclusion of each new tower configuration for duplication before she included it in the set of towers. Together, they found 16 towers, unaware that only 15 towers were distinct. One tower was missing; another had a duplicate. Figure 7.1 shows their initial configuration of towers.

Figure 7.1 Initial configuration of towers red (dark tint) and yellow (light tint)

The girls discussed their solution with the researcher who asked if they were convinced that they found all the towers and whether they could convince others that they found them all. The girls responded by comparing their total number of towers with those of other students working in the classroom.

Meredith: Do you think we could convince her?
Jesse: How much do you have?
Both girls: Sixteen.
Jesse: That's how much we have . . .
Jesse: There's no more.
Jackie: We can't find any more.

The need to convince Jesse that they had all the towers led Jackie to suggest that she and her partner "organize" their towers. Together, they placed towers in pairs of *opposites.* The idea of *opposites* was at first defined differently by each girl (see Figure 7.2). For Meredith, two towers were *opposites* if they had the same pattern, but with alternate colors in each of the tower positions. For Jackie, *opposite* towers had the property that the color configuration was identical when one of the towers in the pair was flipped.[4]

Together, Meredith and Jackie continued to work on the problem. The girls came to use Meredith's definition of *opposite*; they reorganized their towers in pairs and found the duplicate tower, now giving them 15 towers. Jackie realized that one tower did not have an *opposite* because the total number of towers was an odd number. Jackie's discovery encouraged them to search for the missing tower. Meredith identified the missing *opposite* tower, once more giving them a total of 16 towers. Once Jackie and Meredith completed building all of their eight tower pairs, they began to record their solution by drawing a picture of a tower, and then drawing a picture of its *opposite.*

The girls then shared their new organization with the researcher, explaining how it emerged with their identification of a duplicate tower, their discovery of a missing tower, and then their arrangement of the towers in pairs of opposites (see Figure 7.3). Jackie and Meredith enthusiastically explained how they arrived at this new arrangement and how they used a strategy of checking for duplicates.

Jackie: And then, to get . . . to get another one, we did the opposite of it.
Meredith: Went like that, we kept on doing it.

Figure 7.2 Partial organization, based on two ideas about opposites

Figure 7.3 Grouping according to number of yellow cubes

Meredith demonstrated how she checked each tower. The researcher asked if every tower had an opposite. The girls replied:

Meredith: Everyone we did had an opposite.
Jackie: But before we were missing one [tower].
Meredith: Cuz we had a double and then we figured out we shouldn't have put it up there [a tower with one red cube in the second from the top tower position], the other one. We should have put it [the red cube] down there [the second from the bottom position], so we . . . Two yellows, one red, and a yellow.

Meredith then explained that they were recording their tower model on a poster. The researcher asked if their results would be convincing to others.

Meredith: Well, we're making a chart.
Researcher: Very nice. Okay, you're making a chart. How are you going to convince people, though, that maybe there aren't 17 towers, or 18 towers? How do you know you have them all? Do you feel sure you have them all?

In response to the researcher's question, Meredith started *again* to reorganize the towers. She began by locating the towers with exactly three yellow cubes. Then, together, the girls located towers with four yellow cubes, and four red cubes. They extended their idea of "opposite" towers to "opposite groups" of towers. The girls then grouped together the towers with two yellow cubes, and then one yellow cube.

The researcher acknowledged aspects of the girls' reasoning that were convincing, in particular, the cases with no, one and all red cubes.

Researcher: I like your groups. I think it's a very interesting way of thinking about it. I just want you to be absolutely sure while you're recording these, that there are no more in any of these groups. Okay?

The researcher continued, attending with more specificity, to the particular group of towers with two red cubes and two yellow cubes: "I'm pretty convinced, I mean, I'm pretty convinced about this group [the tower of height four with all red cubes], right?"

She then attended to the towers that were built with all red cubes: "The towers with no yellows, there probably is only one way do to it. Okay." The researcher continued, pointing to the group of towers with exactly one yellow cube: "And I'm pretty convinced about some of these groups" [the group of towers with exactly one yellow cube and the group of towers with exactly three yellow cubes].

The researcher verified the completeness of the categories that were visually convincing and then moved to the group of towers of height four with exactly two yellow cubes:

Researcher: This one, really, I have a very hard time seeing it, maybe because I wasn't here when you actually built them, but I'm wondering how other people are going to react to that. So think about it as you're recording these. Think about how you might convince people about this group, okay? All right. I'm going to let you work and record.

In response to the researcher's request to justify the completeness of the group containing two of each color, the girls reviewed their towers and rearranged them as in Figure 7.4. They then engaged the researcher in conversation about their results.

Researcher: Ah . . . so these are all there are with one yellow?
Jackie: And the last ones are four [of a color] . . . [the tower with four red cubes and the tower with four yellow cubes]
Researcher: Okay, and this is four yellows [the tower with all yellow cubes], and this is how many yellows [the tower with four red cubes]?
Meredith: None.
Researcher: No yellows, right?
Meredith: So we put it down here [She placed the tower with no yellow cubes preceding the towers with one yellow cube.]
Researcher: Oh, that's a good idea. So let's go through this again. These are four yellows, these are three yellows, these are two [yellows] . . .
Meredith: Those are one [yellow] and zero [yellows].
Researcher: One and this is zero . . .
Meredith: So we made sort of like groups.

The researcher again asked the girls about the completeness within categories, starting with the category of exactly one yellow cube:

Figure 7.4 Refined organization by cases

Researcher: This is interesting, these groups. Now within these groups are you sure you have all [towers] with say, one yellow [cube]?
Meredith: Yeah.

As the conversation with the researcher continued, Meredith extended her earlier visual explanation by offering an *argument by contradiction*. She indicated that the condition of the problem required that towers be four tall.

Researcher: Oh. How do you know there aren't any more?
Meredith: You can't go any higher . . .
Researcher: Why not?
Meredith: The step, because we can only do four [four cubes tall] and there's four steps.

The researcher continued, attending to the group of towers with exactly three yellow cubes. Meredith referred to the "stair-like" pattern and Jackie indicated how it was "going up."

Researcher: How do you know that you have all the ones with three yellows, though? It's very hard for me to see that.
Meredith: Cause it goes up like a pattern, like stairs, like just all of them do.
Researcher: Where are the stairs?
Meredith: Like . . .
Jackie: Going up here, going up there.
Researcher: Oh, I see, going up here and going up there. That's what you're saying.
Meredith: There's like sort of like a pattern.

In response to the researcher's request for an explanation that the group of three yellow cubes was complete, Meredith rearranged the set of towers in elevator-like fashion for exactly one red (three yellow) and exactly one yellow (three red) cube.

Researcher: Is there a way that might help me better be sure that we have all the ones [towers] with three . . . three yellows?
Meredith: Wait, this should go . . . [She begins to rearrange her set of towers with exactly three yellow cubes into a staircase pattern where the red cube moves down one position at a time beginning in the top position.]

Figure 7.5 shows the staircase arrangements for towers with exactly one yellow cube and exactly three yellow cubes.

The researcher then focused attention on the six towers with exactly two yellow cubes, asking if they had them all. Jackie responded by indicating that she checked for all possibilities, that each tower had an opposite, and that they could not find any more.

Figure 7.5 Staircases for towers with one yellow cube and three yellow cubes

Figure 7.6 Organization of towers with two yellow cubes

Researcher: Interesting. This one really gives me a headache though [the group of six towers of height four with exactly two yellow cubes]. This one's really dizzy. I am just wondering if we have every single one, if we can't think up one more? Or two more?

Jackie: Before I was checking it and for the one [the group of four towers of height four with exactly three yellow cubes], for the one, for the three yellows, and I was going "Is there a way we can figure it out?" and I checked every one, when it was like this, I think, and I couldn't find any more for this one.

Researcher: Why is that? Why do you think that is, Jackie?

Jackie: Cause I think we did all the "opposites," and I think that's all we could really find.

Researcher: So you think so. So you're pretty convinced aren't you?

Jackie: Ah hmm.

To account for all towers that had exactly two yellow cubes, Meredith introduced a recursive strategy. She combined this with the visual display of a staircase for two of a contiguous color, finally organizing them in opposite triples as indicated in Figure 7.6. As she pointed to the pattern with two yellow cubes, she said: "It would go up and over." After grouping the towers now focusing on red, Meredith indicated: "It sort of has the same pattern as this [the yellow cube arrangement]. Each one is sort of like steps."

The researcher then referred to her four-tall tower with all yellow cubes. Meredith identified this group as "four reds" or "zero yellows," indicating ease in reversing notations.

Researcher: This was a different one [the tower of height four with four yellow cubes], right? This was four yellows.
Meredith: Yeah.
Researcher: Okay, so we had, what was this [towers with four red cubes] again?
Meredith: [Towers with] Four reds, or I mean zero yellows.

The researcher asked again about towers with exactly one yellow cube. In response, the girls referred to all of the groupings that emerged to produce an argument by cases.

Researcher: [Towers with] Zero yellows, and these [towers with exactly one yellow cube]?
Meredith: [Towers with] One yellow.
Researcher: And these were [pointing to the respective categories]?
Both girls: Two yellows, three yellows, four yellows.
Researcher: Okay, and you're convinced that you found all the towers for each of these groups.
Meredith: Cuz they have patterns, see . . .

We see from this exchange that a reference is made to patterns. Earlier, the girls worked with patterns of pairs of towers and focused on pairs of opposites. Then they attended to patterns within groups of towers, such as staircases and elevators. Now we see the girls extend their original use of *within* group patterns to patterns *across* groups.[5]

Classroom Sharing

After pairs of students worked on developing methods for justifying that they had found all the possible towers of height four when selecting from two colors, they prepared to share their results with their classmates. Sharing their findings with the entire class gave the children further opportunity to explain their reasoning. Also, it provided for others to offer their justifications, and follow the ideas of others. At the sharing time, all groups had arrived at a total of 16 towers. Now they were invited to offer their arguments to others.

When it was Meredith and Jackie's turn, they brought their configuration of towers to the front of the classroom. What is interesting in this session is that they not only presented their organization by cases, but they also attended to accounting for all possible towers within a category:

Meredith: Well, we did sort of like patterns, and we did a pattern for each one, we made groups of [towers with] one yellows, two yellows, then three yellows. And we made patterns with the one yellows, and the two yellows, and the three yellow, and we made also patterns with the four. Just went like that . . .
Researcher: Four of a color. Four yellows, and what did you call this?

Meredith: Zero yellows. We went one up so we went one up on this [the tower with one yellow cube in the bottom position of the tower] and put a yellow there. [She referred to moving the yellow cube from the bottom tower position to the second to the bottom tower position.] We went one up and put a yellow there and we put one yellow on the top, and the rest we put red. And then we put, on the two we put one there, one there to start off. Then just like the one and then we put them on top. We put three on the top, three reds on the top, and we put three yellows on the bottom of the reds, the two reds and one yellow. Then we put the same pattern over here [referred to the group of towers with the two yellow cubes]. And then we put just like the pattern of this right over here. We put this at the end.

Researcher: Okay, so these are [towers with] one yellow and [towers with] two yellows.

Meredith: And three, and I did the same thing on this [towers with three yellow cubes] as that [towers with one yellow cube], but the yellows were the reds instead.

Previously, the researcher challenged the girls to find a convincing argument for the six, four-tall towers with exactly two of a color. For the category consisting of two of each color, they extended their visual pattern display and for further support offered a *recursive* argument.

The researcher asked the class if it were possible to make another four-tall tower with exactly one yellow cube and three red cubes that was different than the set of four presented by Jackie and Meredith. Andrew, offering an argument by *contradiction*, responded that Jackie and Meredith's method was exhaustive, indicating, first, that you could make more towers only if you violated the condition of making four-tall towers. He backed it up with a *recursive* argument to account for all possible four-tall towers with exactly one yellow cube.

Andrew: Well, if you had like . . . you could . . . you couldn't because you have three reds and you would need a five [cube] tower to make another just one having one yellow [cube].

Researcher: That's interesting, but why is that?

Andrew: Well, because if just had . . . you take away the yellow on the last one [the tower of height four with the yellow cube in the bottom tower position] and put it up one more, then it would be the same as the last one on that side [the tower of height four with the yellow cube, in the second to the bottom tower position].

Researcher: This guy?

Andrew: If you move that yellow [cube] up one more [position] to the second floor that would be the same as the second one [tower].

Researcher: As this guy, right? [indicating the tower]

Andrew: Yeah, and if you move that one [the tower with the yellow cube in the second from the bottom tower position], one higher, it would be the same as the third [the tower with the yellow cube in the third from the bottom tower position].

Researcher: Yeah, you mean like this one?

Andrew: Yeah, if you moved that one it would be the same as the last [the tower with the yellow cube in the top tower position].

Researcher: Moving up one at a time, you say. That's neat! Okay, are there any more, any more? Then you're saying there aren't any more.

Andrew: Yes.

Researcher: Do you agree with this, disagree with this?

Meredith: I agree with that.

Discussion

In a natural way, the girls began by making towers. Their initial strategy was to build a tower and then check to see if it was new. When asked how they knew they had all the towers, they used certain visual patterns to develop an organizational scheme that made it easier to account for missing towers and for finding duplicates. In the process of organizing and reorganizing, new methods of argumentation were developed. They defended their claims that they had all the towers with arguments by contradiction and recursion.

Researcher questioning resulted in more articulate explanations both among groups and within groups of students. Researcher questions motivated the students to develop more efficient organizational schemes, which in turn suggested more sophisticated arguments for why they had all the towers. From the girls, and again later from Andrew during the classroom sharing session, we see a variety of forms of proof ideas. Triggered by further researcher questioning, the visual displays gave way to more elaborated justifications by the 8 year olds who made convincing arguments by cases [Jackie and Meredith], contradiction [Meredith and Andrew], and recursion [Meredith and Andrew].[6]

Conclusion

Towers do not come organized; children bring organization to them. Diverse readings of the towers resulted in the various different organizations such as towers randomly built, grouped by cases, and viewed recursively. The process by which this occurred in the sessions described here is a careful mixture of free play, collaboration, questioning, revisiting the problem, and a call for justification. The crucial point we wish to make is that the types of deductive argumentation that the students ultimately engaged in—including case-based reasoning, proof by contradiction, and recursion—neither developed spontaneously in the students nor were suggested by a teacher. Rather, they emerged naturally in response to a researcher's request for students to justify their work The problem solving that has been documented in these episodes is remarkable in that the children are only 8 years old.[7]

While there is general agreement that proof learning for students is difficult, our research has shown that, in the process of convincing oneself and others of the validity of a solution, arguments are presented by young children that take the form of mathematical proof. It is interesting to note that the "proof-like" arguments were developed initially without the intent or a charge "to prove." Rather, the forms of argument arose *naturally* in response to a challenge to convince oneself and others of the correctness of a solution. The children expressed interest in the variety of ideas and approaches that were revealed. They questioned and monitored their own and one another's ideas in terms of whether they "made sense."

Under the conditions of our research program, the children worked on the tasks, decided what to look at and how to work with what they chose to observe. They decided to represent their ideas in certain ways. In so doing, they made discoveries about their represented ideas; and because they had different ways of working and imagining, different data and different explanations about their data emerged. Because children worked collaboratively and were the arbiters of what made sense, they attended to one another's arguments and compared the arguments of others with their own. It is important to emphasize that the more elegant verbal justifications emerged in response to researcher questioning. A deep understanding of the problem solution and attention to details in providing justifications made possible the articulate explanations that were produced by the children. The researcher's role was not to lead students to think or reason in particular ways. Rather, it was to elicit and make public their ideas that were expressed in the problem solving activity and the physical representations that were produced.

The component of the longitudinal study reported here gives insight into the processes by which the children built careful arguments that are convincing to themselves and others. We see how the processes, as well as the results of earlier building, were reconsidered and reformulated in a reciprocal interaction with the construction of newer and more general ideas. The children shared their own revisions and interpretations. They shared their work with others, folded back to earlier ideas (Pirie & Kieran, 1994), and revised past work through critical review and a more thorough reconsideration. In this way, they purposefully were able to reshape specific items. The children focused and directed the investigations; their discourse, as we have seen, folded back on itself through their own critical review and purposeful reshaping of specific items of earlier work. Were the children *taught* to prove? This depends on one's definition of teaching. If one regards teaching as providing a context for exploring well-defined, open-ended problems, collaborative work, engagement in conversations about meaning with teachers, then maybe so.

The children's initial intent was to share with the teacher/researcher what they were doing. What did the researcher do? The researcher provided attention to details and elicited from the children explanations that made sense. The researcher's implicit intent, from the start, was to draw out from the children their reasoning and their ideas about what makes an argument convincing. What is clear is that the children *learned* about the *idea* of proof.

Implications

It may not be effective for teachers to require their students to offer proofs before they have built for themselves an understanding of what it means *to prove*. Rather, it may be more effective for teachers to capitalize on students' natural interest in *sense making*. This can be accomplished in classroom environments by providing interesting investigations for students to pursue and by making available appropriate materials and sufficient time to think deeply about problems. What the analysis in this chapter suggests is that by giving children ample opportunity to build and refine their justifications for solutions to problems that are meaningful to them, their modes of argumentation become more sophisticated and proof like. Students need to talk about their ideas to one another and to the teacher, who can model listening and questioning techniques. Teachers can encourage students to explain their thinking in detail and respond by asking others to give explicit feedback as to what aspects of the reasoning make sense and what aspects may require further thought. Sense making and reasoning in problem solving come naturally to students. There are so many untapped opportunities awaiting teachers to foster this growth in their classrooms.

Notes

1. The history of the research program has been described in detail elsewhere (see Maher, 2005).
2. The research was supported, in part, by National Science Foundation grants MDR9053597 and REC-9814846. The opinions expressed are not necessarily those of the sponsoring agency and no endorsement should be inferred.
3. For statements of the mathematical investigations posed, visit the Robert B. Davis Institute for Learning website: http://www.gse.rutgers.edu/rbdil/site/.
4. Stephanie and Dana called this tower configuration "cousins" (see Maher & Martino, 1996a).
5. For other examples, see Maher and Martino, 1996a, 1996b.
6. For other examples, see Maher and Martino, 1996a, 1999.
7. See, also, 14-year-old Stephanie's reasoning with block towers (Maher & Speiser, 1997).

8

Aspects of Teaching Proving in Upper Elementary School[1]

DAVID A. REID AND VICKI ZACK

Reform documents and curricula (e.g., NCTM 2000) advocate that mathematical reasoning-and-proving should receive more emphasis at all grades, including in elementary school. This represents a challenge for teachers who have not in the past been expected to include these processes in their teaching and who may feel unprepared to do so. In this chapter, we will discuss teaching reasoning-and-proving in grade 5 with the aim of identifying aspects of teaching that seem important to the successful development of children's proving. We will do this primarily by analyzing a multiyear teaching experiment in Vicki Zack's classroom.

Research on Children's Reasoning

The body of research on children's reasoning in elementary school contexts is not as extensive as the literature on secondary and university students; however, it is still not possible to review all the relevant research in a short space. For an extensive review see Stylianides and Stylianides (2008). They note that some studies have concluded that elementary school students are not yet old enough to engage in proving. Other studies, however, have found that children can engage in deductive reasoning at much younger ages. For example, Galotti, Komatsu, and Voelz (1997) report:

> [Y]oung children can do more than draw deductive and inductive inferences. Even by second grade, they show the beginnings of implicit recognition that these two types of inferences are different and, as is appropriate, show more consistent answering, and higher confidence, in deductive inferences than in inductive inferences (although confidence in deductive answers is not as high as it should be). (p. 77)

Stylianides and Stylianides also report that some types of deductive reasoning develop more easily than others. Direct deductions (*modus ponens*) are understood and used by younger children more often than indirect deductions (*modus tollens*); however, both types can develop in early elementary school aged children, especially with a teacher's intervention.

While such research suggests elementary school students can learn to prove, they do not offer examples of the teaching of proving. A number of studies deserve mention as they offer some hints to elementary school teachers interested in developing their students' reasoning-and-proving. Lampert (1990, 1998a, 1998b) describes her teaching of a US grade 5 class in which conjecturing and proving are evident. Maher and her colleagues (e.g., 1996a, 1996b) report on a longitudinal study that began with the developing reasoning of students in grade 3 and follows them through their school years. Ball and Bass (2003) examine the evolution of reasoning in Ball's grade 3 class. Finally, Boero, Mariotti, and their colleagues (Boero, Garuti, Lemut, & Mariotti, 1996; Boero, Garuti, & Mariotti, 1996) report on an Italian grade 8 teaching experiment centered around sun shadows that suggests elements that could be incorporated into teaching at earlier grades.

What We Mean by "Proof" and "Proving"

The words *proof* and *proving* are used with a number of different significations in the mathematics education literature (see Reid, 2001, 2005) so it is useful to clarify further what we are referring to. We use the words *prove* and *proving* to refer to the activity of reasoning deductively, regardless of how this reasoning is expressed, and what purpose it serves. We use *proof* to refer to arguments which are the expression of the activity of proving, whether verbal or written, formal or informal.

Proving only becomes evident when it is expressed verbally or in writing. In the data recorded in Vicki Zack's classroom, there are many examples of deductive reasoning. Here we will use one to illustrate the kind of reasoning we are focusing on and calling proving.

Our example is a student's explanation to her peers of her method of solving the problem of determining the number of squares in an n by n grid (the "count the squares" problem, see Figure 8.1). She is explaining why the number of squares of a certain size in a 10 by 10 grid is always a square number, and more specifically, that the square numbers involved are 100, 81, 64, . . ., 4, 1. At the beginning she is gesturing to show how a 2 by 2 square fits across the top of the grid nine times.

Maya: Can everyone see? So you count 1, 2, 3, 4, 5, 6, 7, 8, 9 right? . . . Since a square—this—any square—the square is 10 by 10 no matter how you turn it, it's always going to be . . . the same. So you don't have to measure it again. You can go 9 times 9. Do you understand why? Yeah? OK, So you go 9 times 9 like Gino said, 81. Then you can do 3 by 3, 1, 2, 3, 4, 5, 6, 7, 8—and then again you don't have to measure again you know. It's going to be

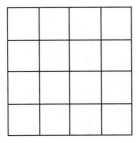

Find all the squares in the figure.
Can you prove you have found them all?
Extension: What if it were a 5 by 5 square? What if it were a 10 by 10 square?
What if it were a 60 by 60 square?

Figure 8.1 The count the squares problem

the same. So 8 times 8—64. And you can keep on going . . . You can do the 7 times 7—49. And 6 times 6—36, 5 times 5—25, 4 times 4—16, 3 times 3—9.

Her argument is proving in that it is based on reference to accepted general principles: "The square is 10 by 10 no matter how you turn it" and established facts (her explicit demonstration of the number of 2 by 2 squares across the top of the grid).

Vicki's students also have the opportunity to record their reasoning in writing, which, if their reasoning is deductive, gives rise to proofs. For example, Micky recorded this proof of a formula for finding the number of diagonals in a polygon:

If you can find all the diagonals possible from one [vertex] you can figure out the whole amount of inside diagonal lines . . . I know that a [vertex] connects with all of the other [vertices] except for 3, itself the [vertex] to the left and right. You subtract 3 from the amount of total sides, . . . here's the rule: (Z = no. of sides) . . . Z − 3 × Z ÷ 2 = no. of diagonal lines in figure. (from Micky's Math Log, May 16, 1994, in Zack, 1995)

Micky's proof has several steps. He is explicit about one: that the number of diagonals from each vertex is Z − 3 (because each vertex is connected by diagonals to all the others, except itself and its neighbors). However, he leaves the justification for another step implicit: every diagonal is counted twice, so the formula ends with dividing by two.

Aspects of Teaching Proving in Vicki Zack's Classroom

In this section, we will illustrate how five key aspects (problem solving, time, conjecturing, expectations, and expertise) contribute to students' learning about proving in the grade 5 classroom of Vicki Zack.

Over a number of years we have collaborated on research on children's reasoning in this context, much of which has been reported elsewhere (e.g., Zack & Reid, 2003, 2004). Here we will shift our focus from the children's reasoning to aspects of teaching. We will present examples of students discussing their solutions for the count the squares problem (see Figure 8.1), specifically for the 10 by 10 and 60 by 60 cases and use these examples as a basis for discussing how several aspects of teaching play out in Vicki's classroom. We also include two "interludes" in which we discuss the reasoning that occurs in the transcripts. (For other examples of students' reasoning in Vicki's classes, see Reid, 2002a, 2002b; Zack, 1999a, 2002; Zack & Graves, 2001; Zack & Reid, 2003, 2004).

Background

Vicki's school is a private school, but draws its students primarily from families with middle-class backgrounds. It is nondenominational, with a population that is ethnically, religiously and linguistically mixed, though most students have English as their first language. Class sizes are in the range 25–27, however in Vicki's classes mathematics was always done in half-groups (12 or 13 children in each group) of heterogeneous ability. While she was working with one group of 12 or 13, the other group was with another teacher (in French, physical education, science, music, or art class).

Problem Solving

Problem solving is at the core of the mathematics curriculum in Vicki's classroom. The school curriculum is based on the reform-oriented Quebec curriculum Défi mathématique/Challenging Mathematics (Lyons & Lyons, 1991, 1996) and has been heavily influenced by the initiatives envisioned by the NCTM Standards (NCTM, 1989, 2000). Non-routine problem solving is central to the mathematics curriculum at all grade levels. It is a problem-solving culture in which the students are expected to support their positions and present arguments for their point of view in most areas of the curriculum.

The problems Vicki chose were not selected according to any criteria except the recommendation of other educators. However, over time, the tasks changed according to what students in previous years had observed. That she used the same problems year after year allowed Vicki to make small adjustments in phrasing or in follow-up tasks that further supported the development of the children's reasoning (Brown, Reid, & Zack, 1998).

The criteria for deciding whether they had solved the problem were in the hands of the students. Specific results were checked by repeating calculations, identifying mistakes, or comparisons with solutions obtained using other methods. The methods themselves were evaluated by the students, according to criteria they developed (see Zack, 2002 for descriptions of counterarguments used by the students to refute an incorrect method of solving a problem).

It should be noted that not all mathematics lessons involved non-routine problem solving. The children learned traditional algorithms and dealt with the language of the mathematics textbook as well, so that they would be familiar with the "formal school textbook" language of mathematics.

The "Problem of the Week" Tasks

Our focus here is on four Problem of the Week tasks, which differ from the other non-routine problem-solving tasks used in the class in that the children were asked to write a detailed explanation when doing these tasks. Over the school year, Vicki assigned approximately 20 Problems of the Week. The sources for these problems were quite varied. The majority came from books on problems solving (e.g., Charles & Lester, 1982; Charles, Lester, & O'Daffer, 1987; Meyer & Sallee, 1983; Moretti, Stephens, Goodnow, & Hoogeboom, 1987). However, two of the tasks (detailed later), Decagon Diagonals and Count the Squares, came from a high school textbook and a teacher's workshop respectively.

Four of the problem of the week tasks will be our focus here: Count the Squares (see Figure 8.1), Prairie Dog Tunnels, Handshakes, and Decagon Diagonals. The prompts for Prairie Dog Tunnels, Handshakes, and Decagon Diagonals are:

Nine prairie dogs need to connect all their burrows to one another in order to be sure they can evade their enemy, the ferret. How many tunnels do they need to build?

If everyone in your class shakes hands with everyone else, how many handshakes would there be?

How many diagonal lines can be drawn inside a figure with 10 sides? [Figures were provided of a triangle, square, pentagon and hexagon, labeled: three sides, zero diagonals; four sides, two diagonals; five sides, five diagonals; six sides, nine diagonals.]

How many diagonal lines would there be in a 25-sided polygon?

How many diagonal lines . . . in a 52-sided polygon?

In these tasks, the children used common techniques such as making organized tables, using diagrams, and searching for patterns in sequences and using differences. Count the squares involves finding the sum for the square numbers from 1 to n^2. The other three problems involve finding the sum for the whole numbers from 1 to n. In them the equivalence of $(1 + 2 + 3 + 4 + \ldots + (n - 1))$ and $n(n - 1)/2$ is discovered through a combination of empirical testing and deduction. Empirical strategies involving drawing all the possible diagonals, or tunnels, or handshakes, are used to establish correct answers when n is small. These give way (as n increases) to reasoning deductively either that each person shakes one less hand than the previous person (the first shakes $n - 1$ hands, the next $n - 2$, and so on) or that each of the n people shakes $n - 1$ hands but that counts each handshake twice so the correct number is $n(n - 1)/2$. The fact that these two methods of reasoning both correctly solve the problem establishes their equivalence. Note that the symbolic formulation used in this paragraph is aimed at an academic audience; it does not represent the language used in the classroom. In Vicki's classroom generalizations are most often expressed as procedures tied to the structure of the problem, for example, "Multiply the number of people by the number of handshakes each person does, which is one less, and divide by two because you counted all the handshakes twice." Vicki then nudges the students to go further, asking them if they can express this idea by using a letter to represent any number of people shaking hands. In some years expressions, like $n(n - 1) \div 2$ are written using variables. Micky's proof (p. 135) shows symbolic formulation of a solution for the Decagon Diagonals problem.

Time

In Vicki's class the children were given a great deal of time to experiment with, think through, discuss and refine their understandings. Each Problem of the Week was assigned on a Monday. The students worked independently on the problem during class time (an extended 90-minute class period) and wrote a detailed description in their math logs. When writing in their math logs, Vicki asked them to "speak" to her in writing as if she were standing at their side. On Monday night, she read the math log entries to see what each child could do independently. She took note of the strategies used, any difficulties and strengths and unique approaches, and the like. On Tuesday, she allowed time for the students to review their logs if they chose to do so and at times prompted some students to reflect on their explanation (for example, to clarify their thinking, elaborate on a diagram or add a diagram).

Wednesday was the day the discussions took place, again in a 90-minute session. The math logs served as an initial basis for the children's discussions. They worked first in pairs (or in a group of three), and then came together in a group of four or five. In these small groups they compared solutions and discussed further, and then reported to the half-class, with more discussion following. On occasion, if the discussions in the small group or the class warranted, class time on Thursday and Friday was also used to allow the discussions to come to a fruitful conclusion. This means that the exploration of the problems occurred over a significant period of time, almost 4 hours per task. Extensive time allotted for work related to proving is also an aspect of other studies in teaching proving at the upper elementary level (for example 10 hours per task in the Italian classrooms described by Boero et al., and the generous time periods in the Maher studies). This allowed for conjectures to be made and explanations sought without being artificially cut short by time constraints. It contrasts sharply with school practices observed by Schoenfeld (1985) involving completing many simple tasks in a short period of time. Schoenfeld sees such practices as leading students to believe that all problems can be solved in just a few minutes. Students who believe that any problem requiring more than 10 minutes to solve is impossible are unlikely to engage in the kind of sustained investigations that gave rise to proving in this context and in the other contexts we have referred to.

Conjecturing

The problems used allowed the students to make hypotheses and discover solutions, which they then proved in order to verify and explain. This example comes from the activity of Maya and Danielle as a pair, discussing their solutions for the count the squares problem, specifically for the 10 by 10 and 60 by 60 cases. In previous classes, they discussed the 4 by 4 and 5 by 5 cases. They solved those cases by systematic counting of the squares, and are now beginning to discover useful patterns. The transcript begins after they have compared their answers for the 10 by 10, found they had different answers, and revisited their calculations. Maya has corrected a small arithmetical error she made and has told Danielle how she attempted a very long calculation to find the answer for the 60 by 60 case, and then made a mistake when she attempted to redo the calculation as a check. Note that Vicki did not intend for them to calculate the exact answer for the 60 by 60 case, only to be provoked by it to think more generally about the problem. However, these students, like many others, were willing to make the effort. (The transcript numbering roughly indicates time elapsed, as the hundreds stand for minutes since recording started.)

Transcript 1

201	*Maya*	So I said, I'm not gonna do it over. I was going to go over here and I said no. I was working on it for 10 minutes, I said, you know what? The important part is the 10 by 10, or that I know how to do it.
202	*Danielle*	See, now—since—I know now how to—use that hunch that I had. I'm going to try—I'm going to try—I'm going to do a quick 60 by 60 and I'm going to see if—I'm not going to count the squares.
203	*Teacher*	[???]
204	*Danielle*	But like I'll draw a 60 by 60—well, I'll try to—well, like the pattern the 7 by 7, the 7 times 7 . . .
205	*Maya*	I did that, 60 by 60 and 9 by 9—you have to go all the way through. I'm sure there is a faster way but that's the way I got it . . .
206	*Danielle*	Never mind.
207	*Maya*	. . . and it works.
208	*Danielle*	Oh.
209	*Maya*	It took me more than a half an hour to add it all up.
210	*Danielle*	But like—did you know when you started you had 60 times 60? Blah, blah, blah.
301	*Maya*	When I started no. I had—I don't know I thought of a pattern. I wasn't sure.
302	*Danielle*	Because now . . .
303	*Maya*	So I started figuring it out and it works for the other ones. That's why I decided to try it. It works for the other ones. Look . . .

In line 202, Danielle refers to a "hunch" that she had about how to do the 60×60 without drawing the squares. In line 301, Maya refers to thinking of a pattern but not being sure. Both are using conjecturing as part of their reasoning. They do not merely conjecture, but also they test their conjectures. Danielle says, "I'm going to try—I'm going to do a quick 60 by 60 and I'm going to see if" (line 202) and Maya checked her pattern by seeing if it applied to the cases she had already worked out and found "it works for the other ones" (line 303).

This kind of conjecturing is typical of the children in Vicki's classes. The processes of conjecturing and proving were intertwined in two ways. Proving made use of insights gained through the explorations that led to conjectures. (For example, cut-out squares of different sizes were used to support the counting of squares, but they were also used in proving. See Maya's proving in the last example and Zack, 2002.) Conjectures were also used as the basis for proving (For example, one child conjectured that the number of squares would always be a multiple of 5, unless the size of the grid was a multiple of 3. He later rejected answers conjectured by his peers if they contradicted his general rule; see Reid, 2002b.)

Expectations

The school is one in which the children are expected to publicly express their thinking, and engage in conjecture, argument, and justification throughout their elementary school life. Vicki used an inquiry-based approach in which the class—students and teacher alike—often pursued questions of personal interest (see, for example, inquiries in the areas of literature and social studies, Zack 1991, 1999b, as well as in mathematics, Zack & Graves, 2001).

The groundwork laid during the year included an expectation that the children would be looking for patterns, and that they could be nudged to think about the mathematical structure underlying the pattern. In addition, there were expectations that everyone's answers should be considered and that answers should not be changed without discussing how they arose and what might be the source of an error. These support the development of beliefs which Lampert (1990) identifies as important to mathematical thinking.

The following transcript illustrates some of the expectations in Vicki's classroom and follows immediately from Transcript 1. In this transcript, the teacher intervenes a number of times in ways that reveal the kinds of communication that were expected in the class. The teacher in this case is not Vicki, who rarely intervened during the small group discussions in the problem solving sessions, but her co-teacher who worked with Vicki throughout the year in support of Vicki's dual role as a teacher–researcher. Vicki's work in establishing expectations was done mainly in whole class discussions and written responses to the math logs.

Transcript 2

| 304 | *Teacher* | Which other ones? Which patterns? |
| 305 | *Maya* | The pattern that I found. |

306	*Teacher*	Which is?

306 *Teacher* Which is?
307 *Danielle* 7 times 7, like . . .
308 *Maya* Yeah, like, did you look at mine?
309 *Teacher* Can you explain it?
310 *Maya* Like over here, over here, for the 60 by 60 where I took them all the way down, um, just that the 1 by 1, you always know that the 1 by 1 is gonna be the side, times the side so that's how I started because you always know that, it's not something you have to figure out. So once I had that I could do 60—and then . . . for the 2 by 2, I did 59 times 59, for the 3 by 3, I did 58 times 58, for the 4 by 4 I did 57 times 57—and so on and so forth.
311 *Danielle* OK, cause like I was thinking, if from the 10 by 10 . . .
401 *Maya* Here, look, can I show you how I think it'll work for this one?
402 *Teacher* Can you just let Danielle finish?
403 *Danielle* If you knew at the beginning, well, from the 10 by 10 you knew you could do 60 × 60, then you'll write it down, just go 60 times 60, . . ., plus the 59, times . . . it would just go down without writing it but if you didn't know that at the beginning you would have to go through all the . . .
404 *Maya* But it takes a while to add it all up.
405 *Danielle* But didn't like um . . .
406 *Teacher* Danielle, you're talking about 60 by 60 or 10 by 10?
407 *Danielle* 60 by 60.
408 *Teacher* Because you said for the 10 by 10 if you knew . . .
409 *Danielle* Oh, because I meant by she did 7 by 7, 6 by 6, 5 by 5, well for 60 by 60, if you knew that it worked for the 10 by 10, maybe you could just go 60 by 60, then 59 times 59, 58 times 58, and those like she knew from the last one. But if you didn't know then you would have to go through all the steps. But you were halfway through, did you know that it would just go, . . . maybe not halfway, but when you started doing the first couple, did you know that it would go 58, 58, 58 times 58?
501 *Maya* Well I did the 10 first. I wouldn't have tried the 60 by 60 until I found a pattern.
502 *Danielle* So you basically knew what you had to do but you just had to write it down.

The teacher's interventions are of three types: she intervenes to request clarification of specific points (lines 304, 406), to ask for explanations (line 309), and to remind Maya to let Danielle finish what she is saying (line 402). These three types of intervention suggest three expectations in Vicki's classroom. Clarity, explanation, and attention to others were valued.

This valuing of clarity, explanation, and attention to others supported the expression of the students' thinking, whether it involved proving or not, but also made that thinking available to others to question. At the close of each session, Vicki distributed a sheet entitled "Helpful explanations/Helpful ideas" and asked the children to note any ideas or explanations they found helpful, to tell why, and to credit the peer(s) who helped. She found that the children became increasingly aware of the contribution others had made to their understanding, and at times could indicate how they have reshaped others' ideas to make them their own (see Zack & Graves, 2001 for further elaboration).

It is possible that Vicki's expectations were informed by her interest in how meaning is made through dialogue. She focused on the quality of the discussions and on the written explanations in the math logs as communicative texts. Her own criteria for judging proofs and other expressions of reasoning were not at the level of the reasoning itself (although that mattered) but more at the level of the clarity in the children's talk about their reasoning. She has observed that on several occasions she only understood a child's argument later, in some cases years later, when she heard it again from another child (Zack, 1995), or when it had been analyzed by a mathematician (Zack, 2002).

INTERLUDE 1: REASONING FROM ONE CASE TO ANOTHER

Before we go on to offer another example of aspects of teaching in Vicki's class, we would like to comment on the reasoning that Maya and Danielle are using. Maya's pattern, which she is using to solve the 60 by 60, is based on her solution of the 10 by 10 (see earlier). This is why: "The important part is the 10 by 10, or that I know how to do it" (line 201). Her comment echoes Lampert's (1990) focus on the method of finding the answer over the answer itself. Her method is to apply a pattern she has seen in the 10 by 10, and which she has checked that "it works for the other ones" (line 303), to solving the 60 by 60. She explains her starting point on the basis of a general rule for all cases, "you always know that the 1 by 1 is gonna be the side, times the side so that's how I started because you always know that, it's not something you have to figure out" (line 310). She then applies her way of solving the 10 by 10 to the 60 by 60, "for the 2 by 2, I did 59 times 59, for the 3 by 3, I did 58 times 58, for the 4 by 4 I did 57 times 57—and so on and so forth" (line 310). She does not express this pattern in general terms (e.g., if she had said "for the 2 by 2s you multiply the number that is one less") but her application of it to another case suggests that it has become a general rule for her, that she can then apply to any specific case.

Her method works, but it is clearly impractical to add up all the square numbers from 1 to 3600. Danielle seems to be looking for a way to make the process more efficient, perhaps by reusing the results from the 10 by 10 case: "If you knew that it worked for the 10 by 10, maybe you could just go 60 by 60, then 59 times 59, 58 times 58, and those like she knew from the last one" (line 409). She may be thinking that after they have calculated the sum from 60 by 60 to 11 by 11, they could then add the result from the 10 by 10, but this idea is never completely articulated. We will see later that a similar idea was used by the other girls in their group of five.

Expertise

In terms of her mathematical background, Vicki could be considered a typical elementary school teacher in that she describes her background in formal mathematics as weak. She is not an expert on the nature and role of mathematical proof, and has not formally proven anything herself since high school. She was a full-time teacher researching her own class, not a university researcher teaching mathematics in an elementary school (like, for example, Lampert and Ball) or a teacher + researcher team (like Maher & Martino and Boero, Mariotti & their colleagues). However, her research interest in how meaning is constructed in dialogue led to a close look at the children's ways of expressing their ideas, and then in turn at issues of convincing and proving.

In our final example we will see how Vicki's expertise fosters the development of her students' mathematical reasoning, even though her expertise is not in mathematics. In it Samantha describes her method of solving the 60 by 60 Count the Squares problem, by making reference to some mathematical content beyond what has been taught in the class. Her terminology is not quite perfect (she says "square root" when she means "square number"), but somewhere she has learned to recognize square numbers.

Transcript 3

1803	*Samantha*	Oh, remember yesterday, when you [Maya] pointed out the pattern, I used that to get it—I did it by the pattern.
1804	*Maya*	What pattern?
1805	*Samantha*	Halfway through—using the pattern, it sort of flipped over . . .
1806	*Maya*	Oh yeah?
1807	*Samantha*	1, 1, 4, 4, can like everybody see? Then 9, 9, and I kept using it and when we got half way through, I just kept using the pattern and I realized that all of the answers are square roots, like in order, like the least to the biggest—the largest I mean.
1808	*Maya*	What do you mean?
1901	*Samantha*	Like, 1 is the first square root, the smallest, 4 is the second, 9, 16, 25 . . . these are all the square roots from smallest to largest.
1902	*Maya*	Oh, of 385?
1903	*Samantha*	No just in general. All the way to 100.
1904	*Maya*	Oh!
1905	*Samantha*	This is for 10 by 10, 9 by 9, 8 by 8, 7 by 7, OK?
1906	*Maya*	Oh yeah.
1907	*Samantha*	What we did for the 60 by 60—can we start talking about the 60 by 60? . . . OK what we were doing for the 60 by 60—we stopped at 10 by 10, so we continued 11 times 11, we got 121, we kept doing all the square roots in order until we get to 60. Then we're going to add, split it up separately and add all our answers together.

Samantha introduces the term "square root" as if the others will understand immediately what she is talking about, but it is not clear that they do. Maya's response "Oh, of 385?" suggests she is thinking of factors. The others are silent. In the moment, it didn't really matter, as they could recognize from her written work that these numbers are generated by multiplying a number by itself, a concept they understood. Their understanding was good enough for their needs in that moment (see Zack & Reid, 2003, 2004 for additional discussion of good-enough understanding).

Vicki's role in connecting students' informal language with standard conventions became evident later, when Samantha presented her method in the half-class discussion. Vicki responded by repeating what Samantha said, changing the terminology to the correct usage, and taking the opportunity to discuss the concept, terminology and notation for square numbers.

Working with the children's ways of making meaning is central to Vicki's teaching. Her expertise lay in listening closely, recognizing potentially fruitful avenues and seizing opportunities to provoke discussion. There was a constant expectation for explanation (e.g., asking "Explain how it works and why it works"), and for generalization ("Will it always work?," "Can you construct a general rule?"). Vicki discovered the richness of the mathematical tasks in large measure due to the children's ways of solving the problem. The tasks lent themselves to algebraic thinking and generalizing and proving. The interest in looking at proving arose from the tasks. There had been no a priori objective to look at proof and proving.

INTERLUDE 2: DIFFERENT PATTERNS OF REASONING

In lines 1803–1807, Samantha refers to a pattern Vicki had pointed out to them at the end of the class discussion of the 4 by 4 case. In that case the sizes of the squares: 1, 4, 9, 16, are the same as the number of squares of each size: 16, 9, 4, 1, but in reverse order. This is a pattern that had been observed in previous years (see Reid, 2002b) and Vicki introduced it into the discussion as it had not emerged spontaneously in this year. From Samantha's comment "remember yesterday, when you pointed out the pattern" (line 1803) it seems that Maya had discussed this pattern with Samantha earlier, but Maya's comments suggests it has not been central to her thinking (lines 1804, 1806). This pattern could have provided another path to Maya's discovery that the 60 by 60 could be solved by adding together the products from 60 by; 60 to 1 by 1. But it was a path not taken.

Although Samantha also did not pursue this pattern further, it did allow her to generate enough data to observe that the numbers are all square numbers (line 1807), which led her to her method for solving the 60 by 60, adding together the products from 1 by 1 to 60 by 60 (line 1907). She was also able, as Danielle had wanted to do earlier (line 409), to reuse the results from the 10 by 10 in her solution to the 60 by 60.

Both Samantha and Maya arrive at methods for solving the 60 by 60 that work. The calculations (except for the order) are the same. But the *reasoning* involved is different. For Maya, it is the process, i.e., starting with the number of 1 by 1 squares and then reducing the numbers to be multiplied by 1 each time, that she generalizes. For Samantha, it is a special feature of the products, being square numbers, that suggests a generalization. Two quite different paths led to their solutions. That they had the opportunity to reason in two different ways is due to the openness of the problem, the opportunities for conjecturing, and the environment of communication in the class. The problem is accessible to all the students as the concepts of squares and counting them are familiar to them. They can all begin, by counting, more or less systematically, the squares. Once they have begun counting, however, many features of the problem can give rise to generalizations. In Maya's case, it was her system of counting that she generalized. In Samantha's case, it was the results of the counting for each size of square. If the problem had not allowed for these different approaches, the reasoning used to solve it might have been less varied. The invitation to conjecture was also important. Conjectures are valued in the class (community expectations include attention to other people's conjectures). Without this support, Maya and Samantha might not have made their conjectures in the first place, let alone offered reasoning to support them. Finally, the culture of this classroom invites reflection and inquiry and requires of the children that they express their reasoning, making it accessible to others and open to requests for clarification and explanation.

Conclusion

The example of Vicki Zack's classroom (and the others cited earlier, as well as psychological research on reasoning) show that students in elementary schools can learn to prove. And if we want their image of the nature of mathematics to be one of a human discipline of knowing and explaining, rather than a set of rules to be memorized, they *should* learn to prove. But this poses a challenge, for many elementary school teachers do not feel they have the mathematical background to teach proving, or even a clear idea of what proving looks like. Can we, on the basis of the aspects of teaching identified here, begin to address this challenge?

We believe the example of Vicki's teaching can be useful here. Teaching experiments like those described by Ball and Bass (2003), Boero, Mariotti, and their colleagues (Boero, Garuti, Lemut & Mariotti, 1996; Boero, Garuti & Mariotti, 1996), Lampert (1990, 1998a, 1998b), and Maher and Martino (1996a, 1996b) may not convince many teachers that they can teach proving because in those cases the teachers had strong mathematics backgrounds or support of a university team of researchers. Most elementary school teachers are not in this position. However, in Vicki's class, her students also learned to prove. It is interesting to consider why this might be so. The combination of three characteristics of her class activities described in this chapter seems central: conjecturing, leaving the criteria for a correct solution up to the students, and expectations for communication. Because the students were making conjectures, they had to have a certain level of understanding of the problem, and they had a personal investment in determining whether or not their conjectures were correct. Because they had to develop their own methods and criteria for verifying their conjectures, they explored ways of reasoning and arguing. Because they had to communicate their arguments to their peers and according to social expectations, they had to be careful in how they formulated their reasoning, they had to be able to explain it clearly, and they had to be willing to reformulate their reasoning if their expression of it was not satisfactory. These three characteristics put an emphasis on reasoning and on arguing. And they seem to be within reach of any elementary school teacher, making teaching proving in elementary schools a possible aim.

Note

1.   This article is based on research conducted as part of two related research projects supported by Social Sciences and Humanities Research Council of Canada, under the direction of the co-authors and funded under grants #410-98-0085, #410-94-1627 and #410-98-0427.

III

Teaching and Learning of Proof in Middle Grades and High School

The chapters presented in Section II provide insight into efforts being made to enrich the nature of elementary schoolchildren's experiences with, and subsequent understandings of, proof. A primary goal of these efforts is to expand and refine students' conceptions of what constitutes evidence and justification in mathematics in ways that build a strong foundation for their experiences with proof in secondary school. Such a goal is important as, traditionally, secondary school and, in particular, middle school, has marked a significant and often abrupt mathematical transition from the concrete, arithmetic reasoning of elementary school mathematics to the development of the increasingly complex, abstract reasoning required for high school mathematics and beyond.

Proof in secondary school has historically found its home in the high school geometry course, and has appeared rarely, if at all, in other high school mathematics courses (and has been virtually nonexistent in middle school mathematics). As noted in the introduction of this volume, however, recent reform efforts have called for significant changes with regard to proof in K–12 mathematics with proof expected to play a much more prominent role throughout the entire school mathematics curriculum. In secondary school, proof is now expected to be a regular and ongoing part of students' mathematics experiences—experiences that are expected to build on students' elementary school experiences with proof. Such experiences should serve to further sharpen and extend students' understandings of proof so that their standards for accepting explanations become more stringent and their methods of reasoning and proof become more sophisticated (NCTM, 2000).

The seven chapters presented in this section highlight aspects of both learning and instruction that provide further insight into students' understandings of proof as well as the kinds of experience with proof that are necessary to strengthen the foundation laid during students' elementary school experiences.

Developing Secondary School Students' Understanding of Proof

Two of the chapters in this section, those by Knuth, Choppin, and Bieda (Chapter 9) and Küchemann and Hoyles (Chapter 10), document patterns in students' mathematical reasoning as they progress through their middle school mathematics education in the United States and England, respectively. In the former chapter, Knuth and his colleagues report results that show modest improvements in students' proving and justifying competencies from the time they begin middle school to the time they complete it. The results also suggest, however, that there is plenty of room for improvement: students demonstrated an overwhelming reliance on the use examples as a means of demonstrating and/or verifying the truth of statements or propositions. To some extent, given traditional school mathematics instruction and curricula, it may be unreasonable to expect more sophisticated justifications from students. It is possible, however, that as the recommendations of reform with respect to proof are more widely implemented—the possibilities of which are demonstrated in many of the chapters in this volume—students will gain more experience with reason giving and begin to develop more advanced means of justifying.

In their chapter, Küchemann and Hoyles focus on student responses to tasks designed to elicit structural reasoning—a form of reasoning that relies on the underlying mathematical structure as a basis for creating an argument (a form of reasoning that also appears, for example, in Deborah Schifter's chapter as students discuss the parity of the sum of two even numbers). They view this form of reasoning as foundational to more sophisticated ways of reasoning (and to proof, in particular). Similar to the level of student progress reported by Knuth and colleagues, Küchemann and Hoyles found modest improvement in students' use of structural reasoning as they progressed from grade 7 to grade 9. They note, however, that such progression was not linear (i.e., in some cases, students "regressed") and that such variability may be due in part to students' lack of familiarity and experience with structural reasoning. Taken together, the Knuth et al. and Küchemann and Hoyles chapters underscore the need for students to have earlier and ongoing opportunities to engage in proving—a message that is reflected in many of the chapters in Sections II and III.

The chapters by Heinze and Reiss (Chapter 11) and McCrone and Martin (Chapter 12) focus on aspects of high school students' understandings of proof; in the former case, on cognitive and affective prerequisites for mathematical reasoning and proof, and, in the latter, on students' perceptions regarding the nature of proof. Heinze and Reiss report results from their study of high school students in Germany (grades 7/8 and 12/13), and take into account not only student performance on proof tasks, but also affective factors such as student motivation and interest (factors typically not considered in much of the research on proof). Although the results show patterns of student performance that are consistent with previous studies, the results also suggest that affective factors may play a role in student performance. In particular, student interest in mathematics was found to be a correlate with performance, a finding that highlights the potential importance of classroom environments in which students are engaged in doing mathematics on a regular and consistent basis.

In their chapter, McCrone and Martin discuss results from their study involving students enrolled in proof-intensive geometry classes in the United States. Through their use of interviews and survey questionnaires, they found that a large number of students judged the validity of proofs by attending to ritualistic aspects of their format (e.g., two columns) rather than attending to the internal logic of the arguments. They also found that students often "appeared to hold some contradictory perspectives on the validity of proof," noting, for example, that although students tended to view general arguments (i.e., proofs) as being "better" than arguments that were based on empirical evidence, they nonetheless also accepted empirically based arguments as valid proofs. The authors conclude that student difficulty might be attributed, at least in part, to students' failure to distinguish between the culture of mathematics and valid arguments in that domain and "everyday" culture and valid arguments in that domain.

The preceding four chapters all raise important concerns about both middle school and high school students' understandings of proof as well as important issues to consider regarding instructional practices and curricula in relation to proof. Although many of the chapters in this volume provide examples of K–16 students engaged meaningfully in proving, it is critical to efforts aimed at enhancing students' understandings of and experiences with proof that such examples become more the rule rather than the exception. In the final three chapters of this section, the authors help us understand classroom practices and teaching in ways that may serve to make the aforementioned examples more the rule than the exception.

Teaching Practice that Supports Proving

Enhancing the role of proof in the classroom requires substantial effort on the part of teachers insofar as they are responsible for ensuring students have the means as well as the opportunities to engage in proving. In Chapter 13, McClain details her efforts to enhance the role of proof in the classroom as she analyzes teaching episodes from her own teaching in a middle school mathematics classroom. She argues that the process of engaging students in significant mathematical discussions can serve as a resource for supporting students' emerging understandings of proof. She further notes that this process of engaging students in significant mathematical discussions "created a setting in which mathematical arguments went through an evolutionary process of refinement leading to more efficient and sophisticated arguments" (Chapter 13, p. 233). McClain's chapter also serves to highlight the critical link between students' opportunities to engage in proving (i.e., the tasks with which students are engaged) and instructional practices that guide what students ultimately learn from such opportunities.

Stylianides and Silver (Chapter 14) also touch on the link between proving opportunities and instructional practices but take a slightly different tack: they focus on pattern identification tasks (a popular type of task particularly in reform-based curricular materials) and middle school teachers' discussion of the implementation of such tasks. In their chapter, Stylianides and Silver analyze teachers' engagement with a pattern identification task in the context of a professional development session. Although the task was one in which it was possible to provide conclusive evidence for the identified pattern, they note that the teachers tended to have difficultly connecting the identified pattern to the underlying structure that made the pattern uniquely determined. The fact that the teachers had difficulty reasoning structurally provides insight into the similar difficulties students displayed in the chapter by Küchemann and Hoyles (Chapter 10)—in order to foster students' abilities to reason structurally, teachers must possess such abilities themselves. This latter point is important given not only the prevalence of such tasks in reform-based curricula, but also their role in promoting students learning to prove.

The final chapter in this section, by Herbst and his colleagues (Chapter 15), focuses on proof in high school geometry courses and, in particular, on the work of proving in geometry from both a historical perspective and from the perspective of geometry teachers. They also provide classroom excerpts of geometry students engaged in proving and discuss their analysis of these excerpts in terms of the systems of norms that regulate students' engagement in proving. They argue that "doing proofs" in geometry classrooms rarely functions to provide opportunities for students to use proof as a tool for learning mathematics, but rather functions primarily as a tool for teaching logical thinking. Their conclusion points to the need to reconceptualize the work of "doing proofs" if conjecturing and proving activities are to play a meaningful role in students' learning of mathematics.

Common Themes and Issues

The seven chapters that comprise this section address and highlight important issues regarding the learning and teaching of proof in secondary school mathematics classrooms. A theme that is clear across the first four chapters is that developing students' understandings of proof will be challenging given students' current difficulties with proof. Each of these chapters provides insight into or suggestions for curriculum and instruction changes that hold promise to both enrich students' experiences with proof and enhance the understandings students develop as a result of such experiences. Knuth and his colleagues stress the importance of helping students understand what constitutes sufficient evidence in mathematics and the need for teachers to develop their "proving eyes and ears," that is, their ability both to recognize opportunities with respect to proof as they arise during instruction and in curricular materials and to capitalize on such opportunities. Küchemann and Hoyles call for curricular and instructional emphases on structural reasoning, a type of reasoning they view as essential to developing an understanding of and the ability to produce proofs. Heinze and Reiss also suggest that affective factors such as motivation and interest play a role in students' learning to prove and need to be taken into account as teachers work to enhance students' understandings of proof. Finally, McCrone and Martin highlight the importance of developing students' understandings regarding the nature of proof itself—what is a proof, what does it mean to prove something, and why is proof even necessary? They argue that teachers need to facilitate students' enculturation into the mathematics community so that students develop an appreciation for and understanding of mathematical ways of thinking and knowing.

The final three chapters in this section all focus on teachers and teaching and illustrate aspects of practice that impact students' learning to prove. McClain provides an account of her own instruction that leads to the evolution of her students' arguments, focusing, in particular, on changes in what counted as a mathematical argument in her classroom community. She argues that such an evolution in students' mathematical arguments is fundamental to the development of their understandings of proof. In McClain's chapter, the role of the teacher is clearly important to the evolution of students' thinking, a role that is also emphasized in the Stylianides and Silver chapter. In their chapter, they step out of the classroom per se, and focus on professional development in which teachers engage in proof-rich patterning tasks. They suggest that if teachers are to help students develop their understandings of proof, then teachers themselves also need support in developing their own understandings of proof. And finally, Herbst and his colleagues challenge mathematics educators to rethink the practice of doing proofs in geometry, suggesting that traditional proving practices in high school geometry classrooms rarely provide students with opportunities to use proving as a means for learning mathematics.

As a collective, the chapters in this section underscore the challenges the mathematics education community faces with regard to enhancing the role of proof in secondary school mathematics. Perhaps as reasoning and proof become a regular and ongoing part of elementary schoolchildren's mathematics experiences, we will see more students enter secondary school with an appreciation for and understanding of proof. Likewise, as we see secondary school students gain further experience with proof, they will likely be better prepared and more successful in proof-intensive courses at the tertiary level.

9
Middle School Students' Production of Mathematical Justifications[1]

ERIC J. KNUTH, JEFFREY M. CHOPPIN, AND KRISTEN N. BIEDA

The nature and role of proof in school mathematics has been receiving increased attention in the mathematics education community with many mathematics educators advocating that proof should be a central part of the mathematics education of students at all grade levels (Ball, Hoyles, Jahnke, & Movshovitz-Hadar, 2002; Knuth, 2002a, 2002b; Schoenfeld, 1994; Sowder & Harel, 1998). Such attention is also reflected in current mathematics education reform initiatives. In contrast to the status of proof in the previous national standards document (National Council of Teachers of Mathematics [NCTM], 1989), its position has been significantly elevated in the most recent document (NCTM, 2000). In particular, the *Principles and Standards for School Mathematics* (NCTM, 2000) recommends that the mathematics education of prekindergarten through grade 12 students enable all students "to recognize reasoning and proof as fundamental aspects of mathematics, make and investigate mathematical conjectures, develop and evaluate mathematical arguments and proofs, and select and use various types of reasoning and methods of proof" (p. 56). These recommendations, however, pose serious challenges for both teachers and their students given that many students have found the study of proof difficult (e.g., Bell, 1976; Chazan, 1993; Healy & Hoyles, 2000; Porteous, 1990; Senk, 1985; Usiskin, 1987).

Thus it comes as no surprise that considerable efforts are being made to enrich the nature of students' experiences with, and subsequent understanding of, proof. In fact, such efforts are reflected in the design of many reform-based curricula—curricula purportedly designed to foster the development of students' justifying and proving competencies. For example, the authors of the *Connected Mathematics Project* (CMP) curriculum assert the following: "Throughout the curriculum, students are encouraged to look for patterns, make conjectures, provide evidence for their conjectures and strategies, ... Informal reasoning evolves into more deductive arguments as students proceed from grade 6 through grade 8" (Lappan et al., 2002, p. 8). Yet, very little research has examined the justifying and proving competencies students in classrooms utilizing reform-based curricula develop. Accordingly, the goal of this chapter is to present and discuss results concerning the nature and range of middle grade (grades 6–8) students' production of mathematical justifications in classrooms implementing CMP.

Proof Production Framework

Researchers have hypothesized that the development of students' competencies in justifying and proving might follow a developmental progression; that is, students' understanding of mathematical justification are "likely to proceed from inductive toward deductive and toward greater generality" (Simon & Blume, 1996, p. 9). Indeed, various proof frameworks have been proposed that reflect such a developmental progression (e.g., Balacheff, 1987; Bell, 1976; van Dormolen, 1977; Waring, 2000). We found Waring's (2000) framework, in particular, compelling for thinking about students' production of mathematical justifications. Based both on our research agenda and analyses of pilot data, we modified her framework and discuss the particulars of our revised framework in what follows.

Proof Production Levels

Proof Production Level 0

Students at this level appear unaware of the need to provide a *mathematical* justification to demonstrate the truth of a proposition or statement. For example, a student might accept a proposition as true because a teacher, parent, or text "says" it's true (cf. Harel & Sowder, 1998); in this case, the justification is "non-mathematical." In other cases, a student might simply state a proposition is true without any reference to why the proposition is true (e.g., "the sum of two even numbers is even because that is just the way it is," "yes the numbers will be equal because they will always be equal").

Proof Production Level 1

Students at this level appear to be aware of the need to provide a mathematical justification, but their justifications are not general; in the majority of cases, students' justifications are empirically based. Among the empirically based justifications, we recognize distinctions (at sublevels) among students who consider checking a few cases, students who consider systematically checking a few cases (e.g., even and odd numbers), students who consider checking extreme cases or "random" cases, and students who consider the use of a generic example (proof for a class of objects) (cf. Balacheff, 1987).

Proof Production Level 2

Students at this level appear to be aware of the need for a general argument, and often attempt to produce such arguments themselves; the arguments, however, fall short of being acceptable proofs. "Falling short" may happen in one of two ways: (1) Students express recognition of the need to provide a general argument and attempt to produce such an argument, however, the argument provided is not a viable argument (i.e., the argument is either incorrect mathematically or it would not lead to an acceptable proof). (2) Students express recognition of the need to provide a general argument and attempt to produce such an argument, however, the argument is incomplete (if completed, the argument would be an acceptable proof). In both situations, the point is that students are attempting to treat the general case. In addition, Level 2 justifications also include responses from those students who demonstrate an awareness that empirical evidence does not suffice as proof—by either expressing recognition of the need to deal with all cases or expressing recognition of the limitation of examples as proof—but who are unable to produce (or attempt) a general argument.

Proof Production Level 3

Students at this level appear to be aware of the need for a general argument, and are able to successfully produce such arguments themselves. We consider the arguments students produce at this level to be acceptable proofs; that is, their arguments demonstrate that a proposition or statement is true in all cases. Arguments categorized at this level typically involve a reference to any assumptions or givens, a chain of deductions used to build the argument, and finally an explicit concluding statement. Although the arguments students produce may lack the rigor or formality typically associated with a proof, their arguments, nonetheless, do prove the general case.

Methods

The data that are the focus of this chapter consist of students' responses to a subset of items from two written assessments that targeted their understandings of various aspects of proving. Both assessments required students to evaluate various mathematical arguments, generate their own arguments, demonstrate their understanding of implication rules and deduction, apply definitions, and discuss the nature of proof. The specific focus of this chapter is on six items for which students were asked to generate justifications regarding the truth of mathematical propositions or statements.

Participants

Approximately 400 middle school students (6th through 8th grades) participated in the study; the students all attended the same middle school. The demographic breakdown of the school's student population is as follows: 29% African American, 8% Hispanic, 12% Asian, and 51% White. The middle school had recently adopted the CMP curriculum and, with the exception of one section of 8th grade algebra, the classes were not tracked (e.g., all 6th grade students were in the same mathematics course). The fact that the school was using CMP is noteworthy given the program's aforementioned emphasis on mathematical reasoning; as such, one might expect "more" from these students in terms of the justifications they produce as compared to their peers in schools implementing more traditional curricula.

Data Collection

Data were collected across 2 school years; each fall, students responded to a different written assessment (the overall focus of which was briefly described earlier). For the first assessment, data were collected from 394 students and, for the second assessment, data were collected from 426 students. As a result of administering the assessments over 2 consecutive school years, there was an overlap of some students who took both assessments; for example, students who took the first assessment as 6th graders, took the second assessment as 7th graders. For this chapter, we present the relevant data from each assessment as the items that required students to produce justifications differed across the two assessments. Moreover, the primary goal of the chapter is to present a picture of middle school students' production of justifications, not to compare the nature of students' justifications from one year to the next.[2]

The six assessment items varied along two dimensions. First, the context for three of the items related to "number tricks" (e.g., "Show that if you take a number, multiply it by 2, add 2 to the product, and divide the sum by 2, the result is always 1 more than the original number"), while the context for the other three items related to number properties (e.g., "Show that the sum of two odd numbers is always even"). Second, the prompts to justify varied among the items, varying from asking students to provide a justification that would convince a classmate to asking students to show that a statement is always true. Although we initially conjectured that we might observe systematic differences in the nature of the justification students provided depending on the particular prompt, this conjecture did not play out in the data.

Data Analysis

The students' responses to the six focus items were analyzed using the proof production framework detailed earlier. In particular, students' responses were coded as representing one of the four proposed proof production levels (as well as sublevels within each level), as non-responses (i.e., items that students left blank), or as non-codable. In the last case, student responses falling into the non-codable category were typically nonsensical (e.g., "an even number plus an even number is even because blue eyes plus blue eyes make blue eyes"), based on a misinterpretation of the item (e.g., student interpreted *consecutive* integers as meaning two *different* integers, such as 5 and 8), or based on computational errors (e.g., a student might conclude that a number trick does not work based on a calculation error).

Results and Discussion

In structuring this section, the results for one set of similar items (i.e., number trick items) are first presented, followed by a brief discussion comparing the items and their results. This structure is then followed for the second set of similar items (i.e., number property items). Last, the primary findings are summarized across all six items. Representative excerpts from students' written responses are provided throughout the presentation of the results in order to illustrate particular findings.

The Number Trick Results

Each number trick item required students to follow a series of calculations and to justify that the ensuing result would always be the same regardless of the starting number. For each item, one or two worked-out examples were provided that illustrated the procedure outlined in the narrative of the problem.

THE THREE NUMBER TRICK

This item was stated as follows: "Mei discovered a number trick. She takes a number and multiplies it by 5, and then adds 12. She then subtracts the starting number and divides the result by 4. She notices the answer she gets is always 3 more than the number she started with. Malaika doesn't think this will happen again, so she tries the trick with another number. Mei and Malaika decide that they will always get a result that is 3 more than the starting number. Do you think they are right? How would you convince a classmate that you would always get a result that is three more than the starting number?"

In this case, there was a predominance of level 1 responses at all grade levels, with examples-based justifications being the primary means of justification within this level. Responses from 6th grade students were categorized at this level 78% of the time, 79% for 7th grade students, and 81% for 8th grade students. The following are typical responses:

> "I would convince them by telling them to give me any number and I would prove to them that it was true." (6th grade student)

> "I would do a few experiments just as Mei and Malaika did to prove my statement." (7th grade student)

> "I would try the strategy a few times to prove it works." (8th grade student)

Given the relatively complex mathematical relationship underlying this number trick, it is not surprising that so many students relied on examples as their means of justification. Indeed, very few students actually attempted a general argument (level 2 or 3)—none of the 6th grade students, fewer than 5% of the 7th grade students, and fewer than 10% of the 8th grade students, with only 2% of the 8th grade students able to successfully produce a proof. The following is an example of a level 2 justification:

> "After multiplying the number by 5, you would add 12, then take away that number. Dividing is taking the number and seeing now many times it goes into it, and you already subtracted one of the original numbers. You would divide a 4 because you originally multiplied by 5 and subtracted one." (7th grade student)

In this case, the student seems to recognize that the result of multiplying the original number by 5 and then subtracting the original number leaves an amount that is four times the original number (i.e., "you would divide a 4"), however, the student's argument is incomplete. In contrast, the following response is an example of a level 3 justification:

> "To convince a classmate I will show them this. It multiplies the number by 5 then subtracts the number, which is also 4 times the number. Then it divides the number by 4, that leaves the starting number. It also adds 12 then divides by 4, 12/4=3. So it has the starting number plus 3." (8th grade student)

Interestingly, of those students who attempted a general argument (successful or not), none of them produced (or even attempted) a symbolic-algebra justification (i.e., $(5x + 12 - x)/4 = x + 3$), arguably the easiest means of justification.

THE 20 NUMBER TRICK

This item (adapted from Bell, 1976) was stated as follows: "Amy and Stephan are trying out a number trick. Amy picks a number between 1 and 10. She adds it to 10 and writes down the answer. She takes the starting number away from 10 and writes down the answer. Then she adds the two answers from the two steps. Stephan picks a number between 1 and 10. He adds it to 10 and writes down the answer. He takes the starting number away from 10 and writes down the answer. Then he adds the two answers from the two steps. What do you notice about the two final answers? Will you always get the same final answer no matter what your starting number is? How would you convince a classmate that you would always get the same answer?"

Level 1 justifications were again the most popular means of justification, although fewer students overall produced such justifications—35% in 6th grade, 51% in 7th grade, and 49% in 8th grade—in comparison to the previous number trick results. Across all three grade levels, there were more attempts to produce general arguments (level 2 or 3); 6% in 6th grade, 14% in 7th grade, and 14% in 8th grade. The following response is a typical level 2 justification: "If you do

$10 - n$ it would equal $10 - n$. If you do $10 + n$ it would equal $10 + n$. So what you're doing is taking away the same amount you are adding" (7th grade student). In this case, the student was clearly on the "right track," however, the argument is incomplete. The following response is similar, however, in this case the student produced a complete argument (a level 3 justification): "You add $10 + x$ with $10 - x$, you can cancel the xs out and you have $10 + 10 = 20$" (8th grade student). Although these two particular students utilized symbolic algebra in their justification, in general, very few level 2 or 3 student justifications involved symbolic algebra; a verbal counterpart of the symbolic-algebra justification was most often produced at these levels.

In contrast to the previous number trick problem, students could prove the invariance of the result by testing the entire set of possibilities, that is, students could try the trick with each number between 1 and 10 in order to show that the result is always 20 (i.e., proof by exhaustion). In fact, a significant number of students (21% in 6th grade, 16% in 7th grade, and 20% in 8th grade) did indeed utilize the method of proof by exhaustion. Typical responses included:

"Go through the procedure step-by-step with every number (1–10)." (6th grade student)

"I would show it to them using every number from 1 to 10." (7th grade student)

"I would try every starting number (1–10) and prove that they all equal the final number of 20." (8th grade student)

Finally, a significant proportion of student responses were categorized as non-codable and, in 6th grade, a significant proportion of responses were also categorized as level 0 justifications. In the former case, there were many students at each grade level who simply responded "I don't know" with regard to how they would convince a classmate that the number trick always produced a sum of 20. In the latter case, most of these 6th grade students simply stated that the number trick would always produce a sum of 20 but did not provide any justification (e.g., "Because it will always come up with the answer of 20").

THE EQUAL NUMBER TRICK

This item (adapted from Porteous, 1990) was stated as follows: "Sarah discovers a cool number trick. She thinks of a number between 1 and 10, she adds 3 to the number, doubles the result, and then she writes this answer down. She goes back to the number she first thought of, she doubles it, she adds 6 to the result, and then she writes this answer down. Will Sarah's two answers always be equal to each other for any number between 1 and 10? Explain your reasoning."

Once again, a significant proportion of students at all three grade levels provided examples-based arguments to justify that the number trick always produced the same result for a given starting number—36% of 6th grade students, 30% of 7th grade students, and 31% of 8th grade students generated such arguments. The results also show that substantially more students (in comparison to the previous two problems) produced or attempted to produce a general argument; in this case, 17% of 6th grade students, 23% of 7th grade students, and 27% of 8th grade students. The following responses are typical of level 2 justifications:

"Yes, because if she always adds 3 then doubles your answer and gets that answer. Then goes back and does it in a different sequence she'll always get the same answer." (6th grade student)

"Yes, when she doubles it plus 3, 3 is doubled, like the original number. If she doubles first, she needs to add double of 3, or 6, to the number." (8th grade student)

Based on these representative responses, it is clear that the students are attempting to justify why the number trick always produces the same number, however, in our view they have not satisfactorily articulated the underlying mathematical relationship for the responses to be coded as level 3 justifications. Students whose justifications were coded as level 3 justifications were better able to articulate the underlying mathematical relationship. The following response are representative:

"Yes, this will always work because in the first step she adds 3 and then doubles. In the second step she doubles and adds 6, but the answer is the same because in step 1 she added and then doubled which adds an extra 3 so that's why we add 6 in the second part instead of 3." (6th grade student)

"Yes, because $(x + 3)2$ and $2x + 6$ are equal. They do the same thing." (8th grade student)

Again, a primary difference between these responses and the responses coded as level 2 is the degree of clarity or completeness of the justification.

Proof by exhaustion was also a possible method of justification, and indeed a number of students used the method (about 2% in 6th grade, 5% in 7th grade, and 10% in 8th grade). The following response was typical:

"Yes they will always be the same. I know this because I checked (see below) [student shows examples for all the numbers between 1 and 10]." (6th grade student)

Interestingly, in comparison with the previous number trick problem, fewer students used this method as a means of justification. (We discuss this further in the next section.)

Finally, it is worth noting the significant proportion of students whose responses were categorized as non-codable: 30% of 6th grade responses, 26% of 7th grade responses, and 24% of 8th grade responses. The majority of these students seemed to misinterpret the problem, in many cases thinking that the end result must always be 20 (the result provided in the worked-out example that accompanied the problem). The following response illustrates such a (mis)-interpretation: "No, because the number comes out differently if you chose a number like 11. It does not come out as 20" (8th grade student).

COMPARING THE NUMBER TRICK ITEMS

There are several differences in the results across these three number trick problems that are worth further discussion. First, students produced many more examples-based justifications in response to the *three number trick* problem (81% overall) than either of the other two number trick problems (46% and 32% overall). We believe this is primarily due to the more complex mathematical relationship underlying the former problem and, as a consequence, students likely had no recourse but to use examples as their means of justification (i.e., a verbal or symbolic argument based on the mathematical relationship was too difficult). Second, the nature of both the *20 number trick* and the *equal number trick* problems afforded students an opportunity to test all of the possible cases to ensure that each number trick always worked—a proof (i.e., proof by exhaustion) was accessible. And in fact, a significant proportion of students used this method of proof for the *20 number trick* problem (19% of the students overall); the method was also used by students for the *equal number trick* problem, although only by 6% of the students overall. The difference in the proportion of students across the two problems may be due to the fact that the mathematics underlying the *equal number trick* problem might have been easier for students to "see" and thus resulted in more attempts to produce a general argument. Not only did more students use proof by exhaustion for the *20 number trick* problem, but they also produced more level 1 justifications, which suggests that they may have had more difficulty attempting a general argument. Finally, a greater proportion of student responses were categorized as non-codable for the *equal number trick* problem (26% overall) as compared to either the *three number trick* problem (11% overall) or the *20 number trick* problem (16% overall). As was discussed in the previous section, this increase was likely due to these students misinterpreting the outcome of the *equal number trick* problem (i.e., that the result must always be 20 as in the worked-out example).

In considering all of the number trick problems, the *three number trick* problem was arguably the most difficult and, not surprising, had the fewest attempts by students to produce a general argument. In contrast, the other two number trick problems may have been more accessible mathematically to students, with the result being an increase in the number of general arguments attempted.

Moreover, if one considers the justifications that were attempts to produce a general argument (levels 2 and 3) as well as those in which the method of proof by exhaustion was used, the results show a significant proportion of students (roughly 30% overall for each of these two problems) moving beyond examples as a means of justification. Interestingly, of those students who attempted to produce a general argument, very few attempted to use symbolic algebra as a means of justification. Porteous (1990) also found "a total absence of algebra" (p. 595) in the justifications produced by students in his study to a problem similar to the *equal number trick* problem. Although the use of symbolic representations (i.e., variables) is introduced in the 7th grade (within the *Connected Mathematics* curriculum), "algebra" has not been a domain of study for most of the middle school students (with the exception of those students enrolled in 8th grade algebra), thus it may not be surprising that very few students used algebra in their justifications. It may also be the case, however, that students lack an "awareness of the usability of algebra in number situations" (Bell, 1976, p. 35), that is, students may not see algebra as a viable means for expressing a general statement.

The Number Property Results

Each of the three number property problems required students to produce a justification regarding the parity of a sum. All of the problems involved mathematical facts with which the students would likely be familiar.

THE CONSECUTIVE NUMBER SUM

This item was worded as follows: "The sum of two consecutive numbers is always an odd number. For example, $5 + 6 = 11$ and $8 + 9 = 17$. Show that the sum of any two consecutive numbers is always an odd number."

Examples-based arguments tended to be the most common approach used by students to show that the sum of two consecutive numbers was odd; this form of argument was used by 54% of 6th grade students, 49% of 7th grade students, and 43% of 8th grade students. A number of students also attempted to produce general arguments (level 2 and 3 justifications), increasing from 12% in 6th grade to 32% in 8th grade. Student responses that were coded as level 2 justifications ranged from those that justified the sum is odd by generalizing (inductively) from a pattern to those that were on the "right track" but were incomplete.

> "[Student shows the following examples: $4 + 5 = 9$; $5 + 6 = 11$; $6 + 7 = 13$; $7 + 8 = 15$] This [i.e., pattern of sums] goes up by 2 every time, and on an odd number if it goes up by two it will stay odd because the odd is every other number." (7th grade student)

> "Whenever you add an even number and an odd number the sum is always odd and any two consecutive numbers is always going to have one odd number in it." (8th grade student)

In the case of level 3 justifications, students typically first defined a particular aspect of consecutive numbers, then used an accepted truth (for middle school students)—the sum of an even number and an odd number is always odd—to deduce the conclusion. The following are representative of this type of justification:

"Consecutive numbers are always one odd and one even number no matter whether I started with one even or one odd. Sums of an even number and an odd number are always odd." (6th grade student)

"If you add two consecutive numbers that means you're adding an odd and even number. If you add one odd number and one even number, it ends up always being odd." (8th grade student)

Interestingly, none of the 7th grade students produced a level 3 justification.

Finally, there were a significant number of responses that were categorized as non-codable (about 28% in 6th grade, 26% in 7th grade, and 25% in 8th grade). In the majority of these cases, we believe that a lack of understanding regarding consecutive numbers led to students attempting to provide a justification to the "wrong" problem. In the wording of the problem, two examples were included as a means of "defining" consecutive numbers; however, as it turned out, for many students this may have been inadequate. Many of the students whose responses were categorized as non-codable seemed to interpret consecutive numbers as meaning two different numbers (e.g., 5 and 8 are "consecutive").

THE EVEN NUMBER SUM

This item was worded as follows: "Show that when you add any two even numbers, your answer is always even. Provide an explanation that would convince a classmate that the answer is always even."

Once again, the most common justification students provided was examples based; 69% of 6th grade responses, 50% of 7th grade responses, and 62% of 8th grade responses. It is interesting to note the increase in level 1 justifications in the 8th grade. We speculate that this increase in level 1 justifications, coupled with the significant proportion of level 0 justifications (about 22%) and the relatively small proportion of level 2 and 3 justifications (about 8%) in 8th grade, may be a result of students not seeing any need to justify such a well ingrained mathematical fact. The following student's response lends support to this speculation: "When you add two even numbers you always get an even because it is a simple fact that that happens" (8th grade student).

Interestingly, we did not code any of the student responses as level 3 justifications. Most attempts at producing a general argument focused on the unit's digit and often used the proposition itself as a part of the argument.

"The reason why if you add two even numbers, it is even is because if the number ends with two even numbers it will always be even." (6th grade student)

"No matter what, whenever you add two even numbers together you get the answer as an even number because the last number in the problem is always even." (7th grade student)

In each of these responses, the students are attempting to articulate an exhaustive proof by examples—the unit digit of any even number must end with 0, 2, 4, 6, or 8, adding any two of these numbers produces a sum whose unit digit is also one of these five numbers (thus an even number itself), therefore the sum of any two even numbers must be even. Healy and Hoyles (2000) noted that this method of proof was also attempted by students in their study. Although this method could certainly be used to construct a level 3 justification, none of the students was able to successfully create such an argument.

We were somewhat surprised that very few arguments students produced were based on the structure of an even number (e.g., all even numbers can be partitioned into groups of two), particularly given that such arguments have been noted in the literature at the elementary school level (Ball & Bass, 2003; Carpenter, Franke, & Levi, 2003) and that the CMP curriculum emphasizes such arguments in a lesson addressing the sums of even and odd numbers. For many students, their working conceptions of "evenness" may have constrained the nature of arguments they could produce (Miyakawa, 2002). If students' conceptions of an even number are based on the parity of the unit's digit, then they are somewhat limited in terms of the nature of arguments they can produce.[3]

THE THREE ODD NUMBER SUM

This item was worded as follows: "If you add any three odd numbers together, is your answer always odd? Provide an explanation that would convince your teacher that the answer is always odd."

Level 1 justifications comprise the largest proportion of responses (as one might expect at this point), with such justifications produced by 41% of 6th grade students, 49% of 7th grade students, and 37% of 8th grade students. The results also show a relatively significant proportion of students whose responses were categorized as level 0 justifications, particularly in 7th grade. In many of the cases, students either restated the proposition or accepted it as fact: "You will always make an odd number" (7th grade student).

The proportion of students who attempted to produce a general argument (level 2 or 3) increased from 20% of 6th grade students to 29% of 8th grade students. Students who were able to successfully produce a proof based their arguments on two accepted truths (for middle school students)—the sum of two odd numbers is even, and the sum of an even number and an odd number is odd.

"If you add two odds, the result is even. An even plus one more odd is odd. So three odds added together always result in odd." (6th grade student)

"We know that odd and odd equal even. So even (two odds) added together with odd equals odd. This shows us that no matter what three odd numbers you add together, the sum will always be an odd number." (7th grade student)

"The reason that when you add three odd numbers together you always get an odd, is that odd + odd = even. That gets rid of two odds. The answer is even, plus another odd would always be odd. This is because odd + even = odd." (8th grade student)

In constructing their deductive arguments, these students implicitly employed the associative property to partition the three addends, determined the parity of the partial sum (the first two addends) using the accepted truth that the sum of two odd numbers is even, and then determined the parity of the final sum using the accepted truth that the sum of an even number and an odd number is odd.

COMPARING THE NUMBER PROPERTY ITEMS

A contrast among the three number property items themselves concerns the role of accepted truths: utilizing an accepted truth in constructing an argument (the *consecutive number sum* problem and the *three odd number sum* problem) versus proving an accepted truth (the *even number sum* problem). In the *consecutive number sum* problem, students could attempt to produce a deductive argument by applying an aspect of the definition for consecutive numbers (i.e., consist of one even and one odd number), using the accepted truth that the sum of an odd number and an even number is odd, and then deduce the conclusion. The *three odd number sum* problem also afforded students a similar means of producing a deductive argument: using two accepted truths—the sum of two odd numbers is even and the sum of an even number and an odd number is odd—students could then deduce the conclusion. In contrast, in the *even number sum* problem, students were asked to prove something (i.e., the sum of two even numbers is even) that was at the level of the accepted truths in the previous two problems. Consequently, the *even number sum* problem did not provide students the same type of affordance for attempting to produce a deductive argument as that afforded by the previous two problems. And as a result, almost one-fourth of the students overall attempted to produce a deductive argument for both the *consecutive number sum* and *three odd number sum* problems, whereas the majority of those students who attempted to produce a general argument for the *even number sum* problem based their arguments on the parity of the unit's digit.

A second contrast among the three problems concerns students' recognition of the truth of each proposition: are students already convinced regarding the truth of a proposition? In the case of the *even number sum* problem, the proposition that the sum of two even numbers is even is a statement of which all of the students were already likely familiar. In the other two problems, however, students may not already be convinced regarding the truth of the propositions as the propositions are relatively novel (for middle school students). Healy and Hoyles (2000) asserted that "students were more likely to assess empirical arguments as general—to believe them to be proofs—if they were already convinced of the truth of the statement" (p. 412). This assertion may also explain why there were more empirical arguments produced in the *even number sum* problem (59% overall) as compared to the *consecutive number sum* and *three odd number sum* problems (48% and 42% overall). In the former case, students may have already been convinced of the truth of the proposition and provided empirical arguments simply to demonstrate that the proposition was indeed true (versus being unsure of the truth of the proposition and thus feeling compelled to prove that the proposition is true).

Middle School Students' Production of Justifications: Patterns, Questions, and Implications

The results reveal several patterns regarding the students' production of justifications; in addition, the results also raise questions regarding students' understandings of proof and also suggest implications for the teaching and learning of proof. These patterns, questions, and implications are discussed in the following sections.

Patterns in Students' Production of Justifications

Similar to previous research regarding students' production of mathematical justifications, the predominant means of justification used by students in this study were empirical arguments. To some extent, given typical school mathematics instruction and curricula, it may be unreasonable to expect more sophisticated justifications from students. Although the following statement from Harel and Sowder (1998) was in the context of discussing the state of proving in undergraduate mathematics education, it applies in the case of secondary school mathematics education as well:

> We may, for example, be fostering the empirical proof schemes *through* our teaching: During instruction, empirical justifications themselves serve as examples of arguments given by mathematicians [and secondary school mathematics teachers], and may inadvertently sanction the empirical proof scheme as a mode of justification fully acceptable in the mathematical context. (p. 278)

It is possible, however, that as the recommendations of reform are more widely implemented students will gain more experience with reason giving and begin to develop more advanced means of justifying.

In looking at the results for all six items across the grade levels, there was an overall increase in attempts to justify as students progressed from 6th to 8th grade. The proportion of students who provided mathematical justifications (levels 1, 2, or 3) versus those students who did not provide mathematical justifications (level 0) increased from 6th grade to 8th grade (with the exception of the even number sum problem). Additionally, there was an overall increase in attempts to produce a general argument as students progressed from 6th to

8th grade. In particular, the proportion of students who attempted to produce a general argument (level 2 or 3) increased from 6th grade to 8th grade (again, with the exception of the even number sum problem). To some degree, this might be a positive indication regarding the influence of reform; by 8th grade, the students in this study had entered their third year of experiencing a reform-based curricular program with its explicit emphasis on reasoning and proof. Thus, as such curricula provide students with opportunities to engage in proving activities, for some students, their awareness of the need to justify as well as their awareness of the need to treat the general case seems to develop as they progress through their middle school education.

Finally, the problems themselves differed in terms of the nature of their *proving affordances*, which, in turn, may have influenced the nature of the justifications that students produced. Both the three number trick problem and the equal number trick problem afforded students the opportunity to use proof by exhaustion and, in fact, students did use this method for both problems. A second type of proving affordance involved the use of accepted truths: both the consecutive number sum problem and the three odd number sum problem afforded students the opportunity to construct a deductive chain based on accepted truths and, again, students did construct their arguments in this manner. There was also a third type of proving affordance evident in the problems, however, it was not one that tended to be taken up by students. In this case, students could have used symbolic algebra as a means of justification for all six problems (although more easily, perhaps, for the 20 number trick and equal number trick problems). In short, the idea of proving affordances speaks to what students have at their disposal with which to reason and, thus, the nature of justifications students produce. In many problem situations in which students are expected to produce a justification (including those discussed here), it may be, for example, that many students are aware of the need to treat the general case (i.e., recognize the limitation of empirical evidence as a means of justification), however, they may not have at their disposal (perhaps due to issues related to content knowledge or to the nature of the proposition they are asked to justify) what they need to produce a general argument, and thus examples-based arguments are their only recourse in attempting to justify. There is some support for this idea in the literature: Healy and Hoyles (2000) found that although students relied on empirical arguments for the justifications they themselves produced, the students did seem to be aware that such arguments were not general.

Questions Concerning Students' Understandings of Proof

Although students' difficulties with proving have been attributed to a variety of factors, one factor, an understanding of generality, seems to be critical to developing an understanding of the concept of proof. One aspect of generality concerns the idea that a proof offers an absolute guarantee regarding the truth of a statement or result. Previous research suggests that many students do not have an adequate understanding of this aspect of generality (e.g., Chazan, 1993; Fischbein

& Kedem, 1982; Porteous, 1990). We also noted evidence that supports a similar conclusion; there were a number of students who attempted to produce general arguments (including students who were successful), and then supplemented their arguments with examples. In the case of the previous research, students reportedly used examples to verify or check arguments constructed by someone else; in our work, students used examples to verify or check arguments that they themselves constructed. The following student responses are representative:

"Yes, your answer is always odd because an odd plus an odd is an even number, so if you add one more odd number in it's like adding an odd and an even and that's always odd. $97 + 95 + 37 = 229$ odd." (8th grade student's response to the three number sum problem)

"Yes I would [get the same result] because she's doing almost the same thing to the numbers. Only because in the 1st one she adds first and multiplies last. 2nd she multiplies 1st and adds last but because she multiplies first she needs to add 6 instead of 3. It would work with any number [student proceeds to work out five examples using 1, 2, 3, 4, and 5 as the starting numbers]." (8th grade student's response to the equal number trick problem)

The questions that arise from such responses are whether students are simply using examples to illustrate the truth of their argument or whether they are simply not aware that their argument is indeed general (or at least is an attempt to be) and thus requires no further evidence. In the former case, students may be taking into account their audience—someone will be reviewing their responses and they use the examples to illustrate their argument for the benefit of the reader (much the way teachers often illustrate proofs with examples). The latter case is more problematic as it suggests that students may not fully understand what it means to treat the general case (see Knuth & Sutherland, 2004).

A second (related) aspect of generality concerns the idea that empirical evidence does not suffice as proof. Again, for many students this aspect of generality appears to be one that they do not adequately understand—a finding that predominates the results of many studies is students' reliance on the use of examples to prove the truth of a statement or result (e.g., Healy & Hoyles, 2000; Porteous, 1990). The results of this study are similar: empirical evidence does seem to suffice as proof for a significant proportion of students. A question that arises is whether students actually believe that empirical evidence suffices as proof or if they simply are unable to produce any other type of justification. Healy and Hoyles found that although the majority of students produced empirical arguments in their study, many of these same students also seemed to be aware that such arguments are not general and, moreover, could in fact recognize correct proofs (constructed by others) if asked to do so. Such results not only suggest that students may not actually believe that examples constitute proof, but also that there is a difference between students' proof production competencies and their proof comprehension competencies.

In sum, this study raises several questions regarding students' understandings of proof that seem worthy of continued research:

1. To what extent do students recognize that a proof treats the general case?
2. To what extent do students think that examples suffice as proof?
3. What role do examples play in students' understandings of proof?
4. What is the relationship between students' proof production competencies and their proof comprehension competencies?

Implications Concerning the Teaching and Learning of Proof

In general, the results of this study suggest that students have difficulty with proof and, in particular, with producing general arguments. If students are to develop their competencies in proving, then proof must play a more meaningful role in their school mathematics experiences. As van Dormolen (1977) aptly stated, "not until we manage to teach our students what giving a proof really means can we expect them to give deductive arguments with understanding and insight" (p. 33). Enhancing the role of proof in the classroom, however, requires substantial effort on the part of teachers insofar as they are responsible for ensuring students have the means as well as the opportunities to engage in proving (Herbst, 2002). This will be no easy task given that previous research has found that many current and future teachers have inadequate conceptions of proof (e.g., Harel & Sowder, 1998; Jones, 1997; Knuth, 2002a; Martin & Harel, 1989) and that they have limited views regarding the nature and role of proof in school mathematics (Knuth, 2002b). One avenue for increasing the likelihood of teachers' success in enhancing the role of proof in the classroom is to help teachers develop better conceptions of proof during their own mathematics education (see Knuth, 2002a, for thoughts on this issue).

Given the inextricable link between teachers' instructional practices and the curricula they implement, another avenue is to design curricular materials that support teachers' efforts to enhance the role of proof in the classroom. Curricular materials not only must provide opportunities for students to engage in proving activities, but must also support teachers' efforts to facilitate such engagement. As an example, consider the following problem from a 6th grade lesson in the *Connected Mathematics* curriculum:

> Which of the following statements are *always true*, which are *never true*, and which are *sometimes true*? Explain your reasoning.
>
> If a number is greater than a second number, then the first number has more factors than the second number.
>
> The sum of two odd numbers is even.

This problem clearly engages students in a meaningful proving activity: it provides an excellent opportunity for a teacher and students to discuss cases in which an example is enough to prove a statement is false and cases in which examples do not suffice as proof. Yet given what we know about students' reliance on examples as proof, the curriculum support materials (e.g., teacher's guide) do not emphasize the importance of such a discussion in this lesson. Moreover, teachers themselves may not recognize this lesson as an opportunity to engage students in conversation about the use of examples as a means of justification—highlighting the need for explicit attention to be given in the design of curricular support materials.

Although many reform-based curricular programs are purportedly designed to support teachers' efforts to enhance the role of proof in the classroom, few studies have explicitly examined the nature of such support. If the second avenue is to be successful, such studies are needed so that, if necessary, curricular support materials can be designed that better support teachers in their effort to promote the development of their students' competencies in justifying and proving (see Carpenter et al., 2003, for an example of such curricular support materials at the elementary school level).

Concluding Remarks

At the very least, this research contributes to our understanding of middle school students' competencies in justifying and proving; it is our hope, of course, that this research will also influence the teaching and learning of proof in school mathematics. There is still a great deal more to learn and we hope our work has raised questions worth further research, particularly research that explores students' understandings of generality as well as research that explores the treatment of proof in school mathematics curricula. Finally, the more we learn about the details of student thinking with regard to proof, the better able we, as mathematics educators, will be able to support the teachers with whom we work in their efforts to enhance their students' understandings of proof.

Notes

1. This research was supported in part by the National Science Foundation under grant No. REC-0092746. The opinions expressed herein are those of the authors and do not necessarily reflect the views of the National Science Foundation.
2. One primary goal of the research project is to study the development of students' conceptions of proof as they progress through middle school—a longitudinal study. The data reported here, however, are essentially data from two cross-sectional studies.
3. On a subsequent assessment, students were asked to define even and odd numbers and, not surprisingly given the results discussed here, the majority of students' definitions focused on the parity of the unit's digit. Interestingly, the curriculum provides a visual definition based on the structure of an even or odd number, yet, very few students used this definition as a basis for their arguments.

10
From Empirical to Structural Reasoning in Mathematics
Tracking Changes Over Time

DIETMAR KÜCHEMANN AND CELIA HOYLES

A major challenge in mathematics education is to develop students' abilities to reason mathematically, that is to make inferences and deductions from a basis of mathematical structures, henceforth referred to as structural reasoning, rather than by arguing, for example, from perception, the assertion of authority, or, in particular, from empirical cases—henceforth referred to as empirical reasoning (for a comprehensive perspective on proof that takes account of all its cognitive, social as well as mathematical constraints, see Harel & Sowder, 1998). Lampert (1990) argues that a common view of mathematics, both in the world at large and in most mathematics classrooms, is one "in which *doing* mathematics means following the rules laid down by the teacher; *knowing* mathematics means remembering and applying the correct rule when the teacher asks a question; and mathematical *truth is determined* when the answer is ratified by the teacher" (emphasis in original). Formal proofs and consistent mathematical argument both require the ability to reason by appealing to the logical structure of the system, that is to engage in structural reasoning. This is a core component of being able to prove mathematically and of developing mathematical understanding.

If, therefore, it is the case that most students have "never learned what counts as a mathematical argument" (Dreyfus, 1999), it is perhaps not surprising that a considerable body of research has accumulated which indicates that school students tend to argue at an empirical level rather than on the basis of mathematical structure (Balacheff, 1988; Bell, 1976; Bills & Rowland, 1999; Coe & Ruthven, 1994; Healy & Hoyles, 2000).

This chapter describes patterns in high-attaining students' mathematical reasoning in the domain of number/algebra and traces development over time in their use of structural reasoning. The analysis presented forms part of the Longitudinal Proof Project (Hoyles & Küchemann: http://www.ioe.ac.uk/proof), which analyzed the development of students' mathematical reasoning over 3 years. Before describing the research and its findings we briefly summarize the approach to mathematical reasoning adopted in England, which is rather different from that of other countries.

Learning to Prove: A Perspective From the English Curriculum

In the 1950s and 60s, academic students, i.e., the 20–30% of secondary school students who were in selective (grammar) schools, met proof mainly in the context of classic Euclidean geometry. However, this systematic treatment of proof more or less disappeared from mathematics curricula for 11–16 year olds during the 1970s and 80s. Thus, for example, Pythagoras' theorem became known as Pythagoras' rule, which students were no longer required to prove but only to apply, perhaps after verifying it through examples drawn on squared paper. In this century, proof has started to make a comeback. Thus, for example, the current National Curriculum for students in English schools requires that most Year 8 students (12+ year olds) should:

> Understand a proof that the sum of the angles of a triangle is 180° and of a quadrilateral is 360°, and that the exterior angle of a triangle equals the sum of the two interior opposite angles. (DfEE, 2001, p. 183)

Here the suggested approach is exploratory, and the proof construction informal. Thus, for angle sum of a triangle, it is suggested that students "consider relationships between three lines meeting at a point and a fourth line parallel to one of them," and that they "explain using diagrams." There is no requirement that the components of a proof be made explicit or that the argument is set out in a formal way, although more recent government-sponsored (but non-statutory) support material is encouraging teachers to do so. Despite the increasing emphasis on proof in the English national curriculum for 11–16 year olds, it is still likely to remain very different from tightly regulated activities in high school geometry in the USA, as described in Chapter 12 (McCrone & Martin) of this volume. Reference, for example, the kind of formal proof shown in "Linda's answer" in Appendix A of that chapter. It is inconceivable that English 11–16 year olds will be required to construct or even consider proofs with statements as formal as "All right angles are congruent" and "By the reflexive property, XY = YX."

As well as being less formal, proving in England tends first to be encountered in the number/algebra domain, rather than in geometry as in most other countries (Hoyles, 1997). In fact, even after a systematic treatment of proof had virtually disappeared from the English school curriculum, students did have the opportunity (potentially at least) to engage in explanation and (informal) proof through extended "investigations," usually in number/algebra. During the 1990s investigational work was incorporated into the national mathematics examination at age 16. Marking schemes were devised to describe the characteristics of a "good investigation," and, as a consequence, the task of "doing" investigations became increasingly procedural and routinized, with an emphasis on generating data and looking for number patterns, even at the upper end of secondary school (Morgan, 1997). Of course, inductive reasoning based on specific cases can be important and fruitful in mathematics (Pólya, 1954a). It can help students develop a feel for a mathematical situation and to form conjectures. It also provides a test for the validity of a general proof, especially where students are uncertain about the scope and logic of their argument (Jahnke, 2005), which is an issue we return to later in this chapter. Cuoco et al. (1996), in their discussion of mathematical "habits of mind," suggest that students should learn to become "pattern sniffers," "experimenters," and "describers," among other things. However, they make the point that the most important habit is to understand "when to use what." We suggest that a crucial habit is to look for mathematical structure, or, as Dreyfus (1999) puts it, to move "from a computational view of mathematics to a view that conceives of mathematics as a field of intricately related structures." This shift would seem to be particularly important in England where, as we have argued above, the curriculum tends to emphasize data and computation and teaching does not emphasize the importance of structural reasoning (see Healy & Hoyles, 2000). This chapter seeks to throw light on this shift and how it is exemplified in any changing patterns of student response to our proof items.

The Longitudinal Proof Project

The analysis presented here forms part of the Longitudinal Proof Project (Hoyles & Küchemann: http://www.ioe.ac.uk/proof), which analyzed the development of students' mathematical reasoning over 3 years. Data were collected through annual administration of a proof test completed by classes of high-attaining students, initially aged 12/13, in England designated as in Yr 8, and finally aged 14/ 15yrs or in Yr 10. Students in England are setted (or tracked) and we targeted students who would be in top sets in Yr 10 (age 14+ years). The schools were randomly selected from within nine geographically diverse English regions. The items in all three proof tests were devised after reviews of the literature and of the curriculum, followed by extensive discussions with teachers. They ranged over the following proof "categories" (although after piloting some categories were dropped as the items turned out to be unsatisfactory):

> Making conjectures; turning conjectures into conditional statements; making and expressing generalizations by engaging in structural reasoning; using generic examples; crucial experiment; general cases which are then limited; given a statement, find (deduce) the value of an unknown, or derive another statement; logical implication; using definitions and structures; transformational reasoning; specialization after a proof; scrambled proof; reasoning from perception.

Items were piloted with up to about 200 students and, overall, 1512 students from 54 schools completed all three tests. The tests comprised items in number/

174 · Dietmar Küchemann & Celia Hoyles

algebra and in geometry, some in open response format and some multiple choice. Each new proof test included some items that were identical or very similar to items from the previous test (*core* items), together with new items. The project used a combination of quantitative and qualitative methods. The quantitative methods included the identification of trends in hierarchically ordered categorical data obtained by coding students' responses to each item in each proof test, and multilevel analyses of student scores in geometry and in algebra to identify significant predictors of progress. The qualitative methods included analyses of interviews with selected students in schools identified from the multilevel modeling as those in which students performed significantly better than predicted (see Hoyles et al., 2005, for a detailed description of the methodology adopted).

Overall in the area of number/algebra, we found an improvement in the use of algebra, although many in our sample of high-attaining students were quite strongly attracted to pattern spotting and computation (as we shall see in this chapter). Also, we found a large gulf between success in a numerical and an equivalent algebraic task. Thus in Yr 10, 88% of students could solve a problem of calculating angles but only 21% were able to re-express the calculation as an algebraic relationship.

In this chapter, we focus on two questions in number/algebra, A1 (see Figure 10.1) and A4 (see Figure 10.3), that both sought to assess whether students engaged in structural reasoning as opposed to appealing to computation or empirical data. As stated earlier, we regard structural reasoning as a core

---

A1 Lisa has some white square tiles and some grey square tiles.
They are all the same size.

She makes a row
of white tiles.

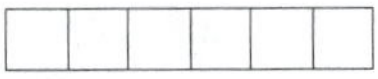

She surrounds the white
tiles by a single layer
of grey tiles.

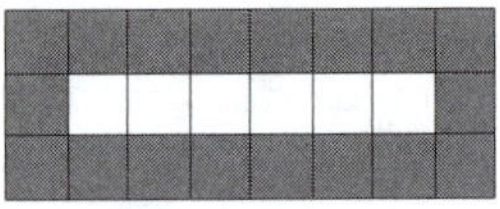

a How many grey tiles does she need to surround a row of 60 white tiles?

Show how you obtained your answer.

b Write an expression for the number of grey tiles
needed to surround a row of *n* white tiles.

Figure 10.1 Yr 8 version (part *a* only) and Y10 version of question A1

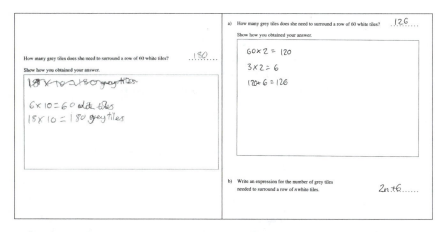

Figure 10.2 Student JG's response to A1 in Yr 8; and Yr 10

component of mathematical proof. Both items featured in more than one proof test so it was possible to identify changes in patterns of response between annual surveys.

Question A1 (Figure 10.1) presents a pattern of white tiles and grey tiles that has to be generalized. It is a familiar type of question in the English school curriculum—apart from the fact that it immediately asks students to make a "far generalisation" (Stacey, 1989), rather than making a series of "near generalizations" (Stacey, 1989) first. It was devised to see whether students could make the generalization on the basis of the pattern's structure or whether, on the basis of spurious number patterns, they would resort to a "function" or "whole-object scaling" strategy (Stacey, 1989) that was inappropriate.

Brown et al. (2002), in their work with student teachers, found that the students quite often chose to "perform computations when reasoning about computations would suffice." Question A4 (Figure 10.3), was based on a suggestion made by Ruthven (1995), and explores this tendency to perform calculations, in a less familiar setting. It uses factorials (a notation that would not be known by students of this age) and concerns divisibility. The first part of the question can be solved by simple computation while in the final part the dividend is too large for this to be a viable strategy, thus (we hoped) forcing a different, more structural approach.

As well as presenting cross-sectional quantitative analyses of responses to both questions, we have gathered together for this chapter a selection of extracts from the interviews with Yr 10 students that involved discussion of A1 or A4, and during which we probed the reasons behind their responses by asking them to compare and evaluate the responses they had made to the same question on different occasions. Unfortunately these interviews could not be systematic, as students were not necessarily selected for their performance on A1 or A4, but analysis of the relevant extracts does throw light on students' thinking on the items and why progress may or may not have been made.

Generalizing a Number Pattern

Question A1 is a standard number/algebra task involving a tile pattern that was familiar to English students. It was designed to test whether students could discern, use and describe a structural relationship between a number of white and grey tiles. A question of this type was included in all the proof tests but took two forms, with the same question used in the proof tests in Yr 8 and Yr 10 (with an additional part in Yr 10), and a different but parallel question used in Yr 9 (not shown here). The Yr 10 version is shown in Figure 10.1 and consisted of two parts, A1*a* and A1*b*. The students were given one example of the relationship (showing six grey tiles and 18 white tiles), and in part *a* asked to generalize this to another number (60) of white tiles and to explain their numerical calculation. In part *b*, the students were asked to write a general relationship involving n white tiles. The Yr 8 version was identical to part *a* but consisted only of this part, as most students of that age would not yet have had much experience of algebraic symbolization. We deliberately built numerical distracters into the item, in the form of simple, but irrelevant, relationships between the numbers of white tiles mentioned in the item (namely, $6 \times 10 = 60$) and between the number of white and grey tiles in the given configuration ($6 \times 3 = 18$).

We recall that an aim of the project was to map out the different kinds of responses to the proof items and to capture progress in reasoning by an analysis of how the frequencies of the different codes changed over time. To achieve this aim, we coded the responses to A1*a* into five broad categories that were based on an a priori analysis of possible response types (using prior research and our own pilot study with over 150 students). Code 1 was given to responses that were incorrect and based solely on spotting number patterns. Code 2 responses showed some recognition of structure but were incomplete or incorrect. Code 3 responses were correct showing a recognition of the structure of the numerical calculation performed. (When we developed the coding scheme, we observed that most students who correctly structured the tile pattern saw the pattern, implicitly or explicitly, as of the form "double and add 6". There are, of course, other equivalent forms, such as "add one at each end, double, add one at each end," but these seemed to be rare and so we did not differentiate between them in our coding scheme.) Code 4 was given to responses that included some explicit description of a general relationship between the different colored tiles and Code 5 was used if this general rule was expressed correctly with variables. Finally, a miscellaneous code was used where students gave no response or responses that did not fit the other codes. Codes 1 to 5 are summarized as follows:

Code 1:   Spotting number patterns ($6 \times 10 = 60$, so there are $18 \times 10 = 180$ grey tiles; or $6 \times 3 = 18$ so there are $60 \times 3 = 180$ grey tiles), no structure

Code 2:   Some recognition of structure (incomplete or draws & counts)

Code 3:   Recognition and use of structure: specific (correct answer, e.g., showing $60 + 60 + 3 + 3$)

Code 4: Recognition and use of structure: general (correct answer and general rule, e.g., $\times 2, + 6$)

Code 5: Recognition and use of structure: general, with use of variables (correct answer and general rule, with naming of variables in words or letters, e.g., multiply the number of white tiles by 2 and add 6, or $2n + 6$)

We judged that responses according to these codes were hierarchically ordered in terms of mathematical "quality," and thus were of the view that, broadly speaking, as students developed mathematically, i.e., became mathematically more capable and aware, they would tend to give higher quality responses. At the same time, a Code 3 response is sufficient to answer the item successfully and thus students who were capable of giving higher quality responses than Code 3 might not have felt the need to do so.

Pattern Spotting and Structural Reasoning

Our first attempt to capture changes in response patterns over time was to record the frequencies of response to A1a classified according to the codes. These frequencies are shown in Table 10.1.

Table 10.1 shows that, despite some improvement between Yrs 8 and 9, a substantial minority of students continued to use "number pattern spotting" strategies giving an incorrect solution of 180 grey tiles. Altogether 35% of the total sample gave such responses in Yr 8. This fell to 21% in Yr 9 but stayed at 21% in Yr 10. Thus the cross-sectional analyses indicated modest improvement followed by plateau. Of course, the fact that the proportions of students giving number pattern responses were the same in Yrs 9 and 10, does not mean these proportions consisted of exactly the same students. We discuss this further later on in the chapter, when we look at the data longitudinally.

Complementing the changes in frequency of pattern spotting responses, the frequency of correct responses (Codes 3, 4 and 5) went up from 48% in Yr 8 to 68% in Yr 9 but only to 70% in Yr 10. However, this small increase from Yr 9 to Yr 10 masks a substantial rise in the appropriate use of variables (expressed in words or with letters) in students' explanations (Code 5 responses), from 16% in Yr 9 to 26% in Yr 10 (and starting from just 9% in Yr 8).

We now turn to our longitudinal analyses of these data. We focus on changes in patterns of response, according to the codes, between Yr 8 and Yr 10 only, since the first part of A1 was identical in those years but not in Yr 9 (the complete longitudinal analyses for item A1 are available in the Year 10 Technical Report of the project—see Küchemann & Hoyles, 2003, pp. 10–18). Table 10.2A shows the code frequencies for A1a longitudinally, in that it cross-tabulates individual students' responses in Yr 8 with their responses in Yr 10. However, as our purpose here is to consider "progress" and as a Code 3 response is sufficient to answer the item successfully, we have grouped the Code 3, 4 and 5 responses. Table 10.2B is derived from Table 10.2A and assumes the validity of the hierarchical ordering of Codes 1, 2, and 3–4–5 combined. Based on this assumption, it shows the

Table 10.1 Code frequencies for item A1*a* for Yrs 8, 9, and 10 (N = 1512)

CODES	Yr 8		Yr 9		Yr 10	
	Number	% of sample	Number	% of sample	Number	% of sample
Incorrect answer						
Miscellaneous non-correct, including no response	163	11	67	4	66	4
Code 1: Spotting number patterns	522	35	317	21	315	21
Code 2: Some recognition of structure	112	7	105	7	75	5
Total incorrect	**797**	**53**	**489**	**32**	**456**	**30**
Correct answer						
Code 3: Correct structure, specific	527	35	572	38	615	41
Code 4: Correct structure, general	53	4	214	14	43	3
Code 5: Correct structure, use of variables	135	9	237	16	398	26
Total correct	**715**	**48**	**1023**	**68**	**1056**	**70**
Total	1512	100	1512	100	1512	100

Table 10.2A Yr 8 by Yr 10 code frequencies for A1*a* with the code 3, 4, and 5 responses combined (N = 1512)

	Yr 10 A1a									
	Code 1		Code 2		Code 3, 4 or 5		Misc.		Total	
Yr 8 A1a	No.	%	No.	%	No.	%	No.	%	No.	%
Code 1	184	12	27	2	283	19	28	2	522	35
Code 2	26	2	9	1	72	5	5	0	112	7
Code 3, 4 or 5	57	4	29	2	612	40	14	1	715	47
Miscellaneous	48	3	10	1	89	6	16	1	163	11
Total	315	21	75	5	1056	70	66	4	1512	100

Table 10.2B Yr 8 by Yr 10 progress for A1*a* with the code 3, 4, and 5 responses combined (N = 1512)

	No.	%
Progress	382	25
Same	805	53
Regress	112	7
Miscellaneous	213	14
Total	1512	100

percentage of students who "progressed" or "regressed" in their responses from Yr 8 to Yr 10 (we have ignored all students who gave a miscellaneous non-correct response in either or both years).

As can be seen from Table 10.2B, the improvement in students' responses from Yr 8 to Yr 10 was not entirely smooth, with 25% of the sample progressing on A1*a* but 7% regressing, giving a "net progress" of 18%. (This is comparable to most of our other core items, where net progress ranged from 2% to 32%.) Focusing on the pattern spotting responses (Code 1) in particular, Table 10.2A shows that over the 2-year period from Yr 8 to Yr 10, well over half of those who gave a number pattern response (Code 1) in Yr 8 progressed to a correct (Code 3, 4 or 5) or partially correct (Code 2) structural response in YR 10, but that one-quarter of those who gave a number pattern response in Yr 10 had given a higher quality response in Yr 8.

Expressing Structure in Algebra

The Yr 9 and 10 versions of question A1 had an added part, A1*b*, where students were asked to write an expression for the number of grey tiles needed for *n* white tiles (see Figure 10.1 for the Yr 10 version). We were interested in whether students were able to express any relationship they discerned in the tiling pattern in algebra, and, indeed, whether this was consistent with their explanations of structure given in words or numbers. We have noted in a previous study (Healy & Hoyles, 2000) that Yr 10 students in England rarely used algebraic symbolization as a language with which to describe mathematical structure, even though they had been taught to do this and indeed they accorded high status to algebraic "proofs."

A1b asks students to map the number of white tiles onto grey, i.e., it requires a function approach ($n \rightarrow 2n + 6$ for the Yr 10 item). Thus we were curious to see whether this would provoke a rethink from some of the students who had used a number pattern approach, namely those whose approach involved scaling (18 grey tiles $\times$ 10 = 180 grey tiles in Yr 10). In the event, in Yr 10 more than half of this subgroup of students ($N = 197$) switched to a function response, although most stuck to a number pattern approach in part b, with 43% giving a response of the form $n \rightarrow 3n$ (which gives 180 when $n = 60$); however, 13% produced a correct algebraic expression, of the form $n \rightarrow 2n + 6$ (with a further 6% producing a partially correct algebraic expression). (Not surprisingly, of the 92 students who in Yr 10 had given a function number pattern response to A1a, i.e. 60 white tiles $\times$ 3 = 180 grey tiles, the vast majority, 87%, gave a response of the form $n \rightarrow 3n$ to part b, with only 1% producing a correct algebraic expression.)

It is also worth noting, that of the 615 students who gave a Code 3 response to A1a in Yr 10 (i.e., a correct, but specific and non-algebraic response), 93% gave a correct algebraic response to part b. Thus part b was effective in prompting students to use algebra where they had not felt compelled to do so in part a, and, as discussed earlier, it provoked a substantial proportion of students to switch from a scalar to a function approach, even if in most cases, probably, this did not lead to a correct restructuring of their answers.

Some Illustrative Interview Extracts

Our analyses of the quantitative data suggests that, for some students at least, there is an element of chance about their responses: rather than being wedded to a particular way of construing such tile patterns (with some going for structural reasoning and some for number pattern spotting), they seemed to hit on one way on one occasion and another way on another occasion; furthermore there is not necessarily any consistency between their non-algebraic and algebraic expressions of the relationship. An examination of individual scripts also shows that some students flipped between approaches on a given occasion.

We now turn to student interviews to help us understand the students' interpretations of the question. The interview extracts used here were all with Yr 10 students who were asked to compare the responses they had made to A1 in Yr 8 and in Yr 10 and to explain any inconsistencies and changes. It is worth recalling that all these interviews were with students whose schools were singled out by our statistical analyses as exceptionally "good" in developing mathematical reasoning. We have selected to report written responses and interview extracts for three students with different response patterns. First, student MS who persisted in making pattern spotting approaches in both Yr 8 and Yr 10, and then students EC and JG who both used pattern spotting in Yr 8 and showed awareness of structure in Yr 10 and had thus apparently made progress, but who differed in their responses to being asked to explain their reasoning. We were not able to interview any student who clearly appeared to regress.

Student MS gave a pattern spotting response in Yr 8 and Yr 10 (and also in Yr 9). However in Yr 8 he used a "×10" scalar strategy while in Yr 10 he used a "×3" function strategy (together with a function response, "$3n$", that was consistent with this in part b).

MS was interviewed a few days after taking the Yr 10 test and was asked about his Yr 8 and Yr 10 responses, which were placed on the desk in front of him. At first he seemed to feel the responses were essentially the same and, as this extract indicates, he remained unperturbed when the interviewer read through the responses again and suggested that they were different:

I   Here (Yr 10) you do something quite different by saying . . . I've got six white tiles and if I multiply by 3 I get 18 grey tiles.

MS   There's 6 there and 18 . . . altogether times it by 3, then I thought it would be the same if you wanted to find out how many grey tiles would be in 60 so I timesed by 3 . . .

I   Did you, did you think about checking it in any way? Or . . .

MS   I was quite confident on this stuff . . . I went onto the next question.

Thus MS seemed confident that the numerical relationships he had found were right, perhaps because of their simplicity, and seemed content to ignore the suggestion that they might be different.

We now turn to student EC who gave a "×10" pattern spotting response in Yr 8 (and Yr 9). Initially, he seems to have embarked on a similar response in Yr 10, in that he has written "6 × 10 = 60". However, this is crossed through and replaced by "60 × 2 = 120" and "+ 3 + 3 = 126", indicating a correct answer based on the geometric structure. EC was interviewed about a week after taking the Yr 10 test and asked to compare his Yr 8 and Yr 10 responses. We were surprised that he chose his incorrect Yr 8 response as the one he now believed was correct:

I   Which one do you think is the right one?

EC   I think this one is [pointing to his Yr 8 response].

I   The Yr 8 one?

EC   Yeah, kind of the first instinct I had.

I   You go by instinct.

EC   Yeah, I think, sort of, in the majority of the time the first instinct is right, so, I think maybe that one looks right.

The interviewer then attempted to probe further and asked EC to explain his Yr 10 response in more detail. EC seemed able to do this but still did not change his mind about the relative merits of his Yr 8 and Yr 10 responses:

I	So, you ended up here in Year 10 with this double thing and then add 6.
EC	Yeah.
I	So, how did that come to you, I mean, why would you have done that?
EC	I think, because we needed 60 and there was 6 along each row, each of the white things, so that means 12, so I just thought that doubling it, and then there's 3 left over, so I just plussed 3 on one, so, I'm not really sure.
I	Okay. I mean, that sounds sensible enough, so, the trouble is, we've still got these two different answers. So are you going to stick with your instinct, your Yr 8 instinct?
EC:	I think so, yes.

It seems surprising that EC was so ready to abandon his correct structural approach in Yr 10 for the simplicity of his "×10" approach in Yr 8, especially when one notes that his Yr 10 (and Yr 9) response to A1b was also correct. Perhaps at this stage, through lack of experience or guidance, EC does not have the meta-knowledge needed to classify his different responses in an appropriate way and to recognize their positive or negative qualities.

We look finally at student JG who made had a similar set of responses to EC, in that she gave a "×10" number pattern response in Yr 8 and a structural response in Yr 10: "$60 \times 2 = 120$, $3 \times 2 = 6$, $120 + 6 = 126$" (see Figure 10.2).

JG was interviewed about a week after taking the Yr 10 test. At first she could make no sense of her Yr 8 response ("I have no idea why I wrote that in Yr 8"), though she comes up with an interpretation eventually:

I	I mean, say someone else had done it, not you . . . could you sort of try and figure out why on earth they did it?
JG	[Long pause] No.
I	No, you can't see any logic in it?
JG	Well . . . yeah. I can now. It's because there's 6 there so, I figured 6 times 10

Yr 8 version of A4	Yr 9 version of A4

A4 a 4! means $4 \times 3 \times 2 \times 1$
     5! means $5 \times 4 \times 3 \times 2 \times 1$
	A4 a 4! means $4 \times 3 \times 2 \times 1$
	     5! means $5 \times 4 \times 3 \times 2 \times 1$

Is 5! exactly divisible by 3?	Is 5! exactly divisible by 3?
Explain your answer	Explain your answer
b What does 100! mean?	b What does 50! mean?
c Is 100! exactly divisible by 31?	c Is 50! exactly divisible by 19?
Explain your answer	Explain your answer

Figure 10.3 Yr 8 and Yr 9 versions of question A4

would be the 60 that they were talking about in the question, and so I just had to times the amount around the outside by the same number. Oh yeah.

However, unlike EC, she prefers her structural Yr 10 response, which she feels makes more sense:

I   Can you say a bit more why it makes more sense?

JG   I don't know . . . just a couple more years' practice of finding patterns and stuff.

I   So how does this answer sort of fit the pattern, the Yr 10 answer fit the pattern better?

JG   Well because, I didn't just times the ones round the outside . . . by the same number as the ones on the inside . . . I worked out sort of a rule for it, rather than just a rule for that, that number.

I   Right. How did you get the rule for the Yr 10 answer?

JG   Well, the three at each end won't change, it's a single row of tiles say . . . you just use the top . . . the grey tiles above and below the white . . .

I   Right, okay . . .

JG's replies here are interesting in several respects. First, she justifies her preference for the structural response with an external reason ("more years' practice"), which, though perfectly plausible, has nothing to do with the quality of the actual response and which is certainly no more valid than EC's quest for simplicity. However, she is able to describe the structure itself very nicely ("three at each end . . . grey tiles above and below") and she does, in fact, say something, albeit in a rather cryptic way, about the quality of this explanation, namely about it being *general*: thus she found "sort of a rule" in contrast to "just a rule for that . . . number." Notice also her statement that "I didn't just times the ones round the outside," which potentially provides a compelling visual test for this number pattern spotting strategy, since the outcome would be a set of grey tiles that no longer fits snugly around the white tiles. Further, she gives a correct response to A1*b*, which shows she is able to express the structure using algebra. All in all, we seem here to be witnessing the beginnings of a meta-knowledge about structural reasoning, even though the concepts and language may not yet be well formed. Unfortunately, we do not know how this knowledge has arisen—although from the limited information we managed to gather about JG's Yr 10 mathematics class, the quest for structural explanations was not a strong feature of its socio-mathematical norms.

Although we cannot say how representative these three interviewees are, their responses do suggest that, for some students at least, the simplicity of number pattern responses may have a stronger appeal than the insight that might be gained from taking a structural approach. The analysis suggests possible discontinuities between modeling with numbers, narrative descriptions of these models and modeling with algebra which in turn might lead some students to re-organize their thinking. It also suggests fragility in appreciating the power of a structural approach, and rather limited ability to describe the characteristics of even correct structural reasoning. This is a phenomenon we have found elsewhere and which may well be widespread even among the highest of our high-attaining students, since they will generally have had little experience of producing mathematical explanations and reflecting on them.

From Calculating to Structural Reasoning

We found a strong tendency for students to work at an empirical level on our proof test items. In the case of A1 discussed earlier, working at an empirical level was largely manifested by pattern spotting responses, which were given by just over one-third of our high attaining sample in Yr 8 and still by just over one-fifth in Yr 10. This empirical tendency was particularly strong, though manifested rather differently in responses to another question, A4, which we used in the Yr 8 and Yr 9 proof tests (but not in the Yr 10 test, for reasons of time and space). A4 had distinctive characteristics in that unlike A1, the mathematical content would not have been familiar to students and a solution strategy that appealed to structure rather than calculation would not have been taught. The question is shown in abbreviated form in Figure 10.3 (the original question was spread over an entire A4-size[1] page, with blank space after each part for students to write their responses). The question has three parts, but our interest here is in parts *a* and *c*. In A4*a* students are asked about the divisibility of 5! by 3 and in A4*c* about the divisibility of 100! by 31 (or 50! by 19 in the Yr 9 version).

Responses to A4*a* were coded into three broad categories, to capture whether students gave incorrect or irrelevant reasons (Code 1), or determined the divisibility of 5! by calculating the value of 5! (Code 3), or gave a correct reason based on the "divisibility principle" (Code 4).

The Codes 1 and 4 were again used in A4*c*, but there was no Code 3 as students did not have the means to calculate the given factorial (100! in Yr 8 or 50! in Yr 9). However, some students (albeit very few) used an inductive reason to argue that the given factorial was divisible by 31 (Yr 8) or 19 (Yr 9), based usually on the observation that in part *a* 120, i.e., 5!, is divisible by 5, 4, 3, 2 and 1. Such responses were given the Code 2.

As with A1, we again judged that the codes were hierarchically ordered in terms of mathematical "quality," and that as students developed mathematically they would give higher quality responses.

The full code frequencies for A4*a* and A4*c* can be found in Küchemann and Hoyles (2003, pp. 23–24). Regarding A4*a*, the Code 3 and 4 frequencies show that most students could correctly determine the divisibility of 5! by 3 (76% in Yr 8, 83% in Yr 9) but the overwhelming majority gave a Code 3 rather than Code 4 response (74% of the total sample in Yr 8, 77% in Yr 9), that is, they did so by calculating 5! and then calculating 120 ÷ 3, and thus essentially by multiplying by 3 and dividing by 3 again. These students showed that they understood what was meant by the notions of factorial and divisibility but it seems they could not put them together to form an explanation. (It is of course possible that some did not give such an explanation because they felt that a demonstration was good enough; however the responses to part *c* suggest that this would have been rare.) Only 2% of the sample in Yr 8 and only 6% in Yr 9 gave a Code 4 response, i.e., based their argument on the fact that 3 was a *given* factor of 5!.

While the latter (Code 4) kind of argument is not required in part *a*, it is essential for part *c*, since students were not allowed calculators. However, as already mentioned, only a slightly larger proportion used such an argument in part *c* (3% of the sample in Yr 8, and 9% in Yr 9). Most students wanted to evaluate the factorial, and had no viable alternative strategy. Some students wrote statements such as: "That would take years to work out and if there is some shortcut I don't know it." Some evaluated 100! as being 2400, on the basis that 100 = 20 × 5 so 100! = 20 × 5!, and so answered "No." Others answered "No," because 100! is even and 31 is odd or a prime.

The response shown in Figure 10.4, given by a student in Yr 8, is a typical response to A4c. Interestingly this student gave a structural response in Yr 9, answering "Yes" to "Is 50! divisible by 19?", because "If you times it by a certain number, you will be able to divide by it." However, we found that very few students managed to move from empirical to structural reasoning on this unfamiliar question. On A4a only 18 students used the divisibility principle in both years, and just 76 students (5% of the sample) shifted from calculating (or a lower level response or no response) to using the divisibility principle, with 19 of the 37 students who had used the divisibility principle successfully in Yr 8 reverting to calculating (or to a lower level response) in Yr 9. On A4c, where the divisibility principle is needed to answer the item successfully, the picture is not much better. Here only 22 students (1% of the sample) used the principle consistently, and 114 students (7%) progressed to using it in Yr 9 having not done so in Yr 8, with 20 of the 42 students who answered part *c* successfully in Yr 8, regressing (or omitting the item) in Yr 9. We found this regression, particularly

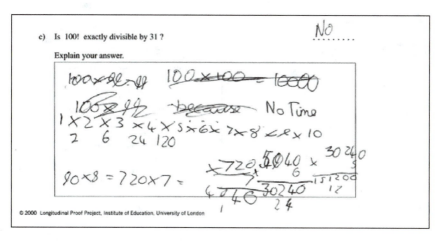

Figure 10.4 Student MH's responses to A4 in Yr 8

on part c, surprising: surely once a student had understood that "if you multiply by a number then the product is divisible by that number" then this would not be "forgotten"?

To gain some insight into these issues, we turn again to our interview data. In the course of our visits to "outlier" schools, we managed to interview 12 students on their Yr 8 and Yr 9 responses to A4. It turned out that none of these was a student who had regressed on part c, which is unfortunate (although perhaps not surprising given that only 15 students regressed out of our total sample of 1512 students). However, several students had successfully progressed from calculating in Yr 8 to using the divisibility principle in Yr 9, and we consider one of these students here.

Student AM was interviewed the day after he had taken the Yr 9 test. His written responses are shown in Figure 10.5. It appears that AM made considerable progress from Yr 8 to Yr 9: he would seem to have a clear understanding of the "divisibility principle" by Yr 9.

As with the A1 interviews, we asked students to compare their responses in different years and to explain why they had written their particular responses.

In AM's Yr 9 written responses he gives an explanation based on the divisibility principle in part a as well as part c, rather than one based on evaluating 5!. Thus, in part a he had written "The number has been multiplied by 3, so it

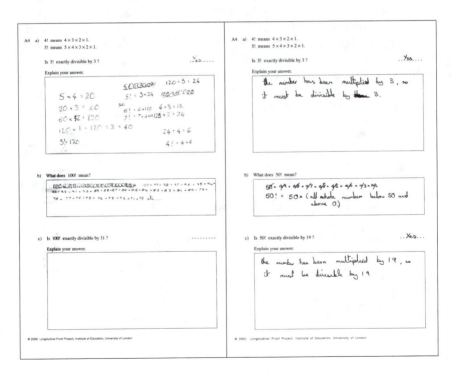

Figure 10.5 Student AM's responses to A4 in Yr 8 and Yr 9

must be divisible by 3." In the interview, he is asked how he arrived at this explanation:

> *AM*   I think I was just thinking about it. I was thinking 5! would be $5 \times 4 \times 3 \times 2 \times 1$ so if it's been timesed by 3 you can almost certainly divide it by 3 and I was just thinking sort of that was that. Also I was saying because it's only then been timesed by 2 and by 1, because 1's obviously not going to change and 2's just going to double it. So you're just going to be able to divide it by 3.

AM does not talk of calculating 5!, but his explanation seems grounded in the steps of the potential calculation. AM feels that it is not enough that $\times 3$ is one element in a string of terms, and he carefully checks that the subsequent terms, $\times 2$ and $\times 1$, will not affect the divisibility by 3. This nicely illustrates the fact that a notion like "divisibility" involves a whole nexus of ideas (see Brown et al., 2002), including some awareness of the associative and commutative properties of multiplication, especially when considering the divisibility of a long string of terms as in the case of 100! and 50!.

Later, AM says he is "almost certain" but not entirely that 5! is divisible by 3. The interviewer explores this further:

> *I*   You said just now it's almost certainly divisible by 3, you're hesitating slightly.
>
> *AM*   I think it is divisible by 3, I think at the time I wasn't completely certain.
>
> *I*   What yesterday, but you're certain today?
>
> *AM*   Yes, after all that I'm almost certain that's good.
>
> *I*   Why is there still this edge of doubt? You say *almost* certain.
>
> *AM*   Don't know. The thing is, I am certain, but not quite . . . I can't see why, I can see slightly why it works but not entirely, I haven't thought "suppose you had a bigger number, would it . . ."

Thus it would seem that for AM, this lack of certainty about the "divisibility principle" (the argument that if you multiply by a certain number the result is divisible by that number) is not because he does not understand the basic argument or appreciate its power, but because of some awareness that other features of the situation (e.g., that in 100! the term $\times 31$ is followed by a long string of other terms) might render the principle invalid. From this (and other interviews) it appears that students who correctly answered part *c* of A4, yet still expressed a need to calculate, may be expressing this need not because they reject or do not value their attempts at a structural explanation, but to *check* that the explanation is valid because they are unsure about parts of their argument. This insecurity about using number relationships because of the possible influence of "other" factors (even when the relationships seem to be understood), might help to explain why some students "regressed" in Yr 9.

Conclusions

We have reported some findings from a longitudinal study of high-attaining students' conceptions of mathematical reasoning. In this study, students' responses to specially designed annual written proof tests were coded and the code frequencies analyzed cross-sectionally and longitudinally. Total scores on each proof test were subjected to multilevel analyses, which were used in part to identify schools in which students performed significantly better than predicted. Interviews were conducted with students from these schools, who were asked to explain their written responses. Although not systematic, extracts from these interviews serve to shed some light on students' thinking on the items and why progress may or may not have been made.

In this chapter, we have focused on two number/algebra questions both of which were designed to investigate patterns in the use of empirical and structural reasoning. One question, A1, involved a familiar task of generalizing a tile pattern (although it was unusual in involving a "far generalization" only); the other question, A4, was less familiar and concerned factorials and divisibility. We found that the use of structural reasoning increased over the years, albeit at a modest rate. This improvement indicates a general cognitive shift from reliance on empirical reasoning to more theoretical reasoning based on the development of meta-knowledge about structures that is doubtless interlinked with the effects of teaching for this high-attaining group. However, despite this effect of teaching, the use of empirical reasoning was still widespread in the form of inappropriate number pattern spotting for A1 and through the desire to perform rather than analyze a calculation for A4.

The quantitative analyses of the longitudinal coded data showed a degree of turbulence in student responses, suggesting that for some students at least, there might be an element of chance about their responses: for example, rather than being wedded to a particular way of construing a tile pattern, they seemed to hit on the underlying structure on one occasion and a superficial number pattern on another occasion. The simplicity of number pattern responses appeared to have a continuing appeal often stronger than the insight that might be gained from taking a structural approach. We also noted inconsistencies in how students construed the tile pattern, most notably when asked to write an algebraic expression where their responses did not necessarily match the approaches used in their earlier reasoning, and though some students might well have been perturbed by this mismatch, there was evidence to suggest that others were not.

Our student interviews lent support to the view that students on the whole showed a lack of confidence in and a fragile grasp of structural reasoning. Initially, we interpreted these findings simply as further evidence of students' lack of appreciation of the power of structural reasoning and of the widespread tendency to use empirical methods—something that is perhaps particularly strong in English schools, because of the curriculum and classroom approaches (see, for example, Morgan, 1997). However, we have modified our views, having reflected on the tentative commitment to different methods exhibited by students in their written responses and interviews. We therefore offer an alternative explanation that distinguishes a more advanced use of empirical reasoning, namely to check the validity of a structural argument.

Jahnke (2005) puts forward the metaphor of "theoretical physicist" as a way of describing students' behavior as they learn to engage in mathematical proof. From this viewpoint, students' recourse to empirical evidence can be seen as a perfectly rational and meaningful attempt to test the validity of a proof argument. Indeed, if proof is seen as something undertaken by "fallible mathematicians . . . as part of a quasi-empirical process" (Reid, 2005; see also, for example, Lakatos, 1976; Lampert, 1990), then this metaphor might usefully be extended beyond the novice student of mathematics. This throws interesting light on an influential study by Fischbein and Kedem (1982), who might be said to have held a "traditional" concept of proof, whereby "a formal proof of a mathematical statement confers on it the attribute of a priori universal validity" (cited in Reid, 2005). They found that even where students agreed that a given proof was correct, many endorsed the idea that further numerical checks would increase their confidence in the theorem. This finding was supported by the study of Healy and Hoyles (2000), who similarly used an item where students were presented with different "proofs" of a statement. They reported that students simultaneously held two different conceptions of proof; those about arguments they considered would receive the best mark and those about arguments they would adopt for themselves. In the latter category, students chose arguments (usually empirical) that they could evaluate and which they found relatively convincing (in the sense that they found them more convincing than did students who chose other arguments) even if they recognized that their scope was limited.

For Fischbein, students who welcomed further numerical checks did not "really understand what a mathematical proof means" (Fischbein, 1982, p.16). Our evidence suggests that though this may be the case for some students, there is an alternative interpretation which may apply to others, namely that they do have some understanding of proof (as, say, a logically ordered set of reasons involving mathematical properties) but they are acknowledging that there might be flaws in the proof (e.g., in the logic or the reasons) which they have not spotted. So, in response to our unfamiliar question about factorials and divisibility, even some students who had shown a basic understanding of the underlying structure and given structural reasons in their written response seemed to need to calculate to achieve full closure. From the perspective of checking the validity of a structural argument, this can be seen as a rational way of coping with a degree of uncertainty about the influence of other features of the situation that might render their reasoning invalid, rather than as a lack of appreciation of such an argument's power. Thus, we suggest that, although recourse to empirical data may in many cases indicate a naive understanding of proof, it need not do so.

We end by summarizing our findings and briefly pointing to some implications for teaching. High-attaining students in our large random sample made progress, albeit modest, in the use of structural reasoning over the 3 years of the project suggesting a positive and cumulative outcome of teaching. However, progress in reasoning from structures was not linear, and was not necessarily retained. Thus despite an overall positive trend, there was unpredictable variation due to issues of interpretation of the task, to changes in curricular emphasis (such as the introduction of algebra), and to an individual student's confidence in their adoption of structural reasoning. We also found that the use of empirical reasoning, in the form of inappropriate number pattern spotting or through the desire to perform rather than analyze a calculation, remained widespread over the 3 years of the project although we identify a more advanced use of empirical reasoning, namely to check the validity of a structural argument.

Given the fragility of student responses, as our longitudinal analyses showed, we conclude that single snapshots of student understanding can be misleading, as students may not have a clear sense of what it means to progress in regard to the quality of a mathematical argument, or they may be trying to express their mathematical ideas in a new representational infrastructure (e.g., algebra). Also, though we do not discuss it in this chapter, we have observed a seeming u-shaped development, reminiscent of "errors of growth" (Bruner, Oliver, & Greenfield, 1966, p. 199), where students appear to regress through applying ideas recently met at school when they are inappropriate, or in ways that are not yet effective.

Our analyses suggest that switching strategies (even between incorrect strategies) might be helpful in catalyzing a new perspective on a problem. This indicates that we can change students' habits of mind (Cuoco et al., 1996). In our recent work on the Proof Materials Project (Küchemann, 2008: see http://www.ioe.ac.uk/proof/PMPintro.html), we collaborated with teachers to see whether this change of habit could be put into effect more widely, in particular by helping students become more aware of different kinds of proof strategies and explanations. We used the student responses to our items as starting points for discussion among teachers, leading them to think about how to encourage students to use different representation for their ideas (e.g., narrative, algebraic and visual), to make connections between them and to justify their reasoning. This helped students focus on structure rather than just on outcomes, as well as to distinguish mathematical from nonmathematical reasons. It is notable that teaching strategies such as speaking about the relationship, using and comparing different representations, and taking a range of particular cases, tend to be used by teachers in lower sets in English schools and not in higher sets, given the twin fears of not covering the curriculum and of students becoming bored if invited to revisit work from a new perspective. A shift in teaching emphasis in this direction for high attainers would seem a useful way forward.

Note

1. A4 is comparable in size to US letter, but with the property that an A4 sheet can be folded in half to produce a rectangular shape that is mathematically similar to the original rectangular shape (the edge lengths are in the ratio 1 to root 2).

11
Developing Argumentation and Proof Competencies in the Mathematics Classroom

AISO HEINZE AND KRISTINA REISS

Do we need proof in school mathematics?
Absolutely. Need I say more? Absolutely.

Alan Schoenfeld

Giving coherent mathematical argumentations and formulating valid proofs are complex tasks for students at all grade levels. These tasks require cognitive as well as non-cognitive prerequisites on the side of the students and, in particular, they presuppose an adequate support in the mathematics classroom. However, little is known about how students' prerequisites emerge and how mathematics instruction may support this process. In the following, we will give an overview on secondary school students' proving competencies and on aspects of its development in German mathematics classrooms. We will present results of a study that focused on describing high school students' mathematical argumentation and proof competencies. These competencies are regarded from a broad perspective and encompass achievement as well as interest, motivation, and emotions related to mathematics.

Students' Competencies in Reasoning and Proof

Mathematical Competency

Mathematics is a proving science and may be distinguished by this aspect from all other scientific domains. In general, it is not at all an easy task to develop a correct mathematical proof, but it asks for a specific mathematical competency. Models of mathematical proof like that of Boero (1999) suggest that this competency encompasses several aspects like generating a hypothesis, identifying possible arguments, or connecting suitable arguments into a valid proof. These aspects are mostly seen from a cognitive point of view but competency is regarded to be a far more complex construct. According to Weinert (2001), competencies may be defined as cognitive abilities and skills, which individuals have or which can be learned by them, however, these abilities and skills will enable them to solve particular problems, and therefore encompass the motivational, volitional, and social readiness and capacity to utilize the solutions successfully and responsibly in variable situations.

Accordingly, being competent in argumentation and proof is more than mastering the technical prerequisites such as using related domain-specific declarative knowledge or knowledge of appropriate procedures. This competency includes the ability to use knowledge in varying contexts under the assumption of an adequate personal disposition. It is a requirement that mathematics educators postulate and it is a challenge for most mathematics curricula that the aspect of application should play a dominant role in the classroom. Thus far, the definition given by Weinert (2001) is in accordance with important objectives of school mathematics. His emphasis on personal disposition goes beyond a mere cognitive view on competency. In addition, mathematical competency depends on the motivational state of the subject, and on his or her willingness and social readiness to adequately perform a specific task.

This notion of competency fits well into patterns of mathematical proving processes as provided by reports of mathematicians about their work. Prominent mathematicians such as B. L. van der Waerden (1903–1996) or Henri Poincaré (1854–1912) who described their struggle with a mathematical proof did not explicitly mention their motivational state but implicitly gave the reader a good idea of their positive attitude towards mathematics (see Heinze & Reiss, 2007). It is self-evident that this positive attitude should also be important for people learning mathematics. In particular, mathematical achievement has to be seen in close relation to aspects like interest in mathematics, motivation to deal with the subject, and the emotional state of a person with respect to mathematics.

Objectives of Reasoning and Proof in the Mathematics Classroom

Developing reasoning and proof competency is a significant objective of mathematics instruction at school. This competency includes the ability to reason coherently and systematically, to identify valid mathematical arguments, and to establish a correct mathematical proof. Obviously, reasoning and proof at school cannot be based on a rigorous mathematical concept of this topic. Reasoning and proof in a school context encompass aspects of logical deduction but do not rely on a well-defined set of basic knowledge and may also use informal elements. In particular, mathematical reasoning in the classroom may be regarded as a local activity. It may be correct only with respect to a specific problem and may lack a global perspective.

In many countries mathematics curricula cover reasoning and proof through all grades and give a specific attention to this topic. Mostly, proving in the classroom is not directed towards rigorous proofs based exclusively on axioms and other (rigorously proven) theorems. The underlying idea is mainly to support students' understanding of proofs and to establish mathematical reasoning as a habit of mind. Developing this habit presupposes a classroom atmosphere that consistently supports reasoning processes in various contexts and at all grades. Even very young students should be encouraged to find arguments supporting or rejecting the validity of mathematical statements. At all grade levels, students should explore mathematical structures and relationships, for example by reasoning inductively from patterns and specific cases. Moreover, students should learn to make use of effective deductive arguments. The aims of reasoning and proof in the mathematics classroom by the end of secondary school comprise that students understand and produce (simple) mathematical proofs, that they are able to perform deductive conclusions, and that they appreciate the value of such arguments (e.g., National Council of Teachers of Mathematics, 2000).

The Empirical Perspective: Students' Performance in Reasoning and Proof

Mathematics instruction is aiming at teaching proof in the classroom, but this is an ambitious aim. Several studies on mathematics achievement and specific studies on students' proof competencies reveal that proving is a complex and particularly demanding mathematical activity.

The ability to reason in a mathematically correct way and to generate a proof demands a sound prior knowledge, including the knowledge of mathematical concepts and heuristic strategies, their application in a problem situation, the use of metacognitive control strategies, and an adequate understanding of the nature of proof in mathematics (Schoenfeld, 1985). Empirical studies from different countries and cultures indicate that many students show deficits in one or more of these facets of proof competency (Healy & Hoyles, 2000; Heinze, Cheng, & Yang, 2004; Klieme, Reiss, & Heinze, 2003; Lin, 2000; Reiss, Hellmich, & Reiss, 2002). In particular, Healy and Hoyles (2000) contributed significantly to the field with a systematic investigation of students' understanding of proofs, ability to construct proofs, and views on the role of proof. Their empirical study was conducted in various types of school spread across England and Wales. Almost 2500 10th grade students, nearly all of them high achievers in mathematics, participated in the study. The results suggested that even these students had great difficulties in generating proofs. The students were far from being proficient in their construction of mathematical proofs, and they were more likely to rely on empirical verification. Nonetheless, most of them were well aware that once a statement had been proven it would cover all cases within its domain of validity. Moreover, the students were frequently able to recognize a correct proof and there was a positive correlation between correctly judging a proof and performance in proof. Their answers were also influenced by considerations that lacked a scientific perspective. An example is that, for many students, it was important to take into account the proof preferences, which they supposed their teacher would have. Students considered that their teacher would be more likely to accept formally presented proofs. When asked for their own preferences, students agreed on proofs that, in their view, had a more explanatory character. Accordingly, the study suggested that students would prefer an explaining style of instruction but regarded their mathematics classrooms as adapting to a more formal side of reasoning and proof.

Most studies on argumentation and proof address the secondary level and geometrical proving as it is performed in many countries at this stage. However, a considerable number of empirical surveys also give evidence that these difficulties are not restricted to students at the secondary level and their first attempts in giving formal mathematical proofs. Research by Senk (1985) revealed wide gaps in respondents' understanding of proofs for North American high school students, moreover, a study by Martin and Harel (1989) showed this for preservice teachers as well. As a consequence one may state that neither mathematics instruction at school nor mathematics instruction at the tertiary level is sufficiently successful in teaching mathematical proof to a majority of students. The elements of a mathematical proof and the process of proving seem to stay incomprehensible even after many years at school.

However, there are significant differences with respect to proving competency within the different groups of students. These findings may help to identify possible reasons for students' difficulties with reasoning and proof problems. Lawson and Chinnapan (2000), for instance, demonstrated that high-attaining students at the upper secondary level were able to activate more knowledge about geometrical shapes and theorems and, above all, to establish more connections between geometrical theorems, examples of use, and drawings than less successful students. This research was supported by results of Weber (2001) which indicated that even domain specific knowledge and a basic understanding of proof principles did not necessarily lead to a correct proof. It was essential for the students to have an adequate knowledge how to combine these elements. A comparison of experts and novices revealed that knowledge concerning the meaning of facts and concepts as well as knowledge of proving schemata for specific types of mathematical problem were important prerequisites for effective and meaningful proofs. These findings corresponded with a study by Koedinger and Anderson (1990), who emphasized that experts constructing geometrical proofs did not merely retrieve declarative knowledge such as definitions, axioms, and theorems from memory and combined these to make logical deductions. In their actual work, experts skipped details of the proving process, and outlined their argumentation in broad terms. Experts used visual models, in which they were able to see properties and connections, as well as pragmatic reasoning schemas in the proving process. Accordingly, geometrical competency is not merely a question of ability, but of specific skills and knowledge, too.

Motivational Aspects and the Learning of Mathematics

The studies cited in the last paragraph concentrate predominantly on cognitive aspects of mathematical reasoning and proof. However, students' learning processes do not exclusively depend on their cognition. Non-cognitive prerequisites like motivation for, interest in, and emotions towards mathematics are aspects of the individual competency as well. Moreover, competency is influenced by the social environment of the individual. The link between individual intrinsic motivation and social environment is the core of modern theories about the impact of motivation for successful learning processes. The subject–object theory of interest may serve as an example. This theoretical approach constitutes a relationship between the individual as a learning person and the learning topic insofar as interest is regarded a long-term characteristic of the relationship which is independent of a singular (present) situation. Within this theoretical framework, Lewalter, Krapp, Schreyer, and Wild (1998) investigated general conditions of classroom instruction that might foster students' individual motivation. They found out that students' motivation substantially depended on the particular teacher responsible for organizing teaching and learning in the classroom. It turned out to be crucial that students experienced the classroom as supportive with respect to their autonomy and competency and that they needed the feeling of being socially integrated. In a meta-analysis, Schiefele, Krapp, and Schreyer (1993) pointed out that there were substantial correlations between subject-specific interest and subject-specific school performance. Their study provides an average correlation of $r = 0.30$ between interest and performance data.

There are some studies that give evidence that emotions are important in learning processes. In particular, motivational dispositions are closely connected to emotions in mathematics learning (Pekrun, 1992). Emotions mediate learning processes and thus influence learning and achievement with respect to motivation, volition, problem-solving strategies, and cognitive resources.

Measuring Proof Competency in the Mathematics Classroom

The process of performing a mathematical proof may be regarded as a complex and challenging task. As proving is a highly important aspect of mathematics, it should nonetheless be positioned in the school curriculum in all grades in order to permanently foster students' understanding of mathematical argumentation and proof. Recent research has significantly added to our understanding of students' performance. We are well aware of deficits that manifest in difficulties of students with mathematical reasoning and proof. More research is needed that is based on a concept of competency reaching beyond the cognitive component. We still lack sufficient knowledge about what constitutes proving competency as a construct that is influenced by cognitive, affective, and social aspects.

Accordingly, there are important research desiderata concerning reasoning and proof in the mathematics classroom. This research should take into account an integrative perspective regarding the individual and his or her personal prerequisites as well as the requirements of mathematics as a subject, which is useful for the individual and the society.

An Empirical Study on Reasoning and Proof in the Mathematics Classroom

Research Questions

Reasoning and proof competency in mathematics classrooms were in the focus of a research project involving German high school students from grades 7/8 and 12/13. This project aimed at identifying students' cognitive and affective prerequisites for mathematical reasoning and proof. In particular, relations between mathematics achievement and personal dispositions were analyzed and described. The data from students of the different age groups should provide information on developments during schooling. In particular, the following research questions were explicitly addressed:

What are adequate personal prerequisites for students' successful proving processes?

In which way are successful argumentation and proof processes correlated with students' interest, motivation, and emotions?

What are the differences between students of various age groups with respect to achievement and motivational factors?

These questions cover diverse perspectives of argumentation and proof processes in the mathematics classroom. Accordingly, they emerged in a series of studies involving students from grades 7 and 8 as well as in grades 12 and 13.

Method and Sample

Part of the research was conducted using a pretest–posttest design. Students at the end of grade 7 took part in achievement tests (see later for an example of a test item) and answered a questionnaire about their motivation, interest, and emotions with respect to mathematics. This questionnaire consisted of different scales concerning interest in mathematics. It was adapted from a more comprehensive questionnaire (Götz, Pekrun, Perry, & Hladkyi, 2001). Subsequently, all students attended instruction on argumentation and proof in their regular grade 8 classrooms according to the mathematics curriculum. This instruction encompassed teacher-guided development of proofs in the classroom setting, classroom discussions on specific geometrical proofs, and work on proof-related exercises from the mathematics textbook. After this teaching unit, the students participated in a posttest on their achievement and completed a questionnaire concerning their motivation, interest, and emotions. The mathematics pretest (13 items, maximum score = 26) and posttest (11 items, maximum score = 22) could be scaled unidimensionally in one latent dimension if the items were rated in a dichotomous way as correct or incorrect (Reiss, Hellmich, & Reiss, 2002). Pretest and posttest used different items, as, the pretest was based on the grade 7 geometry curriculum whereas the posttest took into account aspects of the grade 8 geometry curriculum. There were some items that were used in the pretest as well as in the posttest. In grades 12 and 13, students were asked to solve proof problems (as used in the Third International Mathematics and Science Study; e.g., Baumert & Lehmann, 1997) and to answer a questionnaire on interest and motivation.

Developing Proof Competencies

The Cognitive Perspective: Results from Studies in Grades 7 and 8

Our first study aimed at identifying students' competencies in argumentation and proof tasks. Accordingly, the data from the pretest as well as from the posttest were analyzed not only with respect to the correctness of the tasks but also with respect to the types of error. The results gave evidence, that German students had similar difficulties as their international counterparts. Students tended to show typical mistakes in their proofs like using empirical arguments, accepting incomplete constraints, or engaging in circular reasoning.

However, a detailed analysis of the data suggested that these competencies could be arranged in a hierarchy of levels. A person's proving competency and the characteristics of proof items had a common latent dimension. As already mentioned, this statement was tested by the fit of the data to a unidimensional model. In our study, we grouped the items at three levels according to a theoretical classification. Level I items assumed basic knowledge and the application of simple rules, level II items assumed a simple argumentation, and level III items assumed a more complex argumentation (see Figure 11.1).

The results fitted into the expected pattern, and we were able to show that low-achieving students (according to their test score) were hardly able to deal successfully with items on level III whereas high-achieving students (according to their test score) performed well on level I and level II items and still satisfactorily on level III tasks (see Table 11.1).

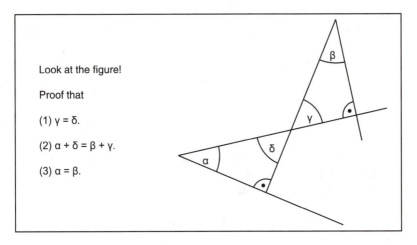

Look at the figure!

Proof that

(1) $\gamma = \delta$.

(2) $\alpha + \delta = \beta + \gamma$.

(3) $\alpha = \beta$.

Figure 11.1 Example for items of competency level II (item 1) and competency level III (items 2 and 3) in the test for grade 7

Table 11.1 Levels of proof competency

Percentages of correct items	Competence level I Application of simple rules	Competence level II Argumentation and reasoning (one step)	Competence level III Argumentation and reasoning (several steps)
N = 659 (grade 7)	M = 69%	M = 56%	M = 24%
Lower third of the students N = 238	51%	22%	0%
Middle third of the students N = 215	72%	61%	18%
Upper third of the students N = 206	85%	89%	50%

The results of the tests in grade 8 confirmed the students' rather low performance. Even after a teaching unit on argumentation and proof most students were only able to solve standard items asking for calculations in a geometrical context. They still had difficulties to identify arguments leading to a proof (Heinze, Reiss, & Rudolph, 2005). Comparing the results students achieved in grade 7 and half a year later in grade 8 indicated that there was hardly a gain in achievement with respect to argumentation skills. Students' performance on argumentation and proof items remained on the same level despite explicit instruction on proof in between. Accordingly, students learned new geometrical content, but could not improve their argumentation competencies.

Moreover, the results indicate that the individual achievement was most prominently influenced by the student's classroom. In this study with students from 27 classrooms, we found significant differences in performance between low-achieving and high-achieving classrooms. In the tests on proving in geometry, the results varied between 22% and 68% correct answers in one test and 12% to 59% in the other (Heinze, Reiss, & Rudolph, 2005). The pretest and posttest class scores had a correlation of $r = 0.604$ ($p < 0.001$), that is, the results on the classroom level were more stable than the results on the individual level ($r = 0.435$; $p < 0.001$). This leads to the question to what extent the classroom level influenced individual performance. For both, pretest and posttest, the scores of the individual performance showed a large variance. The variance on the classroom level was restricted by the higher reliability of the means. Thus, individual students might improve their performance significantly above the average of their classroom.

Based on these findings it can be assumed that classrooms that are low achieving in reasoning and proof at the end of grade 7 can hardly improve above average during regular teaching units with a particular emphasis on this topic.

The Cognitive Perspective: Results from a Study in Grades 12 and 13

A study with 12th and 13th graders revealed that these students had a better mastery of standard items asking for mere declarative knowledge when they had to solve geometry proof tasks that demanded mathematical knowledge at the lower secondary level. The results indicate that the students had the content knowledge that was necessary for solving the problems even had they learned these facts some years ago. However, they were not able to correctly use their declarative knowledge in a mathematical proof. Even mathematically advanced students tended to use one-step statements and were not able to combine them in a chain constituting a correct mathematical proof (Klieme, Reiss, & Heinze, 2003).

Interviews gave even more evidence that the students had the declarative knowledge needed for the single steps, but lacked the procedural knowledge that was a prerequisite for completing the task. It turned out that the students in grade 12 and 13 were able to use simple strategies like starting from the assertion and then working backward. Therefore, in contrast to the 8th graders more students were able to succeed in a two-step argumentation by solving these items with one step forward and one step backward. However, items that required more than these two steps were mastered only by the high achievers.

The data suggest that there is hardly a qualitative gain between the lower secondary and the upper secondary level with respect to procedural knowledge necessary for proofs. Students in grade 12, and, respectively, 13, are approaching the end of their school career. Even then, they seem to use similar heuristic strategies as 7th and 8th graders do and which are more or less of a quite simple nature. Twelfth and 13th graders were better able to use backtracking as strategy but were not able to use this technique on more challenging items. Students from both age groups seem to lack the ability of exploring a problem context in a systematic manner and of generating an idea for a proof. As a result, younger as well as older students have difficulties in using their declarative knowledge as a basis for mathematical proof.

Interest, Motivation, and Emotions

Interest, motivation, and emotions are personal dispositions that may be connected to a specific subject or to a specific topic within this subject. We addressed this topic a small series of studies. The studies were based on data we collected from questionnaires on students' interest, motivation, and emotions.

The students in grades 7/8 received a questionnaire that consisted of 48 items. It aimed at evaluating their interest in mathematics, achievement motivation, self-determined learning as well as their fear of failure, boredom in the mathematics classroom, and fear of mathematics as a classroom subject. The answers were given on a five-point Likert scale reaching from "strongly disagree" to "strongly agree." Sample items are given in Table 11.2.

The analysis of the data on interest and motivation started with a principal component factor analysis, which gave empirical evidence for the theoretically suggested factors. With respect to these factors, the explained variance added up

Table 11.2 Sample items of the interest, motivation, and emotions questionnaire

Scale	Example
Interest in mathematics	"I actively participate in my math class, because I am very interested in mathematics."
Fear of failure	"When taking the math test, I worry I will get a bad grade."
Boredom	"I think the math class is boring."
Achievement motivation	"Because I look forward to getting a good grade, I work hard for the test."
Self-determined learning	"I can estimate and control pretty well myself how I can solve math problems."
Fear of mathematics classroom	"When thinking about my math class, I get nervous."

to 57.5%. The questionnaire was also used for the grade 12/13 students so that results can be compared. Table 11.3 provides these results transformed to values on a scale from -1 ("strongly disagree") to 1 ("strongly agree").

The data described in Table 11.3 indicate that students in grade 8 as well as in grade 12 showed little fear of mathematics as a subject, they showed little fear of failing in their achievement, and they were not bored in the mathematics lessons. However, the results indicate that students were hardly interested in mathematics. Significant differences between grade 8 and grade 12 students could be identified for interest in mathematics, achievement motivation, and self-determined learning. Grade 12 students still reported a low interest in mathematics, but their interest level was slightly higher than that of grade 8 students. The achievement motivation, which was quite high for the 8th graders, was found near to the midpoint of the scale. As could be expected, compared to their counterparts in 8th grade the 12th graders felt that their learning was more self-determined. The different scales correlate with each other. For example, there is a significant correlation between the scales concerning fear of failure and fear of the mathematics classroom ($r = 0.601$, $p < 0.001$; computed for the grade 7/8 population) as well as a negative correlation between interest in mathematics and boredom ($r = -0.439$, $p < 0.001$; computed for the population in grades 7/8).

It should be mentioned that results from these questionnaires are to be interpreted with care. The scales have no absolute point of origin and already small variations in the formulation of the item texts may let students agree or disagree to an item. Furthermore, the problem of social desirability may have an influence.

The questionnaires on interest, motivation, and emotions were administered to complete classrooms and were evaluated on the classroom level as well. Aggregating the individual data from the six scales gave evidence that there were considerable differences between classrooms but much lower differences within the specific classrooms (Heinze, Reiss, & Rudolph, 2005). The data in

Table 11.3 Interest, motivation, and emotions, regarding mathematics

	Grade 8 (N = 417)		Grade 12 (N = 95)	
	Mean*	SD	Mean*	SD
Interest in mathematics	−0.27	0.44	−0.12	0.44
Fear of failure	−0.26	0.51	−0.27	0.51
Boredom	−0.45	0.48	−0.44	0.43
Achievement motivation	0.40	0.44	0.09	0.36
Self-determined learning	0.15	0.42	0.32	0.32
Fear of mathematics classroom	−0.76	0.33	−0.71	0.38

* Values transformed to a scale from −1 to 1.

Table 11.4 are based on 27 grade 7/8 classrooms. They show that all aspects of interest, motivation, and emotion covered by the questionnaire were differently assessed from students in different classrooms. The least range is found for fear of mathematics, however, that is, students are hardly fearful of mathematics at all.

In some classrooms, nearly all students gave extreme values for the scales, whereas other classrooms showed a broad distribution of the individual scores. Nevertheless, many classrooms showed tendencies to agree or disagree as a whole to specific scales. It can be assumed that those tendencies were influenced by their specific mathematics instruction. Probably there is a "classroom climate" not only with respect to cognitive factors but also with respect to the emotional status of the students.

Relations Between Cognitive and Affective Variables in the Mathematics Classroom

Competency is regarded as a construct that encompasses cognitive and affective aspects, accordingly, possible correlations between these aspects should be revealing. Thus, correlations between achievement scores and scores for the six factors extracted from the questionnaire on interest and motivation were determined on the individual level. From these six factors only two were significantly correlated with the achievement tests. It is probably not surprising that the correlation between interest in mathematics and achievement ($r = 0.212$; $p < 0.01$) is highly significant. Students with a relatively high interest level in mathematics showed better results than their less interested counterparts. Moreover, fear of failure ($r = −0.218$; $p < 0.05$) turned out to correlate negatively with students' achievement. This gives rise to a conjecture which is probably not restricted to mathematics, namely that fear negatively influences performance. The study gave also evidence that high-achieving students showed a somewhat higher score for the component self-determined learning, whereas low-achieving students seemed to have more fear of the mathematics classroom (these differences were significant, but not substantial).

Table 11.4 Class means regarding interest and motivation (scale from −1 to 1)

Class means	Min.	Max.	Mean*	SD
Interest in mathematics	−0.62	0.03	−0.27	0.17
Fear of failure	−0.55	0.27	−0.26	0.18
Boredom	−0.71	0.08	−0.46	0.17
Achievement motivation	0.07	0.71	0.40	0.13
Self-determined learning	−0.25	0.37	0.14	0.14
Fear of mathematics classroom	−0.96	−0.53	−0.75	0.11

* Values transformed to a scale from −1 to 1.

Correlations between motivational aspects and achievement do exist; however, it is unclear whether high achievement in mathematics enhances a student's interest or whether high achievement presupposes interest. A similar consideration can be made for the connection between mathematics achievement and fear of mathematics. There is little knowledge about the operating mechanisms here.

Conclusion

Research suggests that students' competencies in argumentation and proof are rather low at all grades. Classroom instruction might foster declarative knowledge but is hardly effective in enhancing procedural skills. Many students in grades 7/8 as well as in grades 12/13 can apply simple rules and concepts and are able to solve problems that require only a one-step argumentation. However, they fail when they have to combine two or more arguments into a chain of arguments. Thus, mathematical reasoning and proof is something obscure for many students during their schooling career. Even if the data presented in this chapter were collected from students in different age groups and lack a longitudinal perspective the problems seem to manifest over the school years (see also Küchemann & Hoyles, Chapter 10, this volume).

It remains an open question as to how cognitive competencies in argumentation and proof can be improved during the years at school. Several intervention studies suggest that it is possible to support specific aspects of these competencies. Mevarech and Kramarski (1997) could show that a mathematics classroom in which metacognitive activities were fostered had a positive effect on students' problem-solving competencies. Reiss and Renkl (2002) developed the concept of heuristic worked-out examples. This is a learning environment focusing on the heuristic aspects of the proving process. It is intended to give students information about how mathematical problem solving is performed. Results of an intervention study with 243 students from grade 8 are promising and suggest that this learning environment might improve students' argumentation in rather complex proofs (Heinze, Reiss, & Groß, 2006).

The approach to learning reasoning and proof presented here presupposes that mathematical competency is not only a matter of cognition. From a theoretical perspective, this is suggested for example by Weinert's (2001) definition given in the first paragraph in this chapter. Moreover, there are reliable and replicated results that document a correlation between individual interest and school achievement as well as a correlation between motivation and school achievement. These results should be regarded as a challenge for mathematics teachers as well as for mathematics educators. In particular, research should try to answer the open questions concerning the complex structure of the effects and the interaction of interest, motivation, emotion, and achievement. There is no sufficient explanation that processes can foster interest in mathematics. More consideration should be given to aspects like the influence of a specific classroom on individual motivation and interest. As pointed out, research at the different grade levels gives evidence for a low interest level in mathematics but students are motivated to achieve in the mathematics classroom and they do not feel bored by mathematics. These somehow conflicting data might be interpreted in a way that students regard the level of difficulty in the classroom as being sufficient but do not feel challenged by mathematics instruction. Further research should answer the questions if and how a better mathematics instruction can promote students' interest and motivation for the subject.

In this respect, it might be of special importance to better take into account the role of emotions. In contrast to interest, which is probably established in a long-term way as an individual disposition, emotions may be experienced in specific situations. Research provides evidence that positive emotions foster a more intensive work and have an effect on the learning effort. Most research in this respect is of a quite general nature. It would be helpful to see more studies on emotions dealing with specific aspects of the mathematics classroom such as argumentation and proof. This research could clarify the influence of emotional aspects and could lead to a more successful teaching and learning.

12

Formal Proof in High School Geometry

Student Perceptions of Structure, Validity, and Purpose[1]

SHARON M. SOUCY MCCRONE AND TAMI S. MARTIN

> The role of proof, I think, is to allow the students an opportunity to organize themselves and get to a point where they can think logically and order their reasoning. But it's not just to prove two lines are parallel or to learn the techniques to prove an angle is a right angle. But to allow them to figure out the process . . . [I want them to] have a good justification for why they think it's true. If they say something is true, they [should] have a good explanation for why.
>
> High school geometry teacher

Our opening quote reflects what most teachers of geometry (and other mathematics courses) want for their students when they study proof. They want students to understand proof as a process and to understand the nature of proof as a tool for not only making a convincing argument, but also for explaining why a mathematical statement is true. With these perspectives in mind and given the earnest efforts by most teachers, why are so few high school students able to produce good mathematical justifications? In this chapter, we take a closer look at how students perceive both the process of proving and the outcomes of that process in order to explore students' ideas of what contributes to good mathematical justifications (or valid proofs). More specifically, we describe high school geometry students' understanding of the nature of proof and their ways of thinking about proof. Our goal was to examine student perceptions in some detail, to ascertain how aspects of instructional and non-instructional contexts may have influenced students' developing notions of proof and the process of proving.

In previous manuscripts, we have discussed *six principles* of proof, student agreement or disagreement with these principles, and implications for instruction. These principles, as first introduced by Dreyfus and Hadas (1987) and revised for our research (McCrone & Martin, 2004), were originally described as common mathematical concepts or obvious statements related to theorems and proving that are "not well understood by most students of average ability" (Dreyfus & Hadas, p. 48). Our revised *six principles of proof understanding* (Table 12.1) were meant to be more comprehensive (although not necessarily exhaustive) statements about proofs and proving with which informed members of a mathematical community would agree.

Table 12.1 Six principles of proof understanding

Principle	Statement and description of subcomponents
Principle 1	**A theorem has no exceptions** a) Knowing that a general statement is true implies that the statement is true for all specific instances. b) Showing that a general statement is not true in one specific instance, proves that the statement is false (i.e., one counterexample disproves a statement).
Principle 2	**Two roles of proof are to convince and to explain** a) One purpose of proof is to establish the validity of a statement. b) Another purpose of proof is to provide insight into why the statement is true.
Principle 3	**A proof must be general** a) One method of proving a general statement is to show that the statement is true for every conceivable instance. b) If a proof is not exhaustive, then a direct proof, indirect proof, or visual proof must be constructed so that all statements or diagrams refer to all classes of figures that satisfy the given conditions. c) Empirical evidence of one or more cases does not constitute a proof. d) Checking a few specific critical cases does not constitute a proof.
Principle 4	**The validity of a proof depends on its internal logic** a) Hypotheses are distinct from conclusions. b) The statements in a proof must be ordered to conform to the laws of reasoning to provide a path from the hypothesis to the conclusion. c) Inherent logic determines the validity of a proof, not ritualistic aspects of the form in which it is presented.
Principle 5	**Statements are logically equivalent to their contrapositives, but not necessarily to their converses or inverses**
Principle 6	**Diagrams that illustrate statements have benefits and limitations** a) Diagrams that illustrate a statement are visual aids. One must keep in mind those aspects of the diagram that reflect the conditions of the statement and those that may vary. b) Focusing on components of the diagrams may be essential for identifying the fundamental relationships that connect the hypothesis to the conclusion.

In this chapter, we take a closer look at students' perceptions regarding themes related to aspects of proof inherent in the six principles including, (a) the purpose of proof (Principle 2), (b) logical structure requirements of proof (Principle 4), and (c) the validity of a proof (Principles 3 and 4). We share evidence from interviews with selected focus students as well as excerpts from students' written responses to a questionnaire in order to delve more deeply into student perceptions of what it means to prove something in a mathematically rigorous way. We also reflect on possible factors contributing to the students' views about the process and products of proving.

Theoretical Considerations

Exploring Understanding and Thinking

Uncovering the understandings of others so that we can scrutinize these understandings and assess the extent to which they match the shared conventions of a community is, by nature, an impossible task. The ideas, conceptions, convictions, and structures in the mind of an individual are, at the same time, intangible, amorphous, and in a constant state of flux and change. As a result, the understandings of others are always elusive. Regardless, teachers who aim to help students learn new ideas or refine previous understanding must continually attempt to gather evidence of students' understandings in order to determine the quality of the learning and to inform instructional decisions. Similarly, researchers, seeking to build models of human understanding to inform practice, must attempt to build a window to students' thinking. Because direct access to another's thinking is impossible, teachers and researchers must rely on indirect sources of evidence from which they can make inferences about students' thoughts, beliefs, perceptions, or mental structures.

Sfard (2000, 2001) addressed some of the issues of capturing a picture of student understanding by characterizing thinking as communication and learning as gaining access to a certain discourse. She also described *object-level rules*, rules governing content of a discipline, and *meta-discursive rules*, rules governing the flow of information exchange within a discipline. Using her description, students' understanding may be assessed by determining the extent to which they are able to follow these discursive rules. Sfard's (2000) research also revealed that differences in classroom environments and expectations can lead to differences in students' perceptions of the meta-discursive rules. This suggests that to make inferences about students' understandings of proof, we must gather evidence by means of verbal or written communication and we should consider the meta-discursive rules of the classrooms in which students participate. If the learning experience is successful, students' communications should reflect the rules of the classroom, or, better yet, the rules and conventions of the discipline. If their communications are at odds with the meta-discursive rules of the learning environment, we should reflect on why this has happened.

Framing Understanding and Thinking

As part of our assessment of students' abilities to participate in proof discourse, we also made use of Harel and Sowder's (1998) proof schemes. Harel and Sowder defined an individual's proof scheme as that "which constitutes ascertaining and persuading for that person" (p. 244). Students can hold more than one kind of proof scheme at a time, indicating that perhaps what constitutes a convincing argument may vary by content or context. Students' proof schemes served as lenses through which they judged the validity or value of given arguments. Thus, by listening to students as they critiqued each other's proofs and as they explained their own thinking through the construction process, we were able to make inferences about their proof schemes. Certain proof schemes are more mathematical than others. Thus, one might expect there to be an overlap between students who embraced both object level rules and the meta-discursive rules for producing proofs and those with more mathematically sophisticated proof schemes.

Methodology and Analysis

To investigate high school students' proof understanding, we provide data from several examples of communication between geometry students and the researchers. In student responses to research instruments, to interview questions, and through classroom observation, we were able to construct a portrait of the students' abilities to participate in the formal discourse related to proof and proving, and more importantly, to explore apparent conflicts between the meta-discursive rules regarding proof in a classroom context and proof in everyday contexts.

Findings reported here come from a 3-year study in which the researchers investigated student understanding of proofs as well as classroom factors that may have contributed to the development of understanding. Details of the larger study are reported elsewhere (Martin & McCrone 2003; Martin, McCrone, Bower, & Dindyal, 2005; McCrone & Martin, 2004). Here, we take a closer look at data from interviews with eight focus students from the second year of the study, along with supporting data from research instruments and classroom observations from the same year. The focus students were enrolled in traditional proof-based geometry classes at two high schools in the mid-western United States. The focus students represented a wide range of ability levels, as determined by the classroom teachers and researchers, and included one female student and one male student from each of four different classrooms. The four geometry classes were observed by researchers on a daily basis for approximately 3 months, during which time the classroom teachers focused on proofs and proof writing along with geometric content that allowed for practice in proof writing (e.g., parallel lines, triangle congruence, similar figures).

After approximately 6 to 8 weeks of proof instruction, students responded to items on a written questionnaire.[2] Student responses to questionnaire items helped us to determine aspects of proof that were troublesome for students or principles that were not well understood. Follow-up interviews with the selected focus students, conducted approximately 2 to 4 weeks after administration of the questionnaire, provided more information for determining their agreement or lack of understanding with regard to the six principles. In particular, students were asked to respond to questions about proof in order to learn more about their understanding of the role of proof (Principle 2), the generality require-ments of proof (Principle 3), and the logic requirements of proof (Principle 4). Students were asked to assess the validity of several proofs as well as to respond to general questions about proof. These interview questions can be found in Appendix A.

Interviews and taped classroom visits were transcribed. Questionnaire data were analyzed using quantitative methods and transcribed data were analyzed qualitatively using methods such as those described by Maykut and Morehouse (1994) and Miles and Huberman (1994). As interviews were reviewed and ana-lyzed, we discovered that students in all four geometry classes had developed similar ways of describing the role and purpose of proof. We also observed that students had some common perceptions about the logic requirements of proof and factors related to establishing a convincing proof.

Using qualitative methods similar to the constant comparative method (Maykut & Morehouse, 1994), interview transcripts were reviewed for recurring patterns in students' responses to the interview questions. There were several themes that emerged from this analysis of the interview data that were not entirely coincident with the original statements of the principles but captured common ideas. These themes were grouped into three categories: structure of proof, validity of proof, purpose of proof. In the sections that follow, we describe each of the themes within their larger category and share excerpts from stu-dent interviews to illustrate and support the themes. We now interpret those findings in relation to students' learning experiences and discuss connections or inconsistencies with respect to responses on the written questionnaire.

Results

Structure of Proof

The first category of proof themes relates to the structure of proof. We define *structure of proof* to mean the essential components of a proof as well as the way in which those components are put together to form a proof. First, we report on student perceptions about the components of a proof. Then, we describe stu-dents' perceptions about the required format in which the components must be presented. The themes presented in this section, and many of the other themes in this report, emerged from students' responses to questions asking them to com-pare a "proof" based on empirical methods with a more general proof written in a paragraph form. Students also addressed issues about structure in their responses to the more general question, "What is a proof?"

THEME 1: THE ESSENTIAL COMPONENTS OF PROOF ARE STATEMENTS AND REASONS

As illustrated in the following sample responses, students repeatedly claimed that statements and reasons are the building blocks of proofs. Although some students were more articulate than others in describing this notion, no students contradicted this claim. For instance, when asked whether Kobi's answer or Linda's answer was a proof (see item 1, Appendix A), students responded in the following way:

HB   This one, Linda's. Um, each . . . well, everything again has a reason for it that's correct.

DW   Linda's answer was a proof because she did take the steps and used a reason why she took that step and then she went through the problem in that order.

RE   I think that Linda's answer is kind of more of a proof because it goes step by step and it leads one thing into another.

When asked, "What is a proof?" students responded:

LC   I think it's just either a statement or you know a flow of, using reason or whatever or what you know, to come to the answer.

AH   I think that a proof is coming to a conclusion through a series of steps supported by facts.

RE   Something we had to do in geometry, because we had to . . . mm, I don't know. I guess that I'd say steps with reasons, slowly telling why something is, I guess.

It is interesting to note that students used "facts" and "reasons" interchangeably, perhaps indicating their understanding that in order for reasons to be used in a proof, they must be considered established facts. Similarly, students mentioned that these statements and reasons must be sequential or linearly ordered. The next theme addresses some specifics of how the statements and reasons must be strung together in order to create a proof in an acceptable format.

THEME 2: PROOFS SHOULD BE STRUCTURED IN A WAY THAT LEADS THE READER FROM GIVEN INFORMATION TO THE STATEMENT THAT MUST BE PROVED

Although some students said that it was acceptable to introduce given information at any point in the proof sequence, most indicated that the given information should really be the first step in building a proof. Of those who described general procedures for ordering statements, all students reported that the statement to be proved, or the conclusion, should be the last statement in the ordered steps. Some of these student perspectives were gleaned from their analysis of sample proofs that did not follow this prescribed format.

    When asked whether Kobi's answer or Linda's answer was a proof, students responded in the following way:

LC   I think a proof is just how you set up to prove things. I think it has to have your conclusion in there and what's given to you in the proof . . . If you had to pick one, I'd say Linda's is more of a proof, but it doesn't have the given in there, but other than that it's good.

HB   This one, Linda's . . . It goes step-by-step. There's, ah, what you have to prove which is at the bottom. I think . . . I don't see the given anywhere, it didn't say given and it didn't say prove . . .

When asked, "What is a proof?" a student responded:

LC   I think you have to have a given and a conclusion for it to be defined a proof.

The format that students seemed to prefer is very rigid, always beginning with given information and leading to the statement to be proved. In fact, if the "givens" and the "prove" statements were not overtly identified, students indicated dissatisfaction with the proof. Much like the step-by-step characterization of proofs described in the first theme, these quotes support the claim that students were unsure if formats (such as paragraphs) that differed from the "two- column" format, which was emphasized in their classes, were acceptable.

SUMMARY: STRUCTURE OF PROOF

It is interesting to note that when asked why a given argument was a proof, students almost exclusively focused on the structure of the argument to support their choice. Similarly, when asked what a proof is, they chose to describe physical characteristics or the organization of a set of statements. Attention to more ritualistic aspects of proof rather than the internal logic was also seen in student responses to items in the proof questionnaire. For example, one questionnaire item asked students whether they agreed or disagreed with the statement that "geometric proofs must list statements and reasons in two separate columns." In two of the four participating classrooms, nearly 50% of the students agreed with this statement (50% agreement in one class and 48% agreement in the other). In the other two classrooms, 26% and 19% of the students agreed with the statement. Such responses hint at a view of proof as algorithm ("this is how we do it") or a product view ("this is how it should look").

Validity of Proof

Although there are many ways to assess the validity of a proof, in this section we focus on the nature of the reasoning (empirical vs. general) and logical ordering of a proof's components. Questions related to interview item 1 (see Appendix A) provided an opportunity for students to express their preferences regarding the nature of the reasoning. Responses to interview item 2 (Appendix A) illustrated students' criteria for assessing the logical ordering of an argument.

THEME 3: NON-EMPIRICAL ARGUMENTS ARE THE PREFERRED METHODS FOR CONSTRUCTING
"VALID" PROOFS, ALTHOUGH EMPIRICAL ARGUMENTS ARE ACCEPTABLE

When asked to compare sample "proofs," one of which relied on an empirical argument and one of which was a general argument, almost all students indicated a preference for the general proof, at least in the sense that their teacher would prefer this proof-writing technique. In contrast, most of the students still clung to the notion that empirical arguments could be counted as proofs.

When students were asked which type of answer (Kobi's or Linda's) their teacher would prefer or assign a higher grade, they all concurred that the teacher would prefer Linda's answer:

LC   I think Linda's answer you'd probably get a better grade on, because that's stuff that we've gone over and we don't usually try to measure things, we try to prove them instead.

JJ   I would probably say Linda's because it uses more . . . like geometry terms. Instead of more like this [points to Kobi's] guess and check. And this [points to Linda's answer] is more like by the facts. So I would say Linda's answer.

GE   Probably Linda's, I would say . . . Because you can do it on your own, and you don't need a resource, and it's more of what we've been learning about proofs and stuff.

LM   That one [Linda's answer]. I could measure everything, but that's not what we're trying to learn. We are trying to learn proofs. I'd use Kobi's answer if we were on that topic. You know, topic of specific answers.

The following responses address whether Kobi's (empirical) demonstration or Linda's paragraph explanation were proofs.

LC   It [Kobi's work] shows that I guess, but I don't know. I guess I'm led to believe that you're supposed to stay away from measuring and things like that. It's not a very consistent way to prove it, but I guess it still proves it.

DW   He [Kobi] did check his answers multiple times, like when he said, I moved point X to different places on the perpendicular bisector and measured segments AX and BX, I found the lengths. So it was basically a guess and check. And I think you could do a proof like that, but I think Linda was a bit more so on the right track.

AH   Um, I think they're both written for proofs, just because they both come to the same conclusion. And they have supporting facts, too.

The students' comments showed that they perceived that general proofs were demanded in their classroom cultures. Although, many of the students did not discount empirical arguments, they were aware that their teachers preferred non-empirical arguments. More specifically, the students seem to imply that one reason their teachers might prefer a general argument is that the format of that style of reasoning enables students to show the teachers what they know about geometry and formal proof writing. In other words, by using theorems or other geometric reasons, the students were able to demonstrate their understanding of the content and the proof "algorithm" they were supposed to be learning in class. This type of rationalization might explain why students can simultaneously accept empirical arguments as valid, while being able to pay lip service to the fact that general arguments are preferred.

THEME 4: THE PURPOSE OF LOGICALLY ORDERING STATEMENTS IN A PROOF IS SO THAT SOMEONE ELSE CAN FOLLOW THE PROOF WRITER'S THINKING

Students were asked to assess the validity of a proof in which the statements were disordered (see item 2, Appendix A). The students' analyses of the validity of this proof and their general responses to questions about the importance of logical ordering, provided insight into students understanding of the role of logic in proofs.

Some of the student comments on an out-of-order proof or how the logical order of a proof affects its validity are as follows.

DW   You could do that [mix up the steps in a valid proof and still call it valid], it would just be harder to understand.

HB   I think it is important because then, um, I guess you don't get lost as easy, but then it just makes sense as you go down. Because it would be all mixed up and you wouldn't know what you were talking about. If it was just like, oh it's supplementary, and then three lines later, it equals 180 degrees, it just wouldn't make sense . . . Yeah, it [a disordered argument] is still a proof. It's probably still valid, I guess to me I like it more in order, but yeah I think it's still valid.

LM   It [a misplaced statement] would have to be before [another statement] because then you don't have the congruent angles. You don't know they are congruent. Then there goes what you are trying to prove right there. Right out the window . . . It's order specific. If you are trying to prove something with another part, that part has to be before it so you already know it. So, you can't just say it later.

AH   Well, yeah, if you take the wrong steps and you still get the same answer, it's not the same as taking the right steps and getting the same answer, because if you don't take the right steps to get there, it kind of makes the property wrong.

In their comments, LM and AH seem to be saying that a proof that is out of order is no longer a proof. However, DW and HB's criteria for accepting an argument as a proof were more lenient. They accepted a disordered collection of statements as a proof, but they indicated that it would not be as easy for the reader to follow the argument as if the statements were ordered more logically.

Students appeared to hold some contradictory perspectives on the validity of proof. On the one hand, they accepted that general arguments are better than arguments based on empirical evidence. On the other, most students did accept those empirical arguments as valid methods of proof. Responses to proof questionnaire items related to the validity of empirical arguments showed similar results. In fact, nearly 70% of the students in three of the four participating classes agreed that an empirical argument is a valid proof as long as representative cases were investigated (e.g., right triangle, equilateral triangle, obtuse triangle). Similarly, they saw the value in having an argument follow a logical order for the purpose of enhancing the communication of ideas. However, some students still accepted a disordered argument as a proof. Perhaps, this classification of disordered arguments as "proofs" is an issue of language. After all, in the schools we observed, students' proof attempts were routinely called "proofs," even if they were only partially correct. This loose terminology may confuse students, or make it more difficult for them to develop a clear sense of what is required for a valid proof.

Purpose of Proof

Several authors have discussed the purpose of proof in the context of describing ways to motivate students to want to learn proof (Chazan, 1993; Hanna, 2000; Hersh, 1993). In our six principles of proof understanding (see Table 12.1), two purposes of proof are identified. Although, there are many purposes of proof (see for example, de Villiers, 1999), our focus was on convincing and explaining. In the proof questionnaire, we asked students to agree or disagree with the statement that the dual role of proof is to convince and explain. Naturally, 100% of the students agreed with this statement. Because the statement itself might be construed as leading, we examined interview transcripts to obtain a more detailed picture of students' perspectives on the purpose of proof.

THEME 5: TWO PURPOSES OF PROOF ARE TO EXPLAIN WHY A STATEMENT IS TRUE AND TO CONVINCE READERS THAT A STATEMENT IS TRUE

This theme was garnered from students' responses to open-ended questions such as "what is proof?" and "what is the purpose of proof?" Although, it wasn't surprising that they echoed well-accepted roles of proof, their reservations about these roles were somewhat more informative.

When asked, "What is a proof?" students said:

RE   Well, it's explanations like I don't know, explanations for something you are trying to prove . . .

HB   Hmm. I guess a proof would be explaining step-by-step how to do a problem.

GE   Something that proves something to be true, I guess . . . You use facts or theorems and I guess to prove that, or apply to the thing, to show that it's true.

When asked to respond to, "What do you think is the purpose of proof?" students said:

AH   Um, to, uh, break down a problem instead of just saying something is this way. Proof is to show why it's that way and how it became that way . . . because, there's always going to be the question of why something is so. And a proof eliminates that question, that doubt, and shows something is absolutely so.

RE   Well, I do not like proofs at all. And I guess they serve a purpose because we are learning them. And I was talking to my social and geographical history teacher, and I asked her what she thought the purpose of a proof was the other day, because I didn't see a purpose to them. And I guess I can kind of understand her answer to it. She said that it helps you later on in life to find out . . . why something is the way it is. Like it helps you to understand . . . life in general [and] why things happen and why things are. And she said it helped her be better organized and have more reasonings [sic] for what she does.

When asked to respond to, "Does a proof really convince you that a statement is true?," one student said:

AH   Um, usually, but there's some instances where it doesn't . . . Um . . . well, I think that even though they've always told us that you can't assume anything by looking at a picture, there is still that assumption that if something looks right then it is. And sometimes a proof just doesn't prove that to me. I still kind of think that I'm right and the proof is wrong.

I   So even when you have proved something, there is still a doubt in your mind?

AH   A little bit, but since it's on paper and there is proof, I really can't contest it.

For the most part, students did have a sense that proving was a way of explaining why something is true. Other students, such as GE, only cited the role of proof to convince the reader of the validity of a statement. It is not entirely clear how to distinguish among students who recognize the role of proof as merely convincing versus those who can see the explanatory power in a proof. Although the excerpts here are meant to illustrate students' perspectives on the purpose of proof, we have uncovered further evidence of students' persistent reliance on empirical arguments or visual cues. In the brief exchange between AH and the interviewer, AH mentioned that although he recognized that proofs he had constructed were meant to be uncontestable, in some situations, he still felt that the visual representation of a relationship was more convincing than the proof. As a result, we can see another sense in which what students perceive they "ought" to be learning about proof, is in conflict with their instincts or natural tendencies. In particular, students have learned that proof ought to be convincing, but sometimes it is not.

THEME 6: THERE IS A DISTINCTION BETWEEN ASCERTAINING AND CONVINCING

Although formal proof instruction in US schools rarely makes the distinction between *ascertaining*, for yourself, that a mathematical statement is true and *convincing* someone else that the statement is true, some students in our study were able to articulate the difference in standards or methods for meeting those two distinct goals. The following excerpts illustrate students' perspectives on this issue.

When asked whether Kobi's answer or Linda's answer were a proof, students responded in the following way:

DW   Kobi's answer seems more of a way that he did it to find what it was, I think Linda's answer was a proof because she did take the steps and used a reason why she took that step and then she went through the problem in that order.

LM   It's . . . this [Linda's] is a geometry proof, like the proofs that we're using in class. This one [Kobi's] is just a measurement; it's proving to you. I mean depending on how you look at proof, if it's proof to prove . . . This is a convince proof [Kobi's]. This is a prove proof [Linda's].

When asked, "What is a proof?," one student said:

LM   A proof . . . There's two types of proofs, I guess. There's a proof where you're proving enough to convince yourself, and maybe a couple other people, of what you believe. And then there's this proof [pointing to Linda's], which would dispute any other thing that said it was, or it was not true.

These students' perceptions were echoed in their responses to our questions about which argument (Kobi's or Linda's) would you use to convince yourself, a friend, or a teacher. Nearly all students felt that the empirical proof would be better to convince themselves or a friend who was not enrolled in a geometry class. Similarly, all students claimed that the general proof would be more convincing to their teacher and their geometry student friends, i.e., the students believed that different situations or audiences called for different standards for or methods of proof.

THEME 7: THE PURPOSE OF PROOF IS TO APPLY THEOREMS TO THE SOLUTION OF
MATHEMATICAL PROBLEMS

A further nuance of students' perceptions about the role of proof is their generalization about when proofs are used based on their classroom experiences.

Regularly in their geometry classes, students were provided a labeled figure, given information, and a statement to be proved. From these experiences, many students appeared to conclude that a primary purpose of proof is to establish the validity of a statement regarding some aspect of a given diagram.

In the context of a discussion about when proof is used, one student commented:

DW   We use theories that are already proved and we use those to prove what the figure is, certain parts of the figure, that we are given. They give us a given for a figure and then using what we know of theories that have already been proved, we use those to prove that the figure is what it wants us to prove.

In response to "What is a proof?," these students replied:

GE   You use facts or theorems and I guess to prove that, or apply to the thing, to show that it's true.

JJ   Like when you are looking at a triangle, or something. You have something that you are trying to prove. You have to use parts of a picture and you have to use facts to get parts of the picture to prove what someone asked you to prove. Like if someone asked you to prove that like one segment was congruent to the other in two different triangles, then you would look at the triangle and use parts of the triangle to prove certain, that the two triangles are congruent. Or either prove the triangles congruent and then prove the corresponding parts are congruent or whatever you are trying to prove. Like basically using facts to prove something in a picture or from an object.

What is interesting about these student responses is that they seem to link proof to the very specific set of educational circumstances in which they have been called on to respond with a proof. Either their teacher has provided a labeled figure with a statement to prove or they have been asked to respond to textbook exercises that require them to produce a proof that a particular relationship holds within a pre-labeled diagram. These responses are reminiscent of Herbst and Brach's (2006) findings regarding the nature of statements for which students are willing to engage in the process of proving. The students did not seem to understand that proof is used to justify the theorems that are the building blocks of an axiomatic system. The students claimed that they use the theorems to prove relationships in a figure, not that the relationships in the figure may themselves represent theorems within a larger axiomatic system. As a result, the students did not have the sense that a purpose of proof, or their geometry course, was to derive a portion of Euclidean geometry stemming from the first axioms and postulates using certain standards of rigor to establish theorems or relationships within the geometric system.

SUMMARY: PURPOSE OF PROOF

Although students were able to acknowledge or echo the notions that proof can be used to convince a reader of the validity of a statement, or, perhaps, provide an explanation for why a statement is valid, the students indicated that proofs are only convincing to teachers and others who were informed about particular formats and rituals. The students seemed to lack any sense that proof-writing rules were self-evident or that the final products of the proving process were necessarily convincing to those outside the small circle of informed participants. As a result, the standards of rigor that guide the development of a convincing argument, are dependent on whether the audience is inside the circle of geometry proof fluency or outside this circle. These standards of rigor are also subject to whether the purpose of the mathematical activity is to find an answer or to construct a proof.

Some students indicated that proof writing was an exercise in which the student shows the geometry teacher that he or she knows the theorems of geometry and how to apply them. The theorems were applied to unique diagrams in which some aspect of the diagram was to be proved to have a specific geometric relationship (e.g., congruency, similarity, parallelism, etc.) to another aspect of the diagram. In this sense, the purpose of proof is very local. None of the students seemed to see the purpose of proof as a means for developing a structured and interrelated system.

Conclusions

Students' perceptions of the structure of proof, the validity of a proof, and the purpose of proof were investigated through interviews with eight focus students as well as through questionnaire responses from over 100 students in proof-based geometry classes. The analysis of student responses provides insights into student understanding of proof. For example, we found that most students attempt to or are successful at following the meta-discursive rules of the classroom in relation to the general nature of a proof and in analyzing a proof. Herbst and Brach (2006) as well as Sfard (2000, 2001) point out that such hidden or perceived rules often govern how students participate in the practices of the classroom. Thus, it is somewhat expected that most students were able to determine and follow meta-discursive rules such as the fact that it was important to have a diagram accompanying a proof or that proofs should contain a list of statements and accompanying reasons in a two-column format. However, we have shown that, perhaps because these rules are so different from their everyday language and ways of communicating or forming convincing arguments, many students developed ways to rationalize apparent contradictions between their intuitive sense making and the rules of reasoning in their geometry classrooms.

One contradiction that students encountered arose from their acceptance of generality requirements for proofs and their inability to reconcile this with their sense that specific arguments were in many ways more credible. All focus students commented that their geometry teachers preferred proofs that were organized

218 • Sharon M. Soucy McCrone & Tami S. Martin

into lists of statements and reasons, written in a step-by-step format. Although the students acknowledged that these general arguments were incontrovertible, they also noted that empirical arguments could be more convincing. The students rationalized that because deductive proofs and empirical arguments both provide reasons for each step, or lead the reader from given information to the desired conclusion, perhaps it may be acceptable to call both types of argument "proofs." Healy and Hoyles (2000) uncovered similar findings when they presented students with a variety of arguments to support a statement and asked students to identify an argument that would get the best mark from their teacher and an argument would be convincing to a friend. Nearly all participants in their study agreed that the more formal looking deductive arguments would receive highest marks from a teacher.

Another source of conflict for students had to do with the logical ordering of an argument. For example, students voiced a preference for arguments in which statements were logically organized because such arguments were easier to follow. However, arguments that were not logically ordered were still considered proofs by students, as long as all the necessary information was accounted for. This perception fits well with our everyday discourse, for we often claim to understand another's intended meaning, even if his or her explanations are somewhat convoluted. As a result, the students in our study appeared to support the claim that a lack of logical ordering does not necessarily invalidate a proof. This view is also consistent with the assessment of the participating teachers: they often gave partial credit for students' written arguments that contained logical flaws. We do not necessarily disagree with such practices, however, these practices might reinforce and explain why students were willing to accept as valid, those arguments that contained logical flaws.

Another interesting point of conflict for students had to do with the purpose of proof. All of the focus students agreed that geometry proofs were helpful for showing that a statement was true. However, we discovered that most of these students had a rather narrow view of what this meant. For most, proofs were helpful for demonstrating some relationship between components of a diagram or for showing the teacher that they understood the content of the geometry theorems. In a sense, students who were successful in writing valid proofs were demonstrating to the teacher that they could function within the discursive community of the mathematics classroom (Sfard, 2000). Students did not see the broader picture of proofs helping to establish the truth of theorems and, hence, helping to build an axiomatic system.

Because students must negotiate back and forth between everyday language and practices and classroom language and practices, it is not surprising that students would try to resolve apparent contradictions in the two worlds. Perhaps these rationalizations are a component of the process of crossbreeding mathematical discourse with everyday discourse (Sfard, 2000), both of which students participate in simultaneously. The students' rationalizations are exactly that: rational explanations for rules that may appear irrational.

The teacher quoted at the beginning of the chapter said that she hoped her students would get organized. For the most part, they did and they saw some value in a rigid structure for developing arguments. She hoped they would learn to think logically. For the most part, they did. They appreciated the value of coherent, logical arguments. However, they could not always produce logical arguments and they expected to be understood despite a lack of perfect ordering of statements within the arguments they produced. The teacher wanted her students to understand that proofs are supposed to explain why something is true. For the most part, they did. But they were not always convinced. Sometimes the students thought that other types of clue or evidence were more compelling than formal proofs.

Students do seem to learn to follow meta-discursive rules related to proof in their geometry classes. But, because they also live in the world of everyday discourse, trying to get students to dismiss everyday discursive rules may be an overly ambitious goal. Students' experiences in the traditional proof-based geometry class have been largely limited to following patterns or models demonstrated by the classroom teacher. This leads to students learning how to write proofs with no true sense of the need for mathematical justification. Perhaps teachers at this level need to be more explicit in helping students understand how and why mathematical practices are different from those in a loosely structured everyday culture. Although, we make reasoned arguments to convince ourselves and others of various "facts" and positions, the rigor of a mathematical system cannot be approximated by everyday standards. In fact, upfront discussions about the nature of mathematical arguments can begin at the middle school (and perhaps elementary) level.

If teachers hope to help students not only follow and understand the taken-as-shared rules of mathematical engagement related to developing valid arguments, they should consider avenues for helping students at all grade levels learn to distinguish between the culture of mathematics and everyday culture and explore reasons for differentiated expectations.

Notes

1. This manuscript is based upon work partially supported by the National Science Foundation under Grant No. 9980476.
2. The development and content of the questionnaire is described in McCrone and Martin (2004).

Appendix A: Interview Questions

1. Suppose point X is any point on the perpendicular bisector of segment AB. Kobi and Linda were asked to decide whether or not the following statement is true: *Triangle ABX is isosceles.*
   - Which student's answer would you use to convince a friend that triangle ABX is isosceles? Why?
   - Which student's answer would get a better grade from your teacher? Why?

Kobi answers:
I believe the statement is true. I moved point X to different places on the perpendicular bisector and measured segments AX and BX. They were always the same, so the triangle is isosceles.

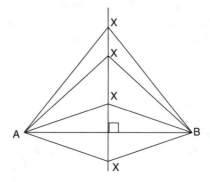

Figure 12A.1 Kobi's diagram

Linda answers:
Since the perpendicular bisector is also a segment bisector, it divides segment AB into two congruent segments, AY = BY. ∠ AYX and ∠ BYX are right angles because XY is perpendicular to segment AB. All right angles are congruent. So, ∠ AYX is congruent to ∠ BYX. By the reflexive property, XY = XY. △AXY = △BXY by the side–angle–side congruence theorem for triangles. Therefore, sides AX and BX are congruent (CPCTC). So △ABX is isosceles.

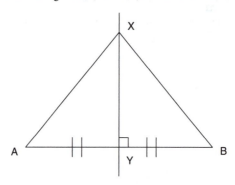

Figure 12A.2 Linda's diagram

2. Consider the following task and response.
   • Are the statements and reasons in the proof written in a logically correct order? Why or why not?
   • How can the logical ordering of a proof affect the validity of the proof?

Given: ∠1 ≅ ∠4, $\overline{PR}$ ≅ $\overline{TS}$, $\overline{RN}$ ≅ $\overline{SN}$
Prove: △*PNS* ≅ △*TNR*.

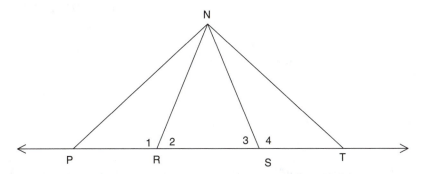

Figure 12A.3 Task diagram

Table 12A.1 Task response

Statements	*Reasons*
1 ∠1 ≅ ∠4	Given
2 ∠1 and ∠2 are supplementary angles, and ∠3 and ∠4 are supplementary angles	If two angles form a straight angle/linear pair, then they are supplementary
3 ∠3 and ∠4 form a straight angle/linear pair	Assumed from the picture (definition of linear pair)
4 ∠1 and ∠2 form a straight angle/linear pair	Assumed from the picture (definition of linear pair)
5 ∠2 ≅ ∠3	Supplements of congruent angles are congruent
6 $\overline{PR}$ ≅ $\overline{TS}$, $\overline{RN}$ ≅ $\overline{SN}$	Given
7 △PNS ≅ △TNR	Side–angle–side congruence postulate

13

When is an Argument Just an Argument?

The Refinement of Mathematical Argumentation[1,2]

KAY MCCLAIN

In order to engage students in the process of elevating mathematical argumentation to more sophisticated levels, the teacher must have a deep understanding of the mathematics under discussion (see McClain, 2004). This is critical in both being able to advance the mathematical agenda and in judging the quality and worth of student arguments. It requires decision making-in-action concerning the pace, sequence and trajectory of discussions in order to ensure that topics under discussion move the mathematical agenda forward.

When focusing on students' offered arguments, the teacher is seen to actively guide the mathematical development of both the classroom community and individual students (Ball, 1993; Cobb, Boufi, McClain, & Whitenack, 1997; Cobb, Wood, & Yackel, 1993; Lampert, 1990). This guiding necessarily requires a sense of *knowing-in-action* on the part of the teacher as he or she attempts to capitalize on opportunities that emerge from students' arguments. With this comes the responsibility of monitoring classroom discussions, engaging in productive mathematical discourse, and providing direction and guidance as judged appropriate. Lampert addresses similar pedagogical issues (1990) in her discussion of the teacher's role in guiding mathematical argumentation as a *zigzag* between conjectures and refutations.

In the analysis in this chapter, I also focus on communication between the students and teacher that leads to the evolution of sophisticated levels of mathematical argumentation. In doing so, I use Toulmin's (1958) scheme for argumentation as a basis for the analysis (see Figure 13.1).

Toulmin describes a process of examining data in order to reach a conclusion. However, he argues that the conclusion must be supported by a warrant and the warrant supported by a backing. In this chapter, I juxtapose the relationship of the backing and the warrant to claim that certain forms of argumentation are more sophisticated than others. This allows me to propose a sequence of phases

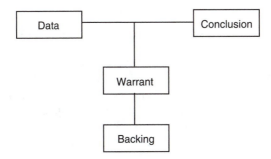

Figure 13.1 Toulmin's scheme for argumentation

or levels that describe the evolution of mathematical arguments over the course of a 12-week classroom design experiment in which I was the classroom teacher.

For the purposes of this chapter, I therefore focus on the sociomathematical norms that became negotiated in the classroom (see McClain & Cobb, 2001; Yackel & Cobb, 1996). Sociomathematical norms include what counts as a *different* mathematical argument, a *sophisticated* mathematical argument, an *efficient* mathematical argument, and an *acceptable* mathematical argument (see Lampert, 1990; McClain & Cobb, 2001; Simon & Blume, 1996; Voigt, 1995). In particular, analyses in this chapter will focus on the evolution of the sociomathematical norm of *what counts as an acceptable mathematical argument* and the norms for justification for these arguments.

Data

The episodes analyzed in this chapter are taken from a classroom design experiment conducted with a group of 29 American 7th grade students (age 12). The primary goal for the experiment was to investigate ways to proactively support middle school students' ability to reason about data while developing statistical understandings related to exploratory data analysis. During the 12 weeks of the experiment, the research team assumed total responsibility for the class sessions, including myself as teacher. It is therefore important to note that my analysis is grounded in both a documentation of my decision-making process as recorded in my daily journal and a retrospective analysis of these decisions.

Analysis of Classroom Episodes

Batteries Episode

Instructional Intent of the Batteries Task

The batteries task involved students analyzing data on the longevity of 10 each of two brands of batteries, *Always Ready* and *Tough Cell.* The goal was to determine which brand to recommend to the school for use in classroom calculators. The recommendation had to be supported by a data-based argument. Data on the longevity of the batteries were presented to the students in the first of the two computer tools that were used in the classroom design experiment. The tool displayed the data values as horizontal bars, the length of which corresponded to the magnitude of the measure, as shown in Figure 13.2.

Although students had been engaging in data analysis activities for 6 days, this was only the second task presented using the first computer tool. A goal for the task was that students find ways to characterize the data that attended to the question (e.g. which brand is better?). Based on analysis of pre-assessment data, the research team conjectured that the students might calculate the mean when comparing two sets of data. The goal was to create a perturbation in this way of thinking and initiate shifts toward ways of reasoning that focused on features of the data distribution including variability. Features on the computer tool that allowed the students to order and partition the data supported this way of reasoning. The goal was then that the students would come to reason about the characteristics of the distribution as they created arguments based on comparisons across the data sets.

Classroom Analysis

The first students to share their argument were Carol and her partner. They had used the range feature on the computer tool to capture the top 10 values by inserting a vertical value bar at 80 hours. I interpreted their use of the range feature as a way to partition the data. In doing so, I assumed that Carol and her partner had reasoned about proportions of each data set that fell within a certain range. However, when Carol gave her explanation, it was clear that I had misinterpreted their activity. This could be due in part to my brief interaction with the students, but also in part to the fact that I had interpreted *their* activity in a

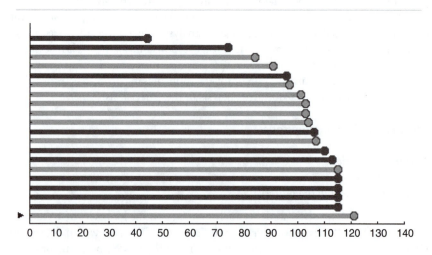

Figure 13.2 Data displayed in the first computer tool

manner that fit with how *I* would attempt to analyze the data. As a result, Carol's explanation was unanticipated.

While it could be argued that Carol and her partner did, in fact, partition the data with the range feature, the particular argument Carol offered did not fit with my interpretation of an argument based on partitioning. Carol stated that she used the range feature to identify the top 10 batteries out of the 20 that were tested and noted that Always Ready were more consistent with "seven out of the top 10." For me, this argument would involve attention to the proportion of each data set that fell within a certain range. However, the choice of the top 10 was, for Carol, a valid basis for selecting Always Ready as the better brand of battery. I, however, was further confused by her use of the term consistent while simultaneously focusing on the top 10. At this point I was wondering if I understood her correctly. For that reason, I asked Janine to explain her understanding of Carol's solution to me.

Janine: I understand.

McClain: You understand? OK, Janine, I'm not sure I do. So could you say it for me?

Janine: She's understanding, I mean she's saying that out of 10 of the batteries that lasted the longest, seven of them are green, and that's the most number, so the Always Ready batteries are better, because more of those batteries lasted longest.

Janine's interpretation of Carol's explanation also did not fit with mine. I was now focused on Carol's use of the term consistent. For me, this implied a smaller range. This was not, however, the explanation that Janine gave of Carol's solution nor did it fit with the notion of consistency. At this point in the discussion, I felt that I was faced with either imposing my interpretation of how to more effectively use the range feature, or simply calling on another student without attending to Carol's solution method. Neither of these moves would support the emergence of mathematical arguments grounded in analysis of the data and would mimic a "show and tell" pattern of discourse.

Fortunately, Juan was able to pose a question that allowed me to explicate my concern. Juan noted that if Carol had chosen to use the range feature to capture the next "bunch of close ones" (e.g. the next four batteries) then there would have been seven of each brand represented within the range feature. At this point, I built up from Juan's question and attempted to get Carol to offer a justification for her choice of the "top 10" instead of, say, 14. However, when she responded that her selection of the top 10 data values was based on "trying to go with half," the students and I were unable to think of a question to challenge her further.

As I deliberated about how to proceed, Brad raised his hand to say that he solved the task a different way. Unable to build from the current discussion, I allowed the shift. Brad had used the value bar to partition the data at 80 hours and had reasoned about the parts of each data set that fell above 80 hours. In doing so, he had judged Tough Cell to be the better brand because he would "rather have a consistent battery that I know that'll get me over 80 hours than one that just try to guess." I followed up by asking Brad why he chose 80 hours as a cut point. He stated that his choice was based on the fact that "most of Tough Cell batteries were all over 80." I interpreted him as viewing 80 hours as a lower limit. Other students supported Brad's conclusion by offering that 80 hours was about 3 days and that that should be "at least" what you expect from a battery.

In retrospectively attempting to tease out important differences between Carol's and Brad's argument, I frame both in terms of Toulmin's scheme for argumentation (see Figure 13.1). Carol and her partner analyzed the data and reached the conclusion that the Always Ready batteries were better. When asked to give a warrant, or to explain how they reached their conclusion, Carol stated that she looked at the top 10 of the 20 batteries. Juan prompted me to ask Carol for a backing, or a justification for her warrant. The backing she offered was that she "was trying to go with half." A similar exercise with Brad's explanation finds his argument different. He chose Tough Cell. Brad's warrant or explanation for his conclusion was that "all the Tough Cell is [sic] above 80." When asked to provide a backing or a justification for his warrant, he explained "I'd rather have a consistent battery that I know that'll get me over 80 hours than one that just try to guess."

A comparison of the two backings offered offers an opportunity to highlight an important shift that occurred in the course of the discussion. Although the students understood Carol's procedure (looked at the top half of the data and noted which brand had the most results there), her choice of the top 10 did not appear to be a valid statistic or backing. In particular, Juan argued that if you modified the position of the range feature and chose the top *14* batteries instead of the top *10*, you would have seven of each brand, as the next four batteries are Tough Cell. Choosing the top 10 was an arbitrary choice not valid for the investigation at hand. Further, Carol did not engage in a discussion with Juan about his proposal. Carol had made her argument and was not interested in engaging in debate or further argumentation. At this point, for Carol any argument that referred back to the data was justified. However, this was not the case for Juan. Juan expected a backing grounded in the context of the situation that would support the initial conclusion (or argument).

Brad, by way of contrast, gave a backing for choosing 80 hours that appeared to be valid for all of the students. He wanted batteries that he could be assured would last a minimum of 80 hours. In particular, his backing was grounded in the situation-specific imagery of the longevity of the batteries. Therefore, his argument was stronger than Carol's.

In this discussion, the students were beginning to negotiate what constituted a sufficient backing or acceptable mathematical argument. Although there is no attempt on my part to clarify the need for a backing that is grounded in the context of the investigation, this does emerge from Juan's challenge. Juan wanted to know what was significant about choosing 10; why not 14? As a result of the students pushing for an adequate backing, the norms for argumentation shifted slightly to include a backing grounded in the question at hand. It was insufficient to describe a procedure. Activity had to be grounded in the question. This exchange constituted a first, small shift in the sociomathematical norm of *what counts as an acceptable argument.*

Transition to AIDS Episode

In the weeks following the batteries episode, the students engaged in a sequence of data analysis tasks. Throughout this process, the norm for *what counts as an acceptable argument* was constantly being renegotiated in the course of whole class discussions. The focus shifted from a *need for* a backing to the *quality of* the backing. As an example, partitioning the data at a cut point became an accepted way of organizing the data. That method was beyond justification. However, the conversations focused on the choice of cut point and the rationale for that particular choice. Students shifted from the *what* to the *why.*

AIDS Episode

Instructional Intent of the AIDS Task

The task for the students in the AIDS episode varied in several ways from the one in the batteries episode. First of all, data were inscribed in the second of the two computer tools. This tool displayed data as line plots and contained numerous features for data structuring. In addition, two data sets could be positioned over one another for comparison (see Figure 13.3).

Over the course of the 12-week teaching experiment, the research team decided that asking the students to choose the better brand or better treatment focused the students' analyses on the choice. By asking the students to think of a way to structure and organize the data so that someone else could make a reasoned decision, the students were asked to investigate patterns in the data and then decide how best to represent those to someone who would not see the data sets in their entirety. For this reason, discussions focused on features of the students' inscriptions that they developed to support their arguments. As students typically agreed on the "better" case, this reformulation of task provided an opportunity to shift the nature of the mathematical arguments from *which brand*

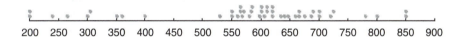

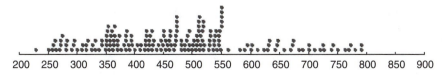

Figure 13.3 Data inscribed in the second computer tool

(or treatment) is better? to *which argument is better?* In this way, students engaged in reflective discourse (see Cobb et al., 1997) about their prior activity. As part of this process, students began to focus on refining their argument in order to strengthen the case instead of simply arguing about the choice. This constituted a significant shift in the norm for *what counts as an acceptable argument.* At this point, an acceptable argument no longer entailed a decision about "which one?" but had to include a strong warrant and backing grounded in the data. This caused a substantial shift in the norms for argumentation in that the correct choice was subjugated to more mathematically significant arguments about the features of the distributions and resulting inscriptions.

In the AIDS task, the students were asked to analyze data on the T-cell counts of 232 AIDS patients and determine which of two treatment protocols was more effective in increasing T-cell counts (see Figure 13.3). The AIDS task was selected because the two data sets contained an unequal number of data points. In particular, one data set was much larger than the other (46 patients compared to 186 patients). In earlier tasks with unequal numbers of data points, students spent a great deal of their time deciding which data points to ignore in order to have equal Ns. One goal of this task was to create a situation where equalizing the sets by eliminating points was not possible. It was hoped that this would create the need for multiplicative ways to structure the data in order to make a comparison since direct additive comparisons would be inadequate.

Classroom Analysis

As the students worked in pairs on their analyses, I again monitored their activity in order to plan for the subsequent whole class discussion. In observing their solutions, I noticed that some students found qualitative ways to make comparisons between the two data sets such as referring to where the "majority" of the data were located. Other students found quantitative ways to make the distinctions by noting the number of data values above and below a certain cut point. Still others had structured the data into four equal groups and reasoned about the percentages of each data set that fell above a certain T-cell value. This range of ways of reasoning from qualitative comparisons based on perceptual patterns in the data to the data structured multiplicatively was a perfect venue for initiating shifts in the quality and sophistication of students' arguments. As an example, just giving the count of data values above 525 (e.g., reasoning additively) would lead one to conclude that the traditional treatment protocol was more effective.

It is only when that number is taken as a proportion or a percentage of the total that an adequate argument can be made. Therefore, I hoped to build from the students' inscriptions and arguments to initiate shifts toward more sophisticated arguments.

As the whole class discussion began, I first asked students to decide if they could understand each report and then decide if they thought it was an "adequate" argument. In doing so, I was attempting to shift the conversations from simply providing explanations of methods, to critiques that raise the standard for argumentation. To facilitate this process, I had taken the students' inscriptions and reproduced them on chart paper. The creation of these artifacts allowed the class to be able to see the reports clearly and removed ownership during the discussion. While the authoring students might have been able to determine which inscription was theirs, it was also the case that other students in the class could have solved the task in the same way. For this reason, the students did not attend to whose solution was being critiqued. In retrospect, it appears to have made it possible for the conversations to focus on refinement of arguments.

Early in the discussion, I introduced the two reports shown in Figure 13.4.

Both of the groups generating these reports had initially structured the data in a similar manner using the computer tool. One had partitioned the data at 525 and the other at 550. Both had then reasoned about the number of data values on either side of the cut point, albeit one in qualitative terms (e.g., majority) and the other in quantitative terms. At this point, Val asked why the one group had chosen to partition at 550. She noted that 550 did not represent the median so she could not determine its importance. Here, Val was asking for a backing for the warrant. Again, using Toulmin's scheme, we note that the students had analyzed the data to reach a conclusion. Their warrant consisted of looking at the number of data values above and below a T-cell count of 550. What was not obvious from the report was the choice of 550. In other words, the report had not included a backing for the warrant. Val wanted to understand why the group had decided to partition the data at a T-cell count of 550 since it did not represent any particular value of the data set such as the median. The norm that had become instantiated in the classroom was that of justifying in terms of the context or the way the data were distributed. Val was making it clear that she could not ascertain the justification from the report. This is evidence that the norms for acceptable mathematical arguments had shifted.

At this point, Meg, who had solved the task in a similar manner, noted that,

Report A

The new drug was better than the old. The majority of the old ones are behind 550 and the majority of the new drug was in front of 550

Report B	
Experimental treatment	Traditional treatment
225–525	225–525
9	130
525–850	525–850
37	56

Figure 13.4 AIDS data as noted in reports A and B

when working on the computer tool, she noticed that the data "lined up straight on 550." I recast this as basing the cut point on a natural break in the data. This was an important concept. One of the goals of the design experiment was that students would view distributions in terms of shape. Here, Meg was offering an explanation based on the shape or breaks in the data as inscribed on the tool. For that reason, I wanted to highlight both the solution and Meg's explanation. This is a significant exchange because it points to another shift in the norms for argumentation. The students would not accept an argument whose backing could not be justified in terms of the analysis of the data.

It is also important to note that Val asked for the backing for the choice of 550 and not 525. In looking at the inscription with a cut point at 525, it could be argued that the data were partitioned on the basis of the midpoint of the extremes (e.g. finding the average of the maximum and minimum value). In previous classes students had attempted to use the midpoint of the two extreme data values as a statistic for describing data sets. They called this value the "middle of the range." I had worked to problematize this way of reasoning, as it has no statistical value in terms of distributions. In this episode, however, the students accepted it as a valid cut point since it coincided with the break in the hills.

Juxtaposing the two reports in Figure 13.4 was a pivotal move. Although I had recast the two inscriptions as partitioning the data around the break in the hills, one group described the results in terms of where the majority of the data fell and the other used the computer tool to determine the number of data values that fell above and below the cut point. Juxtaposing these two reports made it possible to problematize the use of direct additive comparisons. This was facilitated by Missy noting that the second of these two reports would be more confusing, "since the old program has more numbers than the new program." I realized the importance of this contribution and recast it so that all members of the class could focus on the possible problem.

McClain: Oh. So it looks like that there's more. They had 56 that were above 525, and they only had 37?
Missy: So it's like, I guess what I'm trying to say is it's harder to compare them.
McClain: What about what Missy said? She just said there were more people in the old program so if you actually looked at the actual numbers of people, you find out that they had 56 that were in this upper range which is where we want to be and these only had 37. So somebody might say the old program was better because there were more. Janine?

As the discussion continued, students were able to acknowledge the problematic nature of the argument, but could not determine a solution. Finally, at Ken's direction, I created a modified graph of the data, as shown in Figure 13.5.

At this point, students offered a variety of ways to reason about the numbers:

Brad: But then there's more people with the old program than there is with the new program.

Figure 13.5 Graph created from partitioned AIDS data

McClain: Juan?

Juan: Then you see that there's 37 is more than half over 525 and 56 is not more than half of 130 . . . more of them on the bottom than on the top.

Student: I don't understand.

McClain: Juan?

Juan: OK, you see how 37 is more than half of 9 and 37 together? But 56 is not more than half of 30, 130 and 56 put together. There's more on the bottom one than on the top one.

McClain: OK, who can help me out with that, who can say that a different way so that I might understand that? Will, can you say it a different way?

Will: Well, in that situation it wouldn't matter how many people were in there because see like . . .

McClain: Big voice, Will.

Will: On the bottom one you have, see what Juan was saying there's more than there is below 525 and so that means that that one is better because the top one it doesn't even have close to half of what the one below 525 is on that one. So that means that if, if that was the same amount of people it had like, if they both had the same amount of people and, but, and they had the numbers and everything, and this one, the bottom one was a however much more that of . . .

Brad had an intuitive understanding of the problem but was unable to clarify other than to point to the difference in the number of patients in each program. Juan reasoned that since over half of the patients in the experimental treatment protocol had T-cell counts greater than 525 and less than half of the patients in the traditional treatment protocol had T-cell counts that high, the experimental was better.

As the discussion continued, it was suggested that finding the percentage would help. Students then calculated the percentage of patients in each treatment protocol whose T-cell counts were above 525. Once completed, the students were able to confirm their earlier, intuitive arguments with the results. Although this discussion was not based on sophisticated arguments, the students acknowledged the need to clarify the apparent problem in order for the inscription and argument to be acceptable.

The last report to be shared in the whole class discussion involved students structuring the data into four equal groups for each treatment as shown in Figure 13.6.

After I posted the report on the whiteboard, Brad stated that he felt it was adequate because "with the four equal groups, you can tell where the differences in the four groups." Brad noted that the new treatment was better because "the three lines for the equal groups" from the experimental were above 525 compared to "only one of them" for the traditional treatment. Although somewhat difficult to understand, I interpreted Brad's explanation to mean that he noted that 75% of the experimental data (i.e., three lines) was above 525 whereas only 25% of the traditional data (i.e., one line) was above 525. It was my belief that many of the students in the class did not understand Brad's offered explanation. Therefore, teasing out the meaning of the graph was an important aspect of the continuing discussion.

I then asked the students if they knew what percentage of the data was in each interval. In their answer the students were able to clarify that they understood that 25% of the data set fell within each of the four intervals. I then noted on the inscription, at the students' direction, that 75% of the patients in the new treatment program were in the same range of T-cell counts as only 25% of the patients in the old treatment. For many of the students, the process of going back to the percentages was necessary to build an understanding to underlie the inscription. They needed to know where the data were behind the inscription.

A review of the lesson finds the whole class discussion initially focused on reports that structured the data by partitioning around the hills. The first solutions offered a qualitative distinction between the two data sets by focusing on the location of the majority of the people. The nest, however, gave a quantitative distinction that provided the opportunity for a problem to arise from direct

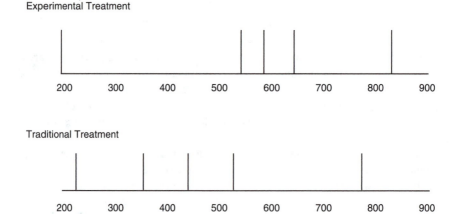

Figure 13.6 Inscription of the AIDS protocol data organized into four equal groups

additive comparisons. This, in turn, offered me the opportunity to problematize the direct additive comparisons and shift the discussion to more sophisticated ways to reason about the data sets. This was the setting for the refinement of the initial arguments.

Throughout the 12 weeks of the teaching experiment, there were continual shifts in the norm of *what counts as an acceptable mathematical argument*. Although there were necessarily cycles of learning during which new inscriptions (e.g. the four equal groups) emerged and required more detailed forms of *explanation of* versus *argumentation with*, nonetheless, reasoning with the inscriptions became normative and created the discourse space for arguments to become refined in the course of discussion.

Discussion

Significant mathematical discussions provide resources for supporting students' emerging understandings while taking their initial understandings as starting points. This process created a setting in which mathematical arguments went through a evolutionary process of refinement leading to more efficient and sophisticated arguments. Concurrently, the students were able to engage in progressively more sophisticated analyses. Their justifications were grounded in the question at hand and over the course of the design experiment they developed, in Toulmin's terms, the need and understanding for mathematically based backings for their warrants. For the students, their discussions and critiques of the varied solutions took the form of refinement.

In Simon's (1997) description of the Mathematics Teaching Cycle he notes the "inherent tension between responding to the students' mathematics and creating purposeful pedagogy based on the teacher's goals for student learning" (p. 76). Purposeful pedagogy in these episodes would entail an explicit focus on mathematical argumentation. This cannot be subjugated to completion of process or finding a solution. It must be an intricate part of the mathematical process.

The analysis provides the basis for a conjectured development of mathematical argument in this classroom. The first phase can be characterized as *argument for defending*. This can be seen in Carol's argument where Juan's question posed no perturbation for Carol. She could not, or chose not, to counter Juan's question with anything other than, "I wanted to go with half." Knowing that this was not acceptable for Juan did not create a problem for Carol. She had given her argument and there was no need for further discussion.

I characterize the next phase, or evolution of the norm as *argument for disagreement*. Although this is a subtle shift, you can see evidence of this when Brad states, "I did it another way." Brad offered his argument as disagreement with Carol's conclusion. At this point in the lesson sequence, students were disagreeing on the better brand or treatment and the primary goal of the argument was to select the better choice. Brad's disagreement was grounded in the fact that he did not agree with Carol's selection of better brand of battery. His argument had more validity with the students in the class since he was able to provide an acceptable backing for his warrant.

The next phase in the evolution was *argument for justification*. In this phase, the students provided an argument to justify their choice or their ways of organizing the data. During this period of time, students had more agreement on the better choice or treatment and discussions shifted toward a better way to argue or organize the data. The justifications (or warrants and backings) then became the object of argument.

The final phase in the evolution of the norm of what counts as an acceptable mathematical argument was *argument for refinement*. This occurred during the AIDS episode when students were working to find a way to best represent the data sets with unequal Ns. All students agreed that the experimental treatment was better. The discussions were focused on which inscription would provide the best argument. In this process, the students worked to refine the offered inscriptions in order for the class to present the best argument.

It is important to note that during this final phase of evolution of the norm, the *best* argument was often not the most sophisticated. As an example, in one task where students were asked to analyze data on study habits to present at parents' night, they selected a somewhat simplistic argument and inscription. When pressed about this, they responded that, "If we do four equal groups, our parents won't understand it." While most likely true, it is important to clarify that the norms were simply that—norms. They were dependent on the particular task, data and audience. This variety points to the students' ability to reason not only *with* the argument, but also *about* it. In this process, the arguments themselves became reified as objects of reflection.

Notes

1.   Support was provided by the National Science Foundation under grant no. REC-0135062.
2.   The episodes featured in this chapter were also the subject of analysis in McClain (2002).

14

Reasoning-and-Proving in School Mathematics

The Case of Pattern Identification [1]

GABRIEL J. STYLIANIDES AND EDWARD A. SILVER

Introduction

One struggle I face [when I teach patterns to my students] is that the kids don't trust themselves, they don't trust the pattern, they don't trust the math involved. Even if they found a formula that seemed to work, they'd only be sure they had the answer if they actually built the whole [pattern] of 50 or 100 [terms].

This excerpt comes from a middle school teacher during a professional development session of the Beyond Implementation: Focusing on Challenge and Learning (BIFOCAL) project (e.g., Silver, Mills, Castro, & Ghousseini, 2006). During the session, this teacher and other middle school teachers engaged in solving and discussing a task that involved pattern identification. Her comment raises important issues related to the teaching and learning of patterns. Why do students have difficulty in trusting patterns? Are there different kinds of pattern? Should students trust all the patterns they find? If not, how can they distinguish between patterns they can trust and patterns they cannot trust? How can students' study of mathematics help them make these distinctions and judgments?

These questions would not necessarily have been considered important in the teaching and learning of mathematics in the middle grades some years ago, but now they are critical. Pattern identification is an essential component of the activity of *reasoning-and-proving*, which has become more visible and prevalent in mathematics curriculum frameworks in recent years not only for high school but also for the elementary and middle grades (see, e.g., National Council of

Teachers of Mathematics [NCTM], 2000). An emphasis on reasoning-and-proving places increased demands on the teachers who, as the teacher comment at the outset of the chapter suggests, many times feel unprepared (in terms of both their mathematical and their pedagogical knowledge) to meet the challenges they face in teaching patterns to their students and in supporting student engagement in complex mathematical activity in the classroom.

Middle school teachers' mastery of patterns is particularly important given that pattern tasks are abundant in the so-called standards-based mathematics textbook series that were funded by the National Science Foundation to reflect the curriculum standards published by NCTM in 1989 and 2000. In an analysis of the *Connected Mathematics Project* (CMP) (Lappan, Fey, Fitzgerald, Friel & Philips, 1998/2004), a popular standards-based middle school (grades 6–8) mathematics textbook series, Stylianides (2005) found that 40% of the tasks in the algebra, number theory, and geometry units were designed to engage students in reasoning-and-proving, and that 25% of these tasks involved pattern identification.

This chapter aims to contribute to the questions we raised earlier and is organized in three sections. In the first section, we situate the activity of pattern identification in the more general context of the activity of reasoning-and-proving. Specifically, we discuss a conceptualization of reasoning-and-proving that derives from the nature of this activity in the discipline of mathematics, and we show how pattern identification connects to reasoning-and-proving. In the second section, we describe briefly parts of an *analytic framework* (Stylianides, 2008) which is based on the same conceptualization of reasoning-and-proving and aims to serve as a useful platform for conducting different kinds of investigations with a focus on reasoning-and-proving.

In the third section, we use relevant parts of this framework to analyze the episode of the professional development session from which the opening excerpt was taken. Our examination of this episode not only provides insight into teachers' mathematical understanding of patterns and the challenges they face when they teach this topic to their students, but also illustrates the utility of the analytic framework in the context of a teacher professional development session. The creation of analytic tools usable in different kinds of investigation in the domain of reasoning-and-proving can provide the means to connect research findings from different investigations, thereby supporting the development of integrated research programs and the accumulation of knowledge across different investigations.

A Conceptualization of Reasoning-and-Proving

In this section, we present the conceptualization of reasoning-and-proving that we use in this chapter, which is based on domains of scholarship that examine this activity from a disciplinary perspective.

Harel and Sowder (1998) note that, "most of the mathematician's work is spent on exploring and conjecturing, not on searching for proofs of well stated propositions, and certainly not on obvious propositions" (p. 236). Lampert (1991) observes:

> In mathematics, new knowledge is produced by testing assertions in a reasoned argument; in a community of discourse, people agree upon a set of assumptions, make generalizations about a given domain, and then explore the boundaries of the domain to which the generalizations apply. (p. 125)

Pólya (1954b) describes mathematical activity as involving two kinds of reasoning: plausible reasoning, through which conjectures are generated, and demonstrative (deductive) reasoning, through which mathematical knowledge is secured. He emphasizes that the discipline of mathematics includes both kinds of reasoning: a mathematical proof is demonstrative reasoning, and mathematics in the making is plausible reasoning. In his own words:

> Mathematics is regarded as a demonstrative science. Yet this is only one of its aspects. Finished mathematics presented in a finished form appears as purely demonstrative consisting of proofs only. Yet mathematics in the making resembles any other human knowledge in the making. You have to guess a mathematical theorem before you prove it; you have to guess the idea of the proof before you carry through the details. You have to combine observations and follow analogies; you have to try and try again. The result of the mathematician's creative work is demonstrative reasoning, a proof; but the proof is discovered by plausible reasoning, by guessing. (Pólya, 1954b, p. vi)

The remarks cited here suggest that the development of mathematical knowledge passes through several stages, ranging from the initial exploration phase and the germ of an idea, to the final stage of a reasoned argument, which is usually a proof (Dubinsky & Tall, 1991). Atiyah (1984/1986) describes eloquently this process:

> In Mathematics, as in the Natural Sciences, there are several stages involved in a discovery, and formal proof is only the last. The earliest stage consists in the identification of significant facts, their arrangement into meaningful patterns and the plausible extraction of some law or formula. Next comes the process of testing this proposed formula against new experimental facts, and only then does one consider the question of proof. (p. 47)

In sum, the activities that typically characterize research mathematicians' work related to proof involve various stages, with *proof* usually being the last. Earlier stages involve what Pólya (1954b) calls "mathematics in the making" and frequently consist of the identification and arrangement of significant facts into meaningful *patterns*; the use of the patterns to formulate *conjectures* and the testing of these conjectures against new empirical evidence; and the effort to understand and provide *arguments* about why things work the way they do. These activities and the connections among them are illustrated nicely in Schoenfeld's (1983) analysis of a problem he worked on with his college students. The problem was about the relationship between the roots of polynomials P(x) and Q(x) of *n*th degree with "reversed" coefficients:

> Observe that the *proof*, like a classical mathematical argument, is quite terse and presents the results of a thought process. But where did the inspiration for the *proof* come from? If you go back over the way that the argument evolved, you'll see two major breakthroughs.
>
> The first had to do with understanding the problem, with getting a feel for it. The problem statement, in its full generality, offered little in the way of assistance. What we did was to *examine special cases* in order to *look for a pattern*. More specifically, our first attempt at special cases—looking at the quadratic formula—didn't provide much insight. We had to get even more specific, as follows: look at a series of straightforward examples that are easy to calculate, in order to see if some sort of *pattern* emerges. With luck, you might be able to *generalize the pattern*. In this case, we were looking for roots of polynomials, so we chose easily factorable ones. Obviously, different circumstances will lead to different choices. But the strategy allowed us to *make a conjecture.*
>
> The second breakthrough came after we *made the conjecture.* Although we had some idea of *why it ought to be true*, the *argument* looked messy, and we stopped to reconsider for a while. What we did at that point was important, and is often overlooked: we went back to the conditions of the problem, explored them, and looked for tangible connections between them and the results we wanted. Questions like "what does it mean for r to be a root of P(x)?," "what does the reciprocal of r look like?," and "what does it mean for (1/r) to be a root of Q(x)?" may seem almost trivial in isolation, but they focused our attention on the very things that gave us a solution [the proof]. (Schoenfeld, 1983, pp. 11–12; some underlines in original were omitted here, emphases added)

Following the position advanced by several scholars that, for students to obtain authentic mathematical experiences in school, these experiences need to be organized in ways that are honest to what it means to do mathematics in the discipline (e.g., Ball, 1993; Bruner, 1960; Lampert, 1992; Schwab, 1978; Silver & Stein, 1996; Stylianides, 2007), we adopt a conceptualization of *reasoning-and-proving* that encompasses the breadth of the activity associated with "identifying patterns," "making conjectures," "providing proofs," and "providing non-proof arguments" (Stylianides, 2008). The first two activities are captured under the notion of "making mathematical generalizations," and the last two under the notion of "providing support to mathematical claims."

Analytic Framework

The conceptualization of reasoning-and-proving that we outlined earlier set the stage for an analytic framework that includes three components: mathematical, psychological, and pedagogical (see Figure 14.1). Stylianides (2008, pp.10–13) provides a detailed discussion and illustrative examples of all parts of the framework. In this section, we describe briefly parts of the framework that relate to our analysis of the focal episode from the professional development session with teachers. Specifically, we first describe the part of the mathematical component of the framework about "identifying a pattern." We then discuss how the framework can support inquiry on reasoning-and-proving from a mathematical, psychological, and pedagogical perspective (each perspective derives from the corresponding component of the framework); in this chapter, we use all three perspectives to analyze the focal episode.

Identifying a Pattern

In this framework, a *pattern* is defined as a general mathematical relation that fits a given set of data. The framework distinguishes between two kinds of pattern— definite and plausible—according to whether or not they are uniquely determined. Specifically, in *definite patterns*, it is possible mathematically for a solver (given the information in a task) to provide conclusive evidence for the selection of a specific pattern. In *plausible patterns*, it is not possible mathematically for a solver (given the information in a task) to provide conclusive evidence for the selection of a specific pattern over other patterns that also fit the data (e.g., one might select the pattern that he/she considers to be the simplest pattern that fits the given set of data).

Table 14.1 illustrates the distinction between definite and plausible patterns. One pattern (expressed in algebraic form) that fits the data in the table is $y = 3x$.

	Reasoning-and-proving			
	Making mathematical generalizations		Providing support to mathematical claims	
Mathematical component	Identifying a pattern	Making a conjecture	Providing a proof	Providing a non-proof argument
	• Plausible pattern • Definite pattern	• Conjecture	• Generic example • Demonstration	• Empirical argument • Rationale
Psychological component	What is the solver's perception of the mathematical nature of a pattern/conjecture/proof/non-proof argument?			
Pedagogical component	How does the mathematical nature of a pattern/conjecture/proof/non-proof argument compare with the solver's perception of this nature? How can the mathematical nature of a pattern/conjecture/proof/non-proof argument become transparent to the solver?			

Figure 14.1 Analytic framework (Stylianides, 2008, p. 10)

Table 14.1 Distinction between definite and plausible patterns

x	0	1	2	3	4
y	0	3	6	?	?

Note: The table shows how two variables relate. Find a pattern in the table and use the pattern to complete the missing entries.

Based on this pattern, the missing entries are 9 and 12. Another pattern that fits the data is $y = x^3 - 3x^2 + 5x$. According to this pattern, the missing entries are 15 and 36. The two pairs (9, 12) and (15, 36) are illustrative of an infinite number of possible solutions that fit the given set of data. For example, any equation of the form $y = ax^3 - 3ax^2 + (2a + 3)x$, where a is a real number, would satisfy the given set of data. From a mathematical standpoint, any answer to the task could be correct. One could simply argue that the missing entries are 0 and 3, taking the first three values of y in the table as the "unit of repeat" of the pattern. It is clear that the pattern in this example is not uniquely determined and, thus, is a plausible pattern. How could the task in Table 14.1 be modified so that the pattern would become definite? One way to make the pattern definite would be to place the task in a situation like this: "A writer has a contract with his publisher to receive \$3 per book that is sold." This situation defines uniquely the pattern $y = 3x$, where y is the amount of dollars received by the writer and x is the number of books sold.

An important goal in mathematics education is to help students understand that not only it is important to find a pattern, but also to see why a generalization holds (e.g., Harel & Sowder, 1998; NCTM, 2000). The previous discussion indicates that whether a pattern is definite or plausible influences the kind of justification one might offer for why the pattern holds. In definite patterns, one can offer conclusive evidence for the pattern, whereas in plausible patterns, this is not possible because the selection of a particular pattern over other possible patterns is not based on mathematical criteria. Yet, this does not suggest that plausible patterns cannot engage students in productive justifications of mathematical generalizations. For example, the students can *assume* a particular structure that transforms a plausible pattern into a definite pattern and then provide conclusive evidence for the specific pattern that fits the data. Let us consider again the pattern in Table 14.1. In most middle school classrooms, the students (and the teacher) would likely assume (implicitly) that the missing pattern is linear and solve the task based on this assumption.

Using the Analytic Framework to Support Inquiry on Reasoning-and-Proving from Mathematical, Psychological, and Pedagogical Perspectives

The *mathematical component* of the framework can support inquiry on reasoning-and-proving activity from a mathematical perspective. In this perspective, the observer (or examiner) of the activity (e.g., classroom activity) is considered to be a mathematically proficient person (e.g., the researcher or the teacher) who analyzes the given activity using mathematical considerations irrespectively of what the solver (e.g., the student) thinks of this activity.

The *psychological component* of the framework focuses on the solver. An inquiry on reasoning-and-proving from a psychological perspective would examine the solver's perception of the mathematical nature of a *mathematical object related to reasoning-and-proving* (e.g., pattern, proof) as this perception is reflected, e.g., in the solver's solution of a task or in the solver's comments about another solver's solution of a task. To illustrate the psychological component of the framework, refer back to the modified version of the task in Table 14.1 (i.e., the version of the task that includes the situation about the writer's contract with the publisher) and consider a solver approaching this task empirically by generalizing the pattern from the numerical values in a table, but believes that her solution uniquely determines the pattern. Although this approach to deriving the pattern is limited from a mathematical standpoint (because it does not connect the pattern to the situation in the task that uniquely determines the pattern, i.e., that the writer receives $3 per book sold), an observer who uses a psychological perspective to examine the solver's work can conclude that this solver appears to perceive the pattern as definite.

The *pedagogical component* of the framework uses both the mathematical and the psychological components. An inquiry on reasoning-and-proving from a pedagogical perspective would focus on how the mathematical nature of a mathematical object related to reasoning-and-proving (as derived by application of a mathematical perspective) compares with the solver's perception of this nature (as derived by application of a psychological perspective). Possible discrepancies emerging from this comparison can help identify potential foci for teachers' pedagogical actions aiming to support the refinement of their students' understandings of the nature of a particular reasoning-and-proving object toward conventional understandings of this nature. The mathematical nature of a reasoning-and-proving object is said to be *transparent* to the solver if the solver's perception of the object coincides (after an instructional intervention, etc.) for valid reasons with the mathematical nature of the object. Otherwise, the mathematical nature of the object is said to be *non-transparent* to the solver. For example, if a solver is convinced that a definite pattern is uniquely determined by generalizing the pattern from the numerical values in a table, then the mathematical nature of the pattern is said to be non-transparent to the solver.

Using the Analytic Framework to Analyze and Discuss an Episode from a Professional Development Session

In this section, we use the analytic framework as a tool to analyze an episode from a BIFOCAL professional development session in which a group of middle school teachers solve and discuss a task, which we call the *hexagon trains task* (see Figure 14.2). This section is organized into two parts. In the first, we situate the episode by providing a brief description of the BIFOCAL project, the structure of

For the pattern shown below, compute the perimeter for the first four trains, determine the perimeter for the tenth train without constructing it, and then write a description that could be used to compute the perimeter of any train in the pattern. The first train in this pattern consists of one regular hexagon. For each subsequent train, one additional hexagon is added. The first four trains in the pattern are shown below.

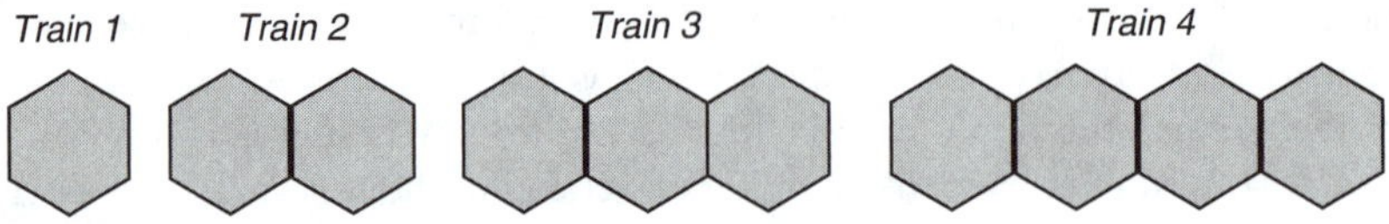

Train 1   Train 2   Train 3   Train 4

Find as many different ways as you can to compute (and justify) the perimeter.

Figure 14.2 The hexagonal trains task

the professional development sessions, and the characteristics of the teacher participants in the project. Finally, we present the focal episode with the teachers. In the second, we analyze and discuss the episode using the analytic framework as a tool.

The Professional Development Program and the Focal Episode

BIFOCAL was a professional development program for in-service middle school teachers (Silver et al., 2006). The teacher participants in the BIFOCAL project were drawn from four small school districts with a total of five middle schools, each of which had adopted CMP; project participants had used this textbook series from 3 to 11 years. The initial cohort was comprised of two or three middle school teachers from each school (a total of 12 teachers); the principals of these schools also participated in some professional development sessions. The teachers in the initial cohort were selected by their districts and were identified as leaders; they had on average 14.5 years of teaching experience and 5 years of experience teaching CMP. In subsequent years of the project, participants in the initial cohort assumed some leadership for school-level and district-level professional development with colleagues; the project also recruited additional participants over time. The episode on which we focus in this chapter was drawn from the first year of the project; in fact, it was part of the first session.

BIFOCAL was built on a foundation of prior work in the Quantitative Understanding: Amplifying Student Achievement and Reasoning (QUASAR) and Cases of Mathematics Instruction to Enhance Teaching (COMET) projects (Silver & Stein, 1996; Smith & Silver, 2000; Stein, Grover, & Henningsen, 1996). In BIFOCAL sessions, COMET cases (Smith, Silver & Stein, 2005a, 2005b; Stein, Smith, Henningsen, & Silver, 2000) were used to support the professional learning of the participants. The COMET cases stimulate reflection, analysis, and inquiry, and they illuminate many of the challenges faced by teachers when they try to move toward more ambitious goals for students in the middle grades. The cases portray the relationships among mathematics, pedagogy, and student learning in the classroom and, thus, permit examination of most components of teaching as a reader analyzes a case. The cases formed a foundation on which we worked to enhance teachers' effectiveness in using the standards-based textbooks to foster students' deep, substantive engagement with challenging mathematical tasks and to enhance their learning.

Before engaging the teachers in reading and discussing a case, there was usually an opening activity in which teachers solved a task that involved mathematical ideas that were similar to the ideas found in the narrative case. This opening experience was viewed as an important precursor to interpreting and analyzing the work of the teacher and his/her students during the class portrayed in the case. The hexagon trains task (see Table 14.2; see also Figure 14.2) constituted the

Table 14.2 The episode

1	As the teachers worked on the hexagon trains task individually and in their small groups, Elliot, the teacher educator, circulated around the room, taking note of their work. After sufficient time, he launched the whole group discussion by noting that everyone appeared to have found an answer to the task, but that there were several different ways used to get the answer. He also emphasized that finding an answer was not the only important thing; it was equally important to explain the thinking used to get to the answer. Betty was the first teacher to come to the overhead and share her solution with the group. She wrote the following:

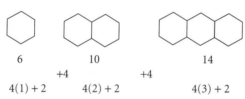

2	

$$6 \qquad 10 \qquad 14$$
$$+4 \qquad\qquad +4$$
$$4(1)+2 \qquad 4(2)+2 \qquad 4(3)+2$$

3	Betty then said that she and her partner, Susan, figured out that the pattern would be $4(n)+2$, where n corresponds to the number of hexagons in the train.
4	*Elliot:* Where does the $4n$ come from? It doesn't seem to be visible in the model though it is visible, perhaps, in the $+4, +4$ you wrote. But would that be obvious to kids?
5	*Susan:* We just measured the sides and found the pattern.
6	Susan continued to explain how she and Betty figured out the formula $4(n)+2$ Specifically, the constant difference of 4 between one train and the next suggested to them that there must be a factor of $4(n)$ somewhere in the formula Once they figured this out, they started looking for the appropriate correction factor. They found this correction factor by trial and error.
7	*Elliot:* We, as adults or more experienced problem solvers, know that we just have to find the proper correction factor to add or subtract in order to get a serviceable formula. How can we get the students to see this, though?

(*Continued Overleaf*)

8 *Nicole*: I teach this unit and my students would struggle with this, and I also struggle with how to help them.

9 *Elliot*: How did you solve the problem Nicole?

10 *Nicole*: I solved it the same way.

11 She then went to the overhead and wrote the following:
1st train: 6
2nd train: $6 + 6 - 2$
3rd train: $6 + 6 + 6 - 2 - 2$
. . . . .
$4n + 2$

12 Elliot reformulated her pattern to $6n - 2(n - 1)$. After this, Nancy, Nicole's partner, joined the discussion.

13 *Nancy*: Oh, I kept thinking it was $6n - 2n - 2$. I'm not sure. Is it $6n - 2(n - 1)$?

14 Elliot then asked about how the pattern could be seen from the figure rather than just from the numerical values.

15 *Susan*: My kids would probably see it as the "seams."

16 Tanya got into the discussion and responded to Nancy's confusion.

17 *Tanya*: It's $6n - 2(n - 1)$ because if you look at Nicole's solution, in the expression $6 + 6 + 6 - 2 - 2$ for Train 3 there are three 6s and two 2s. So, for Train 4 there will be four 6s and three 2s. Train 3 has one 2 less than Train 4.

18 After this remark, Tanya shared her solution to the problem, describing how she figured out immediately that the pattern would be $4n + 2$. She drew the following:

19

20 She then explained that the $4n$ corresponds to the dotted lines (top and bottom) and that the 2 corresponds to the side lines. She also noted that each additional hexagon would add four dotted lines and showed that the new hexagon would be added to the middle of the two hexagons that appear in the figure.

21 Elliot clarified Tanya's solution by saying that every hexagon has two segments at the top, two at the bottom, and two at the sides. The factor of 4 in Tanya's formula corresponds to the two top and the two bottom sides of *each* hexagon. The +2 corresponds to the side segments of the two hexagons that are at the end of the train (these segments correspond to the vertical line segments in the figure). The side segments of all other hexagons are "lost," because they correspond to interior lines and these lines are not part of the perimeter of the figure. He noted that this related to Susan's earlier comment about "seams."

22   Nicole noted that the way the picture was used to find the equation was an eye-opener for her. In her teaching, she would only explain the pattern from numerical values in a table.

23   After this remark, Elliot showed two pieces of student work that illustrated other ways of visualizing the problem, such as adding successive hexagons to the train at the end rather than internally. He led a discussion of how these other methods could be seen in relating the symbolic solution to the train itself. Elliot then posed a question that shifted the discussion to issues of teaching.

24   *Elliot*: Based on your experiences teaching patterns, what are some of the challenging issues that emerge?

25   *Nina*: One struggle I face is that the kids don't trust themselves, they don't trust the pattern, they don't trust the math involved. Even if they found a formula that seemed to work, they'd only be sure they had the answer if they actually built the whole train of 50 or 100.

26   Nancy then suggested that this might relate to some research on brain she had heard about.

27   *Nancy*: Perhaps Nina's sixth-graders' brain cells just hadn't developed enough for such abstraction as was necessary to trust the formula. The students may still be concrete thinkers.

28   *Nicole*: It is so hard not to just do it for them.

29   *Roberta*: We start the year with sequences. I tell my students that if it is going up by a fixed amount then you are going to multiply it by that amount. It seems to help them get on board. It helps so much when I give them that little hint.

30   Elliot noted that this is one way to go. The issue, however, is that the students may not necessarily understand why this approach works.

---

opening activity for the first case we discussed with the initial group of 12 teachers in the opening, day-long professional development session. The task provided an opportunity for teachers to look for the underlying mathematical structure of a pattern and to use that structure to derive a rule that could be used to compute (conclusively) the perimeter of any train.

Analysis and Discussion of the Focal Episode

We begin our discussion of the episode with a mathematical analysis of the hexagon trains task. Using language of the analytic framework and adopting a *mathematical perspective*, we can say that the hexagon task provided an opportunity for teachers to look for the underlying mathematical structure of a *definite pattern* and to use that structure to derive a rule that could be used to compute the perimeter of any train. More specifically, the task gives rise to a definite pattern because it is possible mathematically for a solver to provide conclusive evidence for the selection of the pattern, perimeter of train $n = (4n + 2)$. Of course, there are many equivalent ways (both algebraic and non-algebraic) to represent the pattern. The mathematical structure of the task specifies the process by which each train in the pattern is created: "The first train in this pattern consists of one regular hexagon. For each subsequent train, one additional hexagon is added." Without this structure it would be impossible mathematically for the solver to decide, for example, whether Train 5 would consist of five hexagons.

An analysis of the episode from a *psychological perspective* shows that many teachers approached the pattern in a limited way: they calculated the perimeter of the first few trains and then they identified the pattern in ways predicated on the basis of the regularities found in these terms, without connecting the generalization with the process by which each subsequent train was constructed from the previous one. In essence, these teachers treated numerical values in isolation from the other information in the task that made the pattern definite, thus treating the pattern as if it were plausible. Betty and Susan's solution (lines 2–6) as well as Nicole's solution (line 8–11) constitute examples of numerical approaches to solving the task.

Despite the fact that many of the teachers approached the pattern empirically, most of them seemed to have no doubt about its definite nature. The apparent discrepancy between these teachers' approach to the pattern and the mathematical nature of the pattern suggests, according to the *pedagogical component* of the framework, that the definite nature of the pattern was *non-transparent* to them. Yet, some of the teachers who identified the pattern based solely on numerical values had doubts about whether the formula they found was an accurate representation of the pattern. Nancy, for instance, could not tell whether her formula of $6n - 2n - 2$ described correctly the pattern in the task and how her formula was different (if at all) from the other formula that was being discussed, namely, $6n - 2(n - 1)$ (lines 12–13). Although an algebraic explanation would probably be an effective method to convince Nancy and the other teachers about the equivalence of the two expressions, it might not be an equally effective method to explain the same idea to students because students may not yet know how to make the necessary algebraic manipulations. Also, by simply showing the equivalence of the two expressions would provide no insight into how the expressions were derived or whether they would be true for any train in the sequence. These could be some of the reasons for which the teacher educator pushed the teachers in the episode to think about how the pattern could be connected to the figures drawn in the task. Another reason for which it was important to relate the pattern to the figures was that, by referring only to the numerical values of the perimeters of the first few trains, it was not possible to decide conclusively on a specific pattern. Take for example the values that Betty wrote on the overhead: 6, 10, and 14 (line 2); any formula of the form $\alpha n^3 - 6\alpha n^2 + (11\alpha + 4) \cdot n + (2 - 6\alpha)$, where α is a real number, would fit these values (there is an infinite number of formulas of different forms that also fit the values).

Tanya was the first teacher in the episode who presented a solution that was connecting the pattern to the figures and was providing conclusive evidence that $4n + 2$ was the correct pattern (lines 18–20). The solution she shared with the group, which suggested that the definite nature of the pattern was *transparent* to her, together with Elliot's elaboration on it (line 21), helped make the pattern transparent to some of her colleagues as well. For example, Nicole remarked that Tanya's solution was an eye-opener to her, because she would otherwise only see the pattern from the table, that is, numerically (line 22).

The fact that numerical approaches were short of convincing some of the teachers about the pattern is not surprising because, as we mentioned earlier, numerical data in isolation are not enough to provide conclusive evidence for a particular pattern in the hexagon task. It is reasonable to expect that students, when they engage in tasks that involve pattern identification, will face similar difficulties with the ones that the teachers in the episode faced. Nina's remark on line 25 is indicative:

> One struggle I face [when I teach patterns to my students] is that kids don't trust themselves, they don't trust the pattern, they don't trust the math involved. Even if they found a formula that seemed to work, they'd only be sure they had the answer if they actually built the whole train of 50 or 100.

This excerpt suggests that students have difficulty distinguishing between situations in which one can justify a pattern without the need for further examination of particular cases (*definite patterns*) and situations in which it is not possible to justify (in mathematical terms) the selection of a single pattern (*plausible patterns*).

It is interesting that, although Tanya's solution to the hexagon task seemed to have a significant influence on some of the other teachers' understanding of the pattern (e.g., Nicole's), none of the teachers made the connection that, perhaps, the students in Nina's classroom would become more confident about their solutions to pattern problems if they realized that it is important not only to find a pattern but also to see why a generalization holds. Rather, one of the teachers, Nancy, attributed students' difficulties with patterns to their innate developmental constraints (line 27); her remark suggests that she believed that there is nothing that teachers can do to help their students overcome the problem that Nina described earlier. Roberta's contribution to the discussion (line 29) was also problematic, but in a different way. Roberta presented an algorithmic method that she teaches to her students to help them find the scale factor in problems that involve pattern identification: the scale factor is the constant difference between consecutive terms. There are two fundamental problems with this method. The first is that, as the teacher educator pointed out on line 30, it does not help students assign meaning to what they are doing. The second is that this method would not work with plausible patterns or with non-linear definite patterns, because it assumes that the difference observed in the first few consecutive terms of the sequence remains constant throughout the sequence.

An important question that arises at this point is the following: how can students learn to distinguish between situations in which Roberta's method is applicable and situations in which it is not? Although addressing this question is critical to the development of students' understanding of patterns, in particular, and reasoning-and-proving, more generally, it may be possible for teachers in the middle grades to "get away" from dealing with it in their classrooms. For example, if tasks in which Roberta's method does not apply are underrepresented or are even non-existent in the textbook series used in the classroom, then students will be able to solve correctly tasks that involve pattern identification without knowing the answer to the question raised earlier. Indeed, the analysis of the algebra, number theory, and geometry units in CMP showed that the majority of the tasks in this textbook series are designed to engage students in linear definite patterns (Stylianides, 2005). If there were more tasks in the textbook series where patterns do not generalize in ways suggested by the first few terms, it would probably help students and teachers recognize the limitations of methods like the one that Roberta proposed.

To conclude, in our analysis and discussion of the focal episode the analytic framework helped us in two important ways. First, it offered us useful language to describe the mathematical terrain relevant to the hexagon task. Second, it helped us compare teachers' initial understanding of the pattern in the task with the mathematical nature of the pattern, and examine how this understanding developed during the session. Although the teachers in the session were dealing with a definite pattern, they initially tended to treat the pattern numerically and generalize it without connecting it to the mathematical structure of the task that uniquely determined the pattern. Later in the session, however, the discussion helped make the pattern transparent to some of the teachers. Making the pattern transparent to the teachers helped some of them gain insights into their teaching (see Nicole's comments).

Teachers' understanding of the distinctions between definite and plausible patterns is important for two main reasons. First, such an understanding will help teachers manage part of the complexity when they implement pattern tasks in their classrooms; mastery of this territory is critical given that pattern tasks are abundant in standards-based middle school textbooks. Second, deep knowledge of these distinctions will help teachers promote their students' understanding of reasoning-and-proving: (1) assisting students in recognizing the limits of empirical evidence (sometimes patterns do not generalize in ways suggested by the first few terms); and (2) enhancing students' appreciation of deductive ways of thinking (sometimes we can justify a pattern selection without the need for further examination of cases).

Conclusion

There is widespread agreement that the activity of reasoning-and-proving should be central to all students' mathematical experiences (e.g., Ball et al., 2002; NCTM, 2000; Schoenfeld, 1994; Yackel & Hanna, 2003). Yet, research shows that students at all educational levels (including university students) often face serious difficulties acquiring proficiency in this mathematical activity (e.g., Chazan, 1993; Coe & Ruthven, 1994; Harel & Sowder, 1989; Healy & Hoyles, 2000; Knuth, Choppin, Slaughter & Sunderland, 2002). Research also shows that teachers themselves often face similar difficulties in reasoning-and-proving (e.g., Goetting, 1995; Knuth, 2002a, 2002b; Morris, 2002; Stylianides, Stylianides & Philippou, 2004, 2007). To make matters worse, the existing research knowledge base provides insufficient guidance about the ways in which proficiency with reasoning-and-proving might be developed over time and how to organize effective professional development for teachers in the domain of reasoning-and-proving.

One reason for a lack of progress on these issues has been a lack of analytic tools usable in different kinds of investigations. This deficiency has made it hard for researchers to connect and build on existing findings on reasoning-and-proving and, therefore, to promote in a coherent way this activity in school mathematics. The analytic framework we utilized in this chapter aims to contribute to the research and development endeavor described here. In this chapter, we illustrated how some particular parts of the analytic framework can help researchers analyze professional development episodes related to reasoning-and-proving. Stylianides (2008) discusses other contributions of the analytic framework in the domain of reasoning-and-proving, notably, how the framework can support inquiry on the connections among written curriculum (what is included in the student's textbooks and the teacher's editions), implemented curriculum (how the textbooks are enacted by teachers and their students in the classroom), and the attained curriculum (student learning that results from the dynamic interactions of students and their teacher with textbook content during instruction).

Our analysis of the episode from the BIFOCAL professional development session suggests that even teachers who are identified by their schools as leaders and are experienced users of standards-based mathematics textbooks, may have difficulties understanding the subtle differences and nuances that exist in pattern identification. Understanding distinctions between different kinds of pattern is important because it can help teachers to manage some of the challenges they face when they implement pattern tasks in their classrooms (e.g., in which cases can we trust a pattern without the need for further examination of particular cases?) An explicit treatment of these issues in professional development and teacher preparation programs is essential if we are to come closer to the realization of our collective intention that all students should develop proficiency in pattern identification, in particular, and reasoning-and-proving, more generally.

Note

1.   This chapter is based upon work supported in part by the National Science Foundation under Grant No. 0119790 to the Center for Proficiency in Teaching Mathematics and in part by the Michigan State University Mathematics Education Endowment Fund for support of the BIFOCAL project. Any opinions, findings, and conclusions or recommendations expressed in this material are those of the authors and do not necessarily reflect the views of the National Science Foundation, the Center, or the university. The authors wish to thank two anonymous reviewers for useful comments on an earlier version of the chapter.

15
"Doing Proofs" in Geometry Classrooms

PATRICIO HERBST,[1] CHIALING CHEN, MICHAEL WEISS, AND
GLORIANA GONZÁLEZ, WITH TALLI NACHLIELI,
MARIA HAMLIN, AND CATHERINE BRACH

One of the reasons we teach students how to do proofs is to develop their capacity and disposition to infer necessary conclusions from given possibilities—a particularly valuable tool for mathematical problem solving. But previous research on students' problem solving in geometry (e.g., Schoenfeld, 1986, 1988) has shown that students who have learned to do proofs still fail to make such inferences to solve problems. Likewise, inasmuch as mathematical propositions state relationships between concepts (rather than describe facts about the world), it is by inferring the necessary conclusions that follow from various possibilities that one can understand how mathematical ideas make sense and connect to each other. Yet, research also shows that students fail to use what they have proved to make decisions in problems. An educational system where students learn "how to do proofs" but nevertheless still fail to make use of proof-generated knowledge, or to use heuristics of proving when problem solving is less than desirable. We might be better equipped to improve such system if we understood why that happens, what role instruction might have in perpetuating that state of affairs.

We discuss the teaching of proof in high school geometry, in search of the features of proof instruction that might account for the observed isolation of proof from geometric sense making (Schoenfeld, 1991). Herbst and Brach (2006) interviewed students about how some possible problems would be handled in their classrooms and found that the tasks for which students accept responsibility to engage in proving are only a narrow subset of those in which proving could potentially play a role. And in those tasks for which students accept responsibility for proving, they do only some of the things that a problem solver might do—the share of labor they take is limited. This suggests that proof instruction in high school geometry needs to be better understood.

We account for the place of proof in intact[2] geometry classrooms by inspecting the views of teachers of three different geometry courses offered in a comprehensive high school in regard to the place and purpose of proof. Then we look at classroom episodes in which we observed students engaged in proving. We position those episodes against a descriptive model of what we have called the situation of "doing proofs." We argue that beyond the surface differences across episodes where proofs are done, a common system of implicit norms regulates the division of labor, the organization of time, and the nature of the exchanges between students and teacher. Those norms underscore that students are socialized into a practice of "doing proofs" that is disconnected from "finding out."

Geometry Instruction at Midwest High School

During the years 2002–2004 we collected records of geometry teaching in Midwest High School,[3] a comprehensive high school with over 1500 students and a mathematics department of more than 15 members. The mathematics department at that time offered three versions of a geometry course. An honors geometry course was usually taken by 9th graders who had taken algebra in 8th grade, or by 10th graders who had excelled in 9th grade algebra. The regular, college preparatory course was available to those who had taken a first course in algebra (this geometry course was usually taken by 10th graders and occasionally by 9th graders). There was also an "informal geometry" course, also taken by 10th graders or older students; this course was avowedly at a slower pace and more focused on the needs of students. Teachers of the honors and regular courses used the textbook by Boyd, Burrill, Cummins, Kanold, and Malloy (1998), while teachers of the informal geometry course were expected to use the textbook by Hoffer, Koss, Beckmann, Duren et al. (1998). In spite of these various offerings, the district claimed to be de-tracked and students and their parents avowedly had free choice of which geometry course to elect (see Hamlin, 2006). For each of these courses, we collected yearlong records of instruction, including weekly videotaped, intact lessons, student work, interviews with students, and interviews with teachers.

We collected records from four classes taught by four different teachers: the honors classes taught by Cecilia Marton and Megan Keating, a regular geometry class taught by Lucille Vance, and an informal geometry class taught by Emma Bello. We interviewed each teacher several times. In the first interview, before the start of the school year, we asked, in particular, for their views on the purpose of proof.

Teachers' Views on the Role of Proof in Geometry

All teachers spoke about proof as a skill that was presumably taught for the formal training it could provide. Cecilia, who had taught geometry for more than 20 years, characterized geometry as a way of thinking, and the doing of proofs as an exercise which students could not easily get through by merely memorizing. What propositions were proved was not important to her: "Some of these theorems we prove seem to . . . be pretty useless, it's more an exercise thinking." Later, when asked about the big ideas of the class, she offered "thinking and problem solving" as "the big things, not the little nitty-gritty things like triangles, and things like that." As for the proofs that students do, Cecilia described them as "just prove these triangle congruent . . . they are again just an exercise in organization, figuring something out, proving something, creativity."

Lucille justified what students would learn while doing proofs: "Anything that you do that you have to have some sort of basis for doing, if you can sequentially justify it or explain it, that's an absolutely perfect skill to have." For Lucille, proofs promote "the ability to think logically and reason, I'd like to see more people explore in terms of possible occupations . . . Things that are legal in nature which seem to follow a line of reasoning . . . that seems to be part of geometry." Megan shared that perspective. She would describe geometry to a parent by saying that "Geometry teaches your child how to think in a logical way." She shared that she frequently told kids that they would be:

[G]oing to want your mom or dad to buy you a car. And you have a set of given facts that you know about getting a car and the thing you want to get to is, "I have a car." And what steps do you need to make in between, this is like a proof. I tell kids this all the time, that you do this every day when you're trying to talk your parents into something; you do little proofs every day. You know the givens, what your parents' values are, and you know what you want.

In commenting on why many people don't value the geometry class, Megan also said: "[The content] doesn't show up a lot in the subsequent courses afterwards . . . but the whole thought thing does, but people don't appreciate the proof stuff at the time." On the matter of content, Megan expressed similar views as Cecilia: "I say to kids, no one in your life . . . is ever going to ask you to prove two triangles are congruent . . . But the thought process that you use to get that, you're going to use every day."

Those three teachers thus argued for the importance of proof on formal grounds. Emma Bello underscored the same point when she explained why she would not do proofs in her informal geometry course anymore. Emma said that there had been a time when informal geometry teachers "still taught a little bit of proof," but that they had since "discontinued it, as the National Council of Teachers of Mathematics has moved away from proof." In particular: "[L]ogic, which has always been the reason we've given for teaching proof, [can be] better [learned] by teaching logic directly, and that's the direction we've gone." Thus if students' engagement in proving was to teach them logical thinking, Emma thought the end did not justify the means.

The four teachers all mentioned other issues as concurrent arguments for the support of reasoning and proof in geometry, including providing extra challenge for the elite students, providing a tool for retrieving content from memory, or that knowing why something is true could provide extra motivation to students. Notably absent from their comments was the use of proof as a tool to figure out solutions to geometric problems, or as a way of distinguishing what is necessarily true from what is not.

What Proof Practice Looks Like in Geometry Classrooms

The comments from teachers just noted point to the observation that students' engagement in proving is a form of mathematical work whose exchange value is the skill of organizing statements in a logical way. That is, teachers consider that what is at stake when students engage in proving is their opportunity to learn logical reasoning.

The approach taken by the informal geometry teachers, as Emma Bello told us, was to develop students' capacity for logical reasoning by teaching them logic rather than proof. At the beginning of the year, Emma taught students inductive and deductive reasoning. Emma explained to her class that inductive reasoning is "finding patterns and predicting what will happen next," and had her students produce counterexamples to demonstrate that inductive reasoning could lead to an incorrect prediction. To introduce deductive reasoning, she defined it as "drawing conclusions from information" taken as true. She provided real world examples, such as "No person who lives in our state lives by the ocean. Joe lives in our city, so we can conclude he does not live by the ocean." Students' work included drawing a conclusion from two given premises.

In the three other classes, geometric proofs were frequent. We present a descriptive model of the activities of proving motivated by our study of records of teaching in those classes. We then provide instances in which we observed those classes engaged in doing proofs and use the model to analyze what happened.

The Situation of "Doing Proofs"

The *two-column proof* (sometimes called the *T-form* by the participants) has been "the standard format for writing proofs" (Sekiguchi, 1991, p. 78). In this format:

> [O]ne draws a long horizontal line and a vertical line downward from the middle to form a letter T, creating two columns under the horizontal line. In the left column, one writes a deductive sequence of statements leading to the statement to prove, numbering each statement. For each step of the deduction one has to write in the right column a reason for the deduction with a corresponding number. (pp. 78–79)

In our prior work we have studied a number of features of classroom artifacts and practices associated with the writing of proofs in two-columns. These features include how diagrams are given and used in proofs (Chen & Herbst, 2005; Herbst, 2004), how events in time are sequenced and paced (Herbst, 2002b; Weiss & Herbst, forthcoming), how relevant ideas are activated for use, how premises for a conclusion are identified, how students decide to give a proof (Herbst & Brach, 2006), and how propositions for students to prove have evolved historically (Herbst, 2002a). We argue that the two-column proof form is a metonymy or a sign of an instructional situation, "doing proofs."

The expression "instructional situation" points to a key construct of our theory of instruction, according to which the work a teacher does can be described as organizing and managing transactions between, on the one hand, the work done by the class in their interactions and, on the other hand, claims laid on having learned objects of study at stake. We use the expression *instructional situation* to designate each one of the figurative marketplaces of such economy of exchanges. An instructional situation is a *frame* in the sense that it includes "principles of organization which govern [social] events . . . and [people's] subjective involvement in them" (Goffman, 1974, p. 10). Different events in classroom encounters are instances of the same instructional situation inasmuch as participants appear to refer or defer to similar norms when they organize and manage the work they do. We model those instructional situations as systems of norms[4] that describe the work participants do, the curricular stakes that they lay claim on, the way they divide labor to do that work, and the way they organize events in time. "Doing proofs" in high school geometry is an example of an instructional situation in the sense that while occasions in which students and teacher produce proofs may be superficially different, those occasions appear to be regulated by a common system of norms.

When we say that "doing proofs" is an instructional situation we mean that under the label "proof" are gathered certain kinds of mathematical work, whose unmarked or default organization consists of a collection of actions done over time that include the filling of the two- column form. We contend that those actions are regulated by norms that describe what work is done and what its exchange value is, how labor is shared between teacher and student, and how that work is organized over time (the sequence and duration of events). "Doing proofs" in intact classrooms is, however, not merely the enactment of a routine procedure. Like many recurrent, transactional human activities, such as playing a sport, it is one thing to say that the activity appears to be regulated or normed; it is quite another thing to say that the activity is carried out in a *uniform* manner, that each instance is *predetermined*, or that actors *consciously apply* a norm or a rule to do it. We mean none of that. Some norms may be explicitly known to the actors, but most are tacit. When students are engaged in proving in the geometry class, they do so in the context of an instructional situation, "doing proofs," which defers or defaults to a system of norms. A norm is a reference to which the real instances defer. In the examples that follow, we strive to display a manifold of superficially different episodes in which a proof was done. We then argue that there are deep similarities among those events, similarities that are visible through what the participants said and do when they repaired breaches to norms.

Our provisional model of the situation of "doing proofs" contains the following norms. The work to be done, producing a proof, consists of (1) writing a sequence of steps (each consisting of a "statement" and a "reason"), where (2) the first statement is the assertion of one or more "given" properties of a geometric figure, (3) each other statement asserts a fact about a specific figure in a diagrammatic or generic[5] register, and (4) the last step is the assertion of a property identified earlier as the "prove"; during which (5) each of those asserted statements are tracked on a diagram by way of standard marks, (6) the reasons listed for each of those statements are previously studied definitions, theorems, or postulates, as well as the "given," and (7) each of those reasons are stated in a conceptual (abstract) register. (8) Students' successful production of a proof exchanges for the claim that they know how to do proofs (and can reason logically, justifying their steps). The conclusion proved or implication between the givens and the conclusion is not the object of knowledge to lay claim on by producing the proof.

The division of labor includes the following norms: the teacher (or the textbook) is responsible for (9) stating that the task is one of producing a proof, establishing (10) the givens in terms of properties of a figure represented in a diagram, and (11) what is the conclusion about that diagram to "prove." The teacher is responsible for (12) activating the ideas that will be used in the proof by way of the diagram or the problem statement, and (13) prompting the production of statements and reasons by way of laying out the two-column form or eliciting differentiated prompts ("what will you say now," "what will be your reason"). The teacher is also responsible for (14) providing a diagram that fairly represents the objects to be used in the proof, and (15) the labels that will be used to refer to those objects, but (16) does not include any objects that will not be used in the proof. The students are responsible for (17) producing the statements in their proper sequential order, (18) identifying the reason for each statement after it is made, and (19) adding marks to the diagram to indicate congruent segments or angles, and possibly adding numeric labels to angles. But students are not expected to alter the diagram (20) by adding or (21) by erasing any geometric objects, including labels for points. The norms for the organization of time include: (22) statements are produced in sequence, so that earlier statements in the written proof are uttered earlier, (23) reasons for each statement are produced after their corresponding statement and before any new statement. (24) The duration of the proof production is gauged in terms of the number of steps, whereas (25) every single statement or reason is produced in a handful of seconds.

Many of those norms are tacit while others are explicit. Not all norms are enforced in every instance of "doing proofs." Rather these norms are the defaults of those instances where a proof is produced in class. When people deviate from the norms, they account for that deviation by way of "repair strategies": they mark the occasion or one of its actions as being special (Mehan & Wood, 1975). We use these norms to examine three different occasions of proof in our corpus of video records.

Instances of "Doing Proofs" in Midwest High School

Practicing Proof at Lucille Vance's Regular Geometry Class

By the beginning of November, Lucille was winding down the chapter on triangle congruence, and her students were getting used to writing proofs. In the lesson we examine, students had worked privately on a number of problems that involved proving claims by way of proving triangles congruent. Lucille had asked individuals to come to the board and put their work up for her to review. The statement of the first problem reviewed (see Figure 15.1) shows compliance with norms (9), (10), and (11) listed earlier since the teacher furnished the givens and prove in terms of objects in a diagram. It also complies with (14), in that the diagram includes all objects needed to produce the proof, even when they were not needed to state the problem: note the inclusion of diagonal $\overline{RY}$, which is not needed to state the givens or the conclusion. $\overline{RY}$ is given to suggest triangles △ RLY and △ YAR whose congruency entails that of angles ∠A and ∠L. The role of the diagram in suggesting what needs to be used is likewise visible in a later exercise (see Figure 15.2) when Lucille reviews what Alan had done.

As she reviewed Alan's proof, Lucille came upon the lines where he had asserted angles congruent on account of being corresponding, and reviewed the proof for the class. As she did that she gave evidence of the role that diagrams play in activating the concepts to be used:

I mean if this is parallel [extends segment $\overline{IA}$ to show that there is a line containing it as in the third diagram in Figure 15.2] to this [does same thing with the other line that contains segment $\overline{UL}$] what's the transversal?

And the same thing we want to think about when these are parallel, [completes the lines $\overline{UJ}$ and $\overline{IL}$ through the other points, using a different stroke than with the previous pair of parallel lines, as in the third diagram in Figure 15.2] what's the transversal?

Lucille's alteration of the diagram appears to repair a breach that occurred when students could not promptly identify the transversal line that defines angles 1 and 3 (viz. 2 and 4) as corresponding. In extending the segments that were said

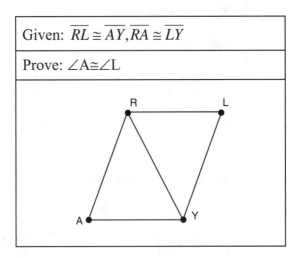

Given: $\overline{RL} \cong \overline{AY}, \overline{RA} \cong \overline{LY}$

Prove: ∠A≅∠L

Figure 15.1 First problem in the worksheet used in Lucille's class

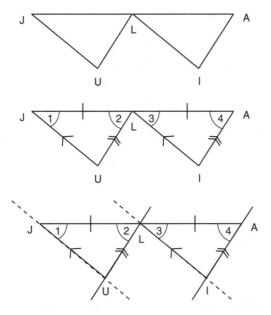

Figure 15.2 Alan's diagram before review starts; Alan's diagram after Lucille has marked the angles; Alan's diagram after the parallel lines have been extended

to be parallel, Lucille showed a diagram closer to the standard used to identify corresponding angles—as if accepting responsibility for students' difficulty seeing corresponding angles. Earlier, Lucille had called attention to the way Alan had done his proof, thus referring to norms for how students are to interact with diagrams when proving: students are expected to mark the diagram (norm 5), and those marks included showing which angles were congruent:

> [B]efore you write things up or whatever you write [points to the two-column proof], mark in your diagram [points to diagram].

Later on, responding to a question by Chloe as to whether they needed to write that angles 1 and 3 were congruent, Lucille stressed that they could not just assume that parallelism meant equal angles, they should make and justify those claims. She added:

Lucille: Now . . . we realize that okay something I want to stress again we realize that . . . we are accepting somewhat erm . . . abbreviated reasons. However if you went to another class even in this building they might want you to be more specific [than writing "corr ∠s" as a justification for why two angles are congruent, which Alan had done].
Alan: Parallel lines?
Lucille: Parallel lines cut by a transversal make corresponding angles congruent.

By marking Alan's departure from the norm (e.g., he had written "corr ∠s" as reasons for two lines of his proof), Lucille deferred to the normative notion that statements in a proof must be justified by a reason, which is a previously accepted theorem, postulate, or definition. Lucille then moved to look at the proof that Lance had put up (see Figure 15.3). As she discussed Lance's proof she held him to the expectation to mark in the diagram the parts that he knew to be congruent.

Even though Lance had already finished his purported proof, and the conclusion of that proof would have entitled him to say that the angles at A are congruent, Lucille contrasted the congruence of the angles at D (which was based on information in the "givens" of the proof) with the congruence of the angles at A (which can only be determined from the conclusion of the entire proof). She continued reading through Lance's proof, and made an issue of the way one must use labels in a proof:

> And, why are we allowed to call those [angles] just by their vertex name? . . . But, is there ambiguity with N and Y?

Then Lucille moved to checking what seems to be the most important piece of the work of proving in high school geometry: making legitimate use of triangle congruency.

The main problem that Lucille noted in Lance's proof was related to not having written all three elements that need to be collected in order to claim that two triangles are congruent. Specifically, both the given and the first line of Lance's proof (see Figure 15.4) included mention that $\overline{DA}$ bisects ∠NDY, which along with the other congruencies stated necessitate the congruence of the two triangles. Yet Lucille insisted that Lance should have further unpacked what it meant to say that $\overline{DA}$ bisects that angle when, as Figure 15.4 shows, she stressed that Lance had missed one statement and its reason. As Lucille recalled, the earlier conversation with Chloe (about Alan's proof) had touched on whether one could use entailments of a statement without writing them. She emphasized the importance of writing out those, since what is at stake is whether students know

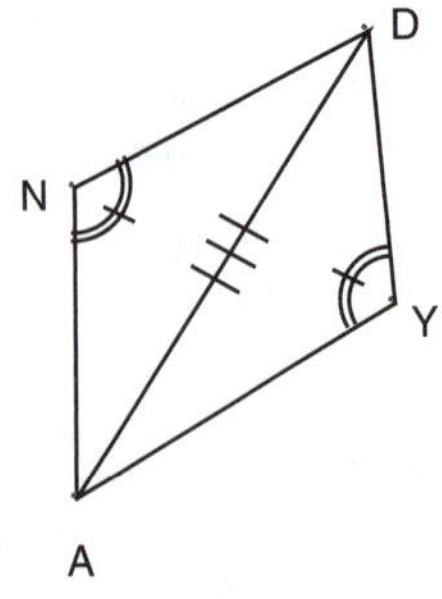

	S	R
	1) $\overline{DA}$ bisects ∠NDY	1) Given
	∠Y≅∠N	
	2) $\overline{AD} \cong \overline{DA}$	2) Same segment
	3) △NDA≅△YDA	3) AAS
	4) $\overline{NA} \cong \overline{YA}$	4)CPCTC

Figure 15.3 Diagram for Lance's proof; Lance's proof

S	R
1) $\overline{DA}$ bisects ∠NDY ∠Y≅∠N	1) Given
2) $\overline{AD} \cong \overline{DA}$	2) Same segment
∠NDA≅∠YDA	Def. bisect angle
3) △NDA≅△YDA	3) AAS
4) $\overline{NA} \cong \overline{YA}$	4)CPCTC

Figure 15.4 Lance's proof after Lucille's markup

how to do a proof. The purpose of the work was not to understand necessary properties of abstract geometric ideas, not even in Lance's problem, which could have led to the claim that in any kite the main diagonal bisects opposite angles. Rather the exchange value of what they had done was to become familiar with doing triangle congruence proofs.

After reading Lance's proof, she said to the class:

You've got to go to get past the . . . that handholding feeling like we have to see them all done, you have to take a position, there's three things you need to show and those three things have to be useful, if you show three angles is that useful? Three angles, angle, angle, angle . . .? No . . . you're looking for things you could use for congruency.

Lucille's case of managing students' engagement in proving, in the context of having them show how they did a typical proof exercise, contrasts with Megan Keating's case of engaging students in proving in a different activity structure, which we show next.

Proving a Theorem in Megan Keating's Accelerated Geometry Class

Megan was teaching different theorems that help decide whether a quadrilateral with specified properties is a parallelogram. Earlier, Megan had written on the board "five ways to prove a quadrilateral is a parallelogram." She then drew the quadrilateral in Figure 15.5 and proposed to prove the last of the five "ways," which stated:

If one pair of opposite sides is both congruent and parallel ⇒ parallelogram

The occasion of proving a theorem allowed Megan to involve her students in setting up of a proof task, something they rarely do (Herbst & Brach, 2006). The charge was to show that if two sides of a quadrilateral are congruent and parallel,

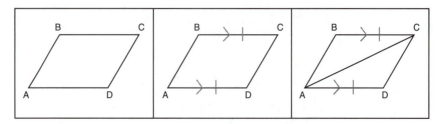

Figure 15.5 Diagram for a way to prove a quadrilateral is a parallelogram

the quadrilateral is a parallelogram. Despite the fact that such a quadrilateral could not, until the proof was completed, be presumed to be a parallelogram, Megan drew a figure that looked like a parallelogram, complying with one of the tacit norms (14) of the situation of "doing proofs": diagrams given are usually accurate, that is, they look like they already have the properties one is to prove they have (Herbst, 2004). Before asking students to start the proof, Megan had students parse the proposition to be proved into *given* and *prove* statements. But she expected more than just an identification of the hypothesis and the conclusion in the succinct statement of the theorem. She expected for "given" not "a quadrilateral has two opposite sides parallel and congruent," but rather "$\overline{AD}$ // $\overline{BC}$ and $\overline{AD} \cong \overline{BC}$." That is, she expected a formulation of the theorem in terms of diagrammatic objects, rather than abstract concepts. As we saw in the previous episode, proof exercises usually confront students not with abstract concepts alone, but with specific diagrammatic objects that instantiate those abstract concepts (Weiss & Herbst, 2007).

Megan maintained control of this moment of problem setup by staying close to the diagram and deciding what would get written on the board and how. In keeping with their usual share of labor in "doing proofs," students' concerns seemed similar to those of Lucille's students—what they should write when they wrote a proof. Jade asked: "If you have to write that out for the homework can we just write it AD, uh, congruent parallel to BC?" The unconventional notation Jade wanted to use—to write $\overline{AD} \cong$ // $\overline{BC}$—does not seem so odd when expressed verbally "*BC* congruent and parallel to *BC*." At stake in such interaction is not only the mathematical correctness of that written statement, but also the importance of distinguishing the properties of parallelism and congruency as bringing separate possibilities about the figure, and the right to decide how abbreviated the filling of a two-column proof with statements and reasons can be.

Once the task had been set up, Megan asked who had an idea of what they could do. Norman raised his hand, and Megan asked him to write the proof at the board. He indicated that he "thought it might be easier if we made it into two different triangles." With Megan's authorization, he drew the diagonal shown in the third part of Figure 15.5. It is usually not expected that students add features to diagrams when they are doing a proof (norm 20). Perhaps for that reason, Megan did not go directly into asking a student to do the proof, but rather asked who had an *idea*. Along those lines, it is important to note that Norman first shared his idea that it would be nice to make the parallelogram into two different triangles, and only after Megan allowed it, Norman drew the diagonal. Note also that Norman's idea to draw the diagonal did not get recorded in the proof itself. One might imagine that the final proof might include a statement to the effect of "Consider the diagonal $\overline{AC}$" justified by the postulate that any two distinct points determine a unique line. On the one hand, the lack of incorporation of such an explicit assertion into the proof could be felt unnecessary on account of the diagram, which made the existence of the diagonal self-evident (once drawn in). The introduction of the auxiliary lines could be argued a part of the pre-proof strategizing, rather than of the proof itself, as it might be if one were not relying on the diagram. On the other hand, although no justification was provided for why one is *legally entitled* to draw in a diagonal, Norman did provide a justification for why it was *strategically useful* (Chazan & Lueke, Chapter 2, this volume) to do so—to get a pair of triangles. In the proof itself, the only attention given to this crucial segment was Norman's allusion to the reflexive property: "AC is congruent to AC."

Like Lance in Lucille's class, Norman added hash marks to a single segment to indicate that it is congruent to itself. The purpose of hash marks in a diagram is ostensibly to indicate that two or more segments are congruent. Thus, placing hash marks on a single segment might be misleading (it might send the reader looking for a second congruent segment that is not to be found) or even nonsensical. But in doing proofs this is quite common (Herbst, 2004); it indicates that the congruence of a segment to itself is one of the three pieces of evidence that will be used to claim the triangles congruent.

The exchanges between Megan and Norman also illustrate how timing usually works as teachers engage students in proving and the role of the teacher in pacing this (norm 13). A prompt like "what else do you know" calls for statements of the proof, whereas a prompt like "why is that" calls for statements of reasons. They tend to alternate in the teacher's speech as ways to pace and prompt students' proof production (norm 23). The episode continues with Megan and Norman collaborating to plan how they would finish the proof. Norman provided the next reason before the corresponding statement, and Megan responded noting to him that he should give the statement first.

Norman: [Norman writes "3." and "3." under "statements" and "reasons," respectively.] Well, you could do alternate interior angles . . .
Megan: Yeah. Show me some. [Norman goes to the diagram and puts numbers in the angles formed by the diagonal and the sides; see Figure 15.6.]
Megan: Okay, Norman, which ones are congruent in that picture?
Norman: Oh, angle 1 is congruent to angle 2 and angle 3 is congruent to angle 4.

Megan corrected Norman, making the class notice that he could not yet say that angles 3 and 4 were congruent. "I know they are alternate interior angles, but I don't have the parallel lines, so I can't say that they are congruent . . . Now we know they're . . . that they are parallel because we know the thing is gonna be a

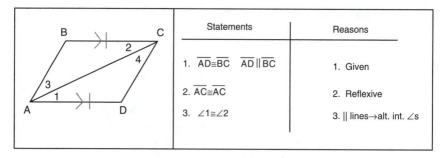

	Statements	Reasons
	1. $\overline{AD} \cong \overline{BC}$ $\overline{AD} \parallel \overline{BC}$	1. Given
	2. $\overline{AC} \cong \overline{AC}$	2. Reflexive
	3. $\angle 1 \cong \angle 2$	3. $\parallel$ lines→alt. int. $\angle$s

Figure 15.6 Parallelogram with angles marked; proof with angle congruence

parallelogram, but we don't know that yet." That last comment discloses one important difficulty a teacher has in managing students' engagement in proving, related to timing: at what moment things can be taken as true and this in turn is related to our prior comment on how diagrams are often drawn. The figure shows (as Megan wanted) that the figure *is* a parallelogram, and, eventually, as Megan said, they would find that out. For students who are still learning to distinguish between what appears to be true and what they can justify as true based on reasons, it is hard to relinquish the perceptual evidence (that angles 3 and 4 are congruent) in favor of what they are given (which only immediately entails that angles 1 and 2 are congruent). But the role of diagrams in proofs is more than to represent what students know at any given moment; it also suggests what they may want to aim for, which is why they are usually drawn accurately. And when students make statements for a proof, they come across as if they are only describing the particular diagram (Herbst, 2004).

As Megan and Norman continued exchanging questions for statements and reasons, the point of the activity seemed to be to demonstrate good use of prior knowledge. Reasons given for statements in a proof are supposed to be previously studied theorems, definitions, or postulates (norm 6). The student has to identify the appropriate reason from the wording and diagram given with the problem (norm 12), retrieve it from memory, and apply it correctly to the particular figure being considered. Accordingly, Norman would say things like "you have that by . . . the definition of parallelogram." The point is not merely whether Norman could justifiably assert that the figure was a parallelogram (which he surely could) but also whether he could identify the correct reason that would entitle that conclusion. Multiple properties (the "five ways") were at that moment listed on the left-hand side of the board, none of which had been proved before, but all of which (except the fifth) were candidate reasons for students to use in the proof. Thus we contend that doing the proof of this particular theorem was not so much motivated on the notion that a theorem may not be used until it has been proven, but rather on the notion that doing a proof gives students an opportunity to practice using the (other) theorems that one is supposed to know. Megan held Norman and the class accountable to actually learning the theorems she had started teaching that day. To have Norman prove one of them was instrumental in making them experience the need for the others; a need that was apparent as he found it necessary to justify his last statement.

The last case we describe, recorded in Cecilia Marton's class, provides yet a different example of how the activities of proving in a geometry class defer to the norms of the situation of "doing proofs" but deviate from merely applying those norms in deterministic fashion.

Reviewing a Homework Problem in Cecilia Marton's Accelerated Geometry Class

In early December, Cecilia's students had a homework problem requiring a proof about an isosceles trapezoid. The problem might be stated conceptually as one of proving that the triangle formed by extending the legs of an isosceles trapezoid is isosceles. The textbook however stated it in terms of diagrammatic objects as shown in Figure 15.7. Cecilia conducted an oral discussion of how students had solved the problem. Rather than surveying the proofs step by step, Cecilia asked students for their strategies.

Cecilia started the review telling that a student in another class had "turned in a 45-step proof." She laughed, recalling her surprise at seeing it, while somebody said "I feel bad for them," and others asked what the problem had been. Everybody else had produced proofs for that problem that were "around 20 steps." Cecilia then asked for the plan for a homework problem on an isosceles trapezoid.

Cecilia asked students to offer the approaches they had taken to solve this problem. Eve suggested the diagram shown in Figure 15.8, proposing that since angles 1 and 2 are congruent and 3 and 4 are supplementary to those they would be congruent—thus Cecilia concluded that would "make the triangle isosceles." Cecilia's interest seemed to be to make clear what the strategy was for Eve's argument, assessing its value as strategy, and devolving this strategy to the rest of the class.[6] The details of Eve's homework could be checked in private. Cecilia, however, was interested in other ways of justifying that the triangle is isosceles. She prompted Kara, "do you have another way of doing that?" Kara proposed looking at the diagram as shown in Figure 15.8. Kara argued that the whole

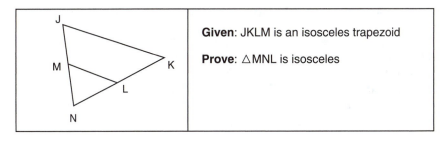

Figure 15.7 Diagram provided by the textbook (drawn according to Boyd et al., 1998, p. 326); statement of problem 40 (from Boyd et al., 1998, p. 326)

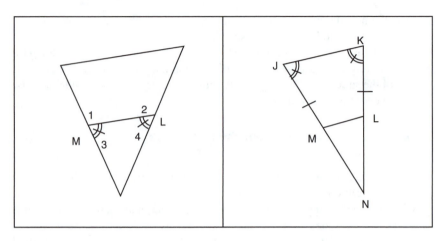

Figure 15.8 Cecilia's diagram of Eve's solution; Cecilia's diagram of Kara's and Erie's solution

triangle JKN was isosceles since the trapezoid was isosceles, and that angles J and K were congruent to angles M and L (respectively). In this case even though a key idea for the proof seemed to have been given by Kara—to look at the whole figure as one big isosceles triangle—Cecilia considered that Kara had to do more to really describe her strategy, and so she pressed for more detail. Kara could not satisfy Cecilia in explaining how she moved from the big triangle to the small triangle and gave up.

Ernest took the discussion again to operating on the angles, offering that "If JKLM is a . . . trapezoid, then $\overline{JK}$ must be parallel to $\overline{ML}$, so angle 3 and angle J are congruent, they're corresponding." Cecilia was satisfied with this explanation: "And since these two are congruent, you can get those two congruent, and get your isosceles triangle." After Ernest's proof, Erie picked up the question Cecilia had previously put to Kara:

Erie: What if you, um, you already know that angle J and angle K are congruent because it's an isosceles trapezoid, so then, can you say that JN is congruent to KN?
Cecilia: Yes.
Erie: And then you can just, subtract JM from JN and then, K . . . L from KN, and then you can say that MN and LN are congruent?
Cecilia: That's right, yeah, that's a good way to do it. You could use the segment addition postulate, and since these two are the same length they are equal to each other, subtract off the equal ones, right Erie?

Thus Erie brought the second big idea that complemented Kara's first idea, and made the proof follow. The consideration of multiple proof strategies had been grounded on attending to students' thinking. Cecilia had said, "I know there are more, okay? This is one where people have quite a few different ways of doing it." After Erie provided her solution, a second value surfaced, namely the importance of pursuing a strategy that would lead to the greatest economy of labor:

Cecilia: Okay, does anyone have a sense, as to . . . [Cecilia numbers the different proof strategies on the board] . . . Anyone have any idea with that, which one might have been the shortest?

Student responses brought to bear what was prior knowledge in the situation (what they knew about isosceles trapezoids, isosceles triangles, supplementary angles) to justify which of the proofs would be shorter. The discussion continued as to which of the three proofs was longer, comparing two of the ideas on account of the number of steps that it took. Cecilia eventually said, laughingly, "We're not making any judgments on these, okay? . . . Wait till you see the 45-step proof I told you about." The discussion about the relative lengths of the different proofs lasted nearly 2 full minutes, longer than the discussion of any one proof (the discussion of Eve's proof lasted only 40 seconds, Kara's lasted 31 seconds, Ernest's 25 seconds, and Erie's 38 seconds.)

At first blush, the reported conversations between Cecilia and four of her students (Eve, Kara, Ernest, and Erie) about the approaches they took to do problem 40 leave the impression that "doing proofs" is very different in Cecilia's class than in Lucille's class. While Lucille had concentrated on inspecting the details of the work students had done, marking the diagram, laying out every step needed, and had acknowledged the importance of appropriate justifications, this episode presents Cecilia as interested only on the different ideas that could be used to make the argument. But the ensuing discussion on the length of the proofs along with the opening comments by Cecilia about the "45-step proof" repair the previous "planning" discussion by indicating what *would need to have been done* for the previous discussion to count as *really* "doing proofs." The work expected of students in Cecilia's class appears to be similar in demands to the work expected of students in Lucille's class, at least in regard to the requisite level of detail and the things that might be written in the "reasons" column.

These differences can be partly accounted for by noting that the two proof exercises were discussed in different activity types (Lemke, 1990): in Cecilia's class, they were discussing a homework problem that could conceivably have taken many steps (certainly many more than any of the proofs done in Lucille's class). In Lucille's class, the work for the day included becoming more comfortable writing proofs and seeing other people's work writing these proofs. Thus whereas the episode in Lucille's class shows compliance with many of the hypothesized norms of "doing proofs" and repairs of particular moments when the norms are being breached, the episode in Cecilia's class consists of one large breach to the situation of doing proofs. Cecilia and her students act as if they were taking time out from what they really should do (survey each proof done) to discuss the ideas of the proofs, but the beginning and ending conversations show that what was really at stake were the formal characteristics of the accomplished proofs.

"Doing Proofs"

The episodes narrated here illustrate three activity types that are quite common in secondary mathematics classrooms: practice problems at the board by students surveyed in public by the teacher in Lucille's class, oral review of a hard homework problem by the teacher with students at their seats in Cecilia's class, and board work by a student in Megan's class. The three episodes are different on the surface. Cecilia did not spend the amount of time per solution that Lucille spent reviewing Lance's work, or the amount of time that Megan spent accompanying Norman in the proof development. But the similarities are more important than the differences. These similarities can be expressed by a common system of implicit norms regulating the events on the surface. We summarize those norms as we show how those norms can be organized.

First of all there are norms that create an exchange structure around "doing proofs" in the geometry class, namely, norms that regulate what counts as valuable work and what value particular pieces of work have at particular times. The use of the two-column form by Megan conveyed the sense that giving a reason for each statement would be valued as work, whereas the lack of use of the two-column form (actually the lack of any writing) by Cecilia pushed students to make more substantive contributions; for example when Erie provided the "segment addition" argument it was not just to fill in a blank space in the "reasons" column but because it was needed as a way to operationalize what it would mean to take away from the legs of an isosceles triangle. Lucille's stated objective to bill the lesson as one that showcased the work of doing proofs gave more value to small details such as how to clearly identify the angle being used to claim congruence. Instead, Cecilia's attempt to showcase different strategies to prove the same conclusion gave more value to quick references to angles and key ideas for the proof.

Second, there are norms that regulate the division of labor between teacher and student when "doing proofs." The teacher (or the textbook) is usually the one who indicates that doing a proof would be the way for students to respond to a mathematical task. The teacher also provides a diagram that includes all geometric objects that will be needed for the argument, and a statement of the given and conclusion in terms of objects in the diagram. The student, by way of contrast, is responsible for producing succinct statements about the diagram, for marking those statements in a diagram, and for justifying statements with clearly spelled out reasons drawn from prior knowledge.

Third, there are other norms that contribute to the existence of a temporal structure for "doing proofs," namely, norms that regulate when things need to be done, in which order, and for how long. The moment in which the proving task is set up (as "given" and "prove") usually precedes the production of the argument. The conclusion, in particular, is always known (and known to be true) from the beginning; the argument is produced after, and done as a sequence of statements, each followed in turn by a reason, with the "given" usually in the first line (justified as "given"), and the "prove" in the last.

Inasmuch as most of those norms are tacit, they are "real" only as hypotheses made by an observer: they exist insofar as they are useful for the observer to understand actual events and contrast them with conceivable ones. A model of the instructional situation of "doing proofs" in terms of a system of norms regulating what classroom participants do when engaging in proving is helpful to investigate what it might mean to create a different place for proof in geometry classrooms. These norms could be considered, and systematically altered, in the design of instructional experiments that depart from custom in order to make possible desired mathematical work and in the design of instructional supports for a teacher to manage such kind of mathematical work.

Conclusion

Over more than a century, American students have learnt to do proofs in high school geometry classes. The justification for the teaching of proof was originally a formal one (learning logical thinking); our teacher interviews show that justification as current. While the informal geometry class has sought students' learning of logical thinking by teaching them logic, other geometry course have continued to teach students to do proofs, allocating students a share of labor, and organizing time and values in such a way that students get involved in logically justifying claims about geometric figures. The examples show that students' main responsibility continues to be the production of statements and reasons in sequence—making statements about a diagram and justifying them on prior theorems, definitions, and postulates. They have little responsibility over fashioning a diagram, making connections to concepts that haven't been activated by the problem or the diagram, establishing what givens might warrant a particular conclusion, or discovering what conclusions might be drawn from a particular set of givens.

The extent to which that training in logic can have an impact on the development of students' capacities for problem solving and to use proof as a tool to *know with*, however, is hampered by the norms that regulate how students engage with proving. The notion that geometry is just a context in which to teach logic works on the presumption that the geometric objects themselves are merely tokens, marks on paper, like the words and sentences used in propositional logic. But the proofs that students do are formulated in terms of given geometric *diagrams* (rather than geometric concepts), and the statements that students make about those diagrams are not always necessary conclusions from previous statements; at times they are descriptions of what is seen to be true in the diagram (Herbst, 2004). Because of this, we argue that geometric objects benefit but also suffer from their representation as geometric diagrams. Diagrams are not as well behaved as the *Ps* and *Qs* of propositional logic, which have no semiotic referent other than that ascribed to them by the formal structure of the logic itself.

Herbst and Brach (2006) have shown how students distinguish between those tasks in which students perceive it relevant to produce a proof, and more open problems the solving of which might benefit from (among other things) the making of initial assumptions or hypotheses about diagrams. Behind that distinction lies the presumption that proving is just a way of organizing logically factual properties of Platonic configurations. It is as if the factual properties of diagrams were just imperfect images of the properties of the abstract object that is to be apprehended, and as if putting restrictions on students' interactions with diagrams would have the effect of turning diagrams into marks on paper, objects amenable to logical reasoning. However, when students arrive into high school geometry, the van Hiele research has told us (Fuys, Geddes, & Tischler, 1988) students are far from having conceptualized geometric figures as abstract, logical objects. For many (perhaps most) of them, the diagram still *is* the object, and acting on diagrams, constructing, measuring, are very much part of their understanding of them.

A view of geometric truths that characterized geometry as a theory of the necessary consequences of possible properties of geometric objects might not only be closest to the heart of creative mathematical activity, but also better able to translate into a pedagogy that could capitalize on students' means of operating in the geometry that they know, and additionally co-opt those means to develop knowing of abstract geometric objects. Yet if conjecturing and proving are to play the role of a tool to know with, successful engagement of students in proving will have to be characterized by a much broader share of labor for students, a more flexible temporal structure, and a more open-ended exchange structure than those observed in the "doing proof" situation.

Notes

1. This work is supported by NSF grant REC 0133619 to Patricio Herbst. Any opinions, findings, and conclusions or recommendations expressed in this material are those of the authors and do not necessarily reflect the views of the National Science Foundation. The authors acknowledge valuable help from Hui-Yu Hsu in locating archival resources.
2. By intact lessons, we mean lessons whose plans did not respond to any intervention that we had participated in designing.
3. All personal and institutional names have been replaced by pseudonyms.
4. There is no assumed connection between our use of the word *norm* and that of Cobb and his colleagues (Cobb & Bauersfeld, 1995). Whereas they have used the word to designate emergent agreements among teacher and students regarding what will be considered appropriate for students to do, the norms we speak about are implicit regulations of the joint activity of teacher and students, as seen from the perspective of the observer. For us, *normative* means not necessarily "correct" or "appropriate" from anybody's perspective but rather "unmarked"; a manner of doing joint activity that is ordinary, in the sense that it goes without saying.
5. For the distinction between diagrammatic, generic, and conceptual registers, see Weiss & Herbst (2007).
6. Students volunteered their different approaches to the problem from their seats and Cecilia recorded some aspects of their ideas on diagrams in the whiteboard. Since the camera was capturing students' faces as they spoke, our records of those diagrams are from occasional shots at the board and they do not permit synchronization of speech and actions on the diagrams.

IV
Teaching and Learning of Proof in College

Arguments that the "essence of mathematics lies in proofs" (Ross, 1998, p. 2) and that "proof is not a thing separable from mathematics . . . [but] is an essential component of doing, communicating, and recording mathematics" (Schoenfeld, 1994, p. 76) reinforce the centrality of proof in mathematical thinking. Indeed, the central role of proof in mathematics becomes more obvious at the college level where the level of mathematical content and students' maturity facilitates the introduction to complex forms of proof. The chapters in Section IV conclude the "proof story across the grades" by shifting our attention to questions and issues specific to teaching and learning proof at the post-secondary level and connecting these questions and issues to those raised in the previous sections.

In Sections II and III, the authors helped conceptualize the role and meaning of proof for young children and high school students respectively. Their questions included: what constitutes an acceptable argument at those levels? What role can proof have in the entire mathematics curriculum? (see Introductions to Sections II and III). They explored students' intuitive notions of proof and how school experiences might nurture and challenge these natural ways of understanding mathematical argumentation towards developing what constitutes a logically valid argument.

At the college level, the questions of what constitutes proof and what role it might have in the curriculum become peripheral. The Committee on the Undergraduate Program in Mathematics (MAA, 2000) maintains that proof should be central to the teaching and learning of mathematics at the post-secondary level. And as Harel and Sowder (Chapter 16) suggest, college students, in particular those who major in mathematics-related areas, are expected to graduate from college with some degree of facility in reading and writing proofs. The questions that guide the teaching and learning of proof at this level concern how to make the type of reasoning that underlies proof central to the *entire* mathematics curriculum.

Furthermore, at lower levels, proof is often given the broader definition of the act of arguing to convince one's community of the truth of an assertion (Balacheff, 1988). As students advance to college courses, an informal argument or explanation is not sufficient (Harel & Sowder, Weber & Alcock in this volume). Students who major in mathematics-centered topics are expected to write *mathematical proofs*, which have a distinct form and well-defined conventions. As Harel and Sowder (2007) argue elsewhere and Weber and Alcock argue in this volume

(Chapter 19), in post-secondary mathematics instruction a sharp shift takes place, and the validity of the proof is dependent not only on its *content* but also on its *form*. This section reflects the attempt to understand this shift and, ultimately, its implications for the teaching and learning of proof in the college classroom. In completing the story of proof across the grades, this section highlights the distinct differences in the character of proof and the common goal to teach students the notion of what constitutes a grade-appropriate logically valid argument.

Developing College Students' Understanding of Proof

Much of the literature on the learning of proof, especially at the college level, has focused on identifying difficulties that students face when reading and writing proofs. This work provides an important starting point in the research base on teaching and learning proof. Indeed, understanding how to support student learning often begins with understanding where students' difficulties lie. Building from this research, this volume uses perspectives that move towards models and frameworks for supporting students in their growth. The authors in this section use new lenses to identify ways in which students' learning processes can be viewed not as an obstacle in instruction but as a basis to inform instruction. Some of the chapters challenge our views by combining perspectives commonly used in mathematics education with perspectives used in other fields as a way of opening a dialogue among researchers and practitioners.

Using the DNR framework developed in their earlier work, Harel and Sowder (Chapter 16) explore the critical issue of how faculty might facilitate the development of student learning of proof and what they know about how students learn proof. They focus on the development of student "proof schemes"—a label for the means that one might use in convincing oneself and others about a matter. In this perspective, each student holds *some* level of understanding and appreciation of proof, so, this examination of students' proof schemes "highlights the student as a learner." Moreover, it focuses instructors' attention in two areas: (1) current student understanding, rather than students' final proof productions and hence their possible shortcomings; (2) the instructional practices that may help in gradually refining student understanding towards the desirable deductive proof scheme, that is, the proof scheme practiced by contemporary mathematicians.

Blanton, Stylianou, and David (Chapter 17) and Smith, Nichols, Yoo, and Oehler (Chapter 18) examine mathematics learning that takes place in classrooms for which social negotiation of mathematical meaning is commonplace (Cobb & Yackel, 1996). Although such classroom environments are rare at the undergraduate level, the perspective used in these chapters partly as a tool to understand students' learning of proof is innovative and, more importantly, their effects on the learning of proof appear to be fruitful.

Blanton, Stylianou, and David (Chapter 17) explore possible ways in which students begin to internalize the language of proof and argumentation as they actively participate in whole class discourse. They maintain that the transfer of responsibility for proving a statement from the teacher to the students supports students' proof learning. They then conjecture that teachers' prompts that encourage student engagement in the discussion towards proof and the development of a critical stance towards statements and claims made publicly can be crucial in students' development of proof construction. They further suggest that students internalize public argumentation in ways that facilitate private proof construction.

Smith, Nichols, Yoo, and Oehler, like Blanton and her colleagues, view proving as an activity that develops within a community of learners and that can be appropriated by assisted participation. They discuss the ways in which a college instructor's choices to assume and relinquish the role of leader in the classroom affect students' participation in the learning community and facilitate their engagement in mathematical discussions about proofs presented during the course of instruction.

Weber and Alcock (Chapter 19) and Selden and Selden (Chapter 20) shift our attention to individual students' learning of proof. They write from the perspective that student learning of proof cannot be separated from the mathematical content in which it occurs. These two chapters consider the character of mathematical proof and its implications for student proof construction.

Weber and Alcock (Chapter 19) frame their discussion around the argument that learning to prove in college classrooms requires students to work within a new *representation system*. Because this process can be simultaneously limiting and empowering, it is important to understand both the challenges and affordances that students face within the (new) representation system of mathematical proof.

Connecting their argument to their earlier work on syntactic and semantic reasoning in proof construction, Weber and Alcock claim that constructing a proof syntactically may allow an undergraduate to produce a valid argument when they could not otherwise do so. However, when students limit themselves to syntactic proof productions, they may not make use of links between the representation system of proof and other informal representations of mathematical ideas, and so may find the proof non-convincing and non-explanatory. By the same token, when students attempt to link the formal representation system in which proofs are produced with their concept images, that is, use semantic reasoning, they may have more choices as to what line of reasoning to pursue. A potential problem with semantic reasoning, however, is that an individual may be able to develop a mathematically correct understanding of why a proposition is true, but be unable to frame their argument within the representation system of proof.

Similar to Weber and Alcock, Selden and Selden shift our attention to mathematical proof as it might be viewed in advanced college courses by proposing a framework for working within proof to facilitate student growth. In particular, they introduce the notion of looking at proofs themselves, because differing types of proof and components of these proofs can correspond to various abilities needed to produce them. They suggest that an analysis of kinds and aspects of proofs should facilitate teaching by coordinating the theorems assigned with student abilities. It should also facilitate the assessment of student abilities and, hence, the study of how students learn about proof.

Teaching Practices that Support Proving

Throughout most of this volume, teaching and learning proof are naturally interwoven. As in the previous sections, the authors in this section draw on their work on student learning of proof to suggest principles for instruction, or use their study of instruction to examine the growth of student learning.

Blanton, Stylianou, and David (Chapter 17) examine teacher practice in an undergraduate mathematics course that embedded the development of proof in the social activity of the classroom. They used selected classroom episodes from this course to design a framework for characterizing whole class discourse on proof, specifically, teacher and student utterances. They then used this framework to analyze classroom discourse, specifically, to study how these utterances impact one another and scaffold student learning of proof.

Smith, Nichols, Yoo, and Oehler (Chapter 18) continue this emphasis on teacher practice. However, while Blanton and her colleagues conduct a fine grain analysis of the teacher's actions and utterances during selected teaching episodes, Smith et al. focus on teacher actions over time. In particular, they look at how one instructor's choices about when to interrupt and direct class discussions of students' work influenced the development of a classroom community of inquiry and encouraged a view of mathematics as a human social activity by engaging students in discourse about mathematics and proof.

These studies examine classrooms that are not common at the college level in that the public negotiation of ideas and collective development of proofs are part of the sociomathematical norms of the classroom. Harel and Sowder (Chapter 16) complement this work by using a wider lens on instructional practices in the teaching of proof and bringing to the forefront the views of a larger group of college instructors who approach the teaching of proof from a variety of perspectives. They study the views of university mathematics faculty of upper division courses, using interviews about students' success and difficulty with proof in the typical university mathematics curriculum. According to Harel and Sowder, the question of critical importance is "what instructional interventions can bring students to see an intellectual need to refine and alter their current proof schemes into deductive proof schemes?" (Chapter 16, p. 279).

After discussing the relation of the structure of the proof to the learning of proof, Selden and Selden make specific suggestions for the teaching of proof, taking into account these features. They describe an approach to teaching in which instruction is integrated into students' construction of proofs, and in which features of proofs are more important than the topics of theorems.

Overall Issues and Perspectives on the Teaching and Learning of Proof at the College Level

This section attempts to bring to the forefront issues of student learning and instructional practice regarding proof at the post-secondary level, while taking into consideration the nature of proof itself, the representation system in which it resides and the norms and culture of mathematics. The studies that are described in this section are rooted in different methodological orientations ranging from the notion that the building of proof should be a part of the sociomathematical norms of the classroom (Blanton et al., Smith et al.) to one based on understanding the instructors' notions of proof (Harel and Sowder), and to orientations rooted in the nature of mathematical nature of proof itself and its implications on instruction and learning (Weber & Alcock, Selden & Selden). But, despite their differences, they all share the goal of understanding ways to facilitate students' growth in reading and writing *mathematical proofs*.

Hence, the chapters in this section examine the shift of students' reasoning from the building of logical "arguments" that may have been the norm in the elementary and secondary grades towards "arguments that obey well-defined conventions that are agreed on by contemporary mathematicians" (Weber & Alcock, Chapter 18, p. 323). Moreover, they provide evidence that, despite the documented difficulties in understanding proof, instruction *can* help students learn to reason deductively. For example, Smith and her colleagues discuss students' "remarkably mature conception of mathematical proof as an essential activity of doing mathematics" (Smith et al., Chapter 18, p. 322) and Selden and Selden report on their success in teaching proof by attending to the special features of various types of proof.

Finally, they pose questions that invite a dialogue between mathematics education and other disciplines and attempt to examine new venues for the teaching and learning of proof which are largely unexplored at the college level. In particular, Harel and Sowder's analysis (Chapter 16) of mathematicians' views and conceptions regarding the teaching of proof and their expectations with respect to student learning and students' growth in deductive reasoning creates an opening for dialogue between college instructors and secondary teachers regarding each community's views and conceptions. Selden and Selden (Chapter 20) also address traditional topics on the teaching of proof and point to disparities between current practices at the college level and the needs of students as suggested by mathematics education research. The sociocultural perspective used by Blanton and her colleagues (Chapter 17) and Smith and her colleagues (Chapter 18) brings to the fore instructional approaches that, as of recently, were largely unexplored at the college level and that invite the college community to further consider these novel approaches and their implications.

Another similar theme across the chapters is the focus (often implicit) on students who are largely within mathematics-related fields (i.e., mathematics, sciences, computer science, engineering). The studies described in this section were implemented in classes aimed towards students with strong mathematics backgrounds and the findings describe these students' success and performance. However, the *Curriculum Guide* (MAA, 2000) maintains that *every* course should help *all* students develop an ability to reason logically and deductively. But, the teaching and learning of proof for students outside of mathematics-related areas remains an open question for us to consider. What, in particular, should the learning goals with respect to proof for these students be, and what types of instructional activities and curricular materials would be appropriate to achieve these goals? Furthermore, what meaning does proof take on in courses for these students and how does that relate to the teaching of proof at earlier levels? If proof is to become a central goal in post-secondary mathematics instruction we need to consider how to incorporate it in all courses in appropriate ways.

This volume seeks to promote discussion on how we can develop students' command of proof at all stages of their education. In our overview of the chapters in Section I, we posed the questions "If children enter secondary grades with an emergent understanding of what counts as a logically valid argument and how to develop one, in what ways can this extend their mathematical understanding in more advanced studies?" And, "What are the implications for curriculum and instruction in secondary grades and beyond when children bring these understandings to the classroom?" (p. 70). As each of the sections of this volume addresses the teaching and learning of proof at different critical junctures in students' development, each points to a different character of proof—an age-appropriate one. Elementary students who currently have opportunities to build and reflect on mathematical arguments will not enter college for several more years. But, before that happens, the college mathematics community will need to think about the kinds of classroom environment and curriculum that will support students in learning its own conceptions of proof.

16

College Instructors' Views of Students Vis-à-Vis Proof

GUERSHON HAREL AND LARRY SOWDER

College instructors who teach mathematics courses to majors during their last 2 or 3 years of undergraduate coursework are naturally a possible source of information about the status of the *proof schemes* of their students. Their views of students' proof schemes and what they say about their own instructional efforts to advance them are an integral part of their *knowledge base*. Their views could be informative in understanding both the obstacles students encounter in developing desirable proof schemes and also the kind of curricular and instructional efforts needed to help students develop such schemes. In particular, we use a general instructional framework (DNR) to help identify likely areas for improvement in instruction on proof. We start with a brief exposure to the DNR framework and later interpret several quotes and impressions from interviews with college instructors in terms of this framework.

Basics in the DNR Framework[1]

DNR-based instruction, or *DNR* for short, can be thought of as a theoretical framework consisting of three interrelated categories of constructs: *premises, concepts*, and *instructional principles*. Three principles—the *duality* principle, the *necessity* principle, and the *repeated reasoning* principle—are foundational in DNR-based instruction. Relevant to this chapter are also several DNR concepts: we begin with the notions of *proof scheme* and *proof*, which are attributes of one mental act—that of *proving*. Following this, we generalize these two notions into the constructs of *way of understanding* and *way of thinking*, respectively, as attributes of any mental act.

Proof Versus Proof Scheme

In Harel and Sowder (1998, 2007), we defined "proving" as the *mental act* employed by a person (or a community) to remove or instill doubts about the truth of an assertion. The proving act is instantiated by a combination of two acts: *ascertaining* and *persuading*. "Ascertaining" is the act an individual employs to remove her or his own doubts about the truth of an assertion (or about its negation), whereas "persuading" is the act an individual employs to remove others' doubts about the truth of an assertion (or about its negation). In this subjective perspective, a *proof* is defined as the particular argument one produces to ascertain for oneself or to convince others about the truth of an assertion, whereas a *proof scheme* is a characteristic of the proving act. For example, asked why 2 is an upper bound for the sequence $\sqrt{2}, \sqrt{2 + \sqrt{2}}, \sqrt{2 + \sqrt{2 + \sqrt{2}}}$, some undergraduate students gave the proof "$\sqrt{2} = 1.41, \sqrt{2 + \sqrt{2}} = 1.84, \sqrt{2 + \sqrt{2 + \sqrt{2}}} = 1.96$ [five more items of the sequence were evaluated] we see that [the results] are always less than 2, . . . Hence, all items of the sequence are less than 2." Other students gave the proof, "The third item is less than 2 because it is the square root of a number that is smaller than 4, this number being the sum of 2 and a number that is smaller than 2. The same relationship exists between any two consecutive terms in the sequence." These two proofs—viewed as such by the students who offered them—are products resulted from carrying out the proving act. They may indicate certain persistent characteristics of these students' mental act of proving. For example, on the basis of a multitude of observations of proofs produced by the first group of students, we may characterize its act of proving as empirical if the proofs they produce are similar in nature to one produced here. Thus, a "proof" is a product of the proving act, whereas "proof scheme" is a persistent cognitive characteristic of that act.

Before we generalize these two notions, we emphasize that despite these subjective, student-centered definitions, the goal of instruction in DNR is unambiguous—namely, to gradually refine current students' proof schemes toward the proof scheme shared and practiced by contemporary mathematicians. This claim is based on the premise that such a shared scheme exists and is part of the ground for scientific advances in mathematics.

Way of Understanding Versus Way of Thinking

Clearly, the proving act itself is never carried out in isolation from other mental acts, such as "interpreting" "symbolizing," "connecting," etc. As with the act of proving, we can—and very often we must—talk about the product and character-istic of such acts. A product of a mental act is referred to as a *way of understand-ing*, whereas a persistent cognitive characteristic of such an act is referred to as a *way of thinking* (Harel, 1998, 2001, 2007).

To illustrate, consider the mental acts of "interpreting." The actual interpret-ation one gives to a term or a string of symbols is a way of understanding because it is a particular cognitive product of her or his act of interpreting. For example, one may interpret the string of symbols $y = \sqrt{6x - 5}$ in different ways: as an equation (a condition on the variables x and y), as a number-valued function (for each number x, there corresponds the number $\sqrt{6x - 5}$), or as a proposition-valued function (for each ordered pair (x, y) there corresponds the value "true" or the value "false."). These ways of understanding manifest certain characteristics of the interpreting act—for example, that "symbols in mathematics represent quantities and quantitative relationships." A person who holds more than one such way of understanding is likely to possess, in addition, the way of thinking that "mathematical symbols can have multiple interpretations." And a person who is able to vary the interpretation of symbols according to the problem at hand is likely to possess the way of thinking that "it is advantageous to attribute different interpretations to a mathematical symbol in the process of solving prob-lems." These are examples of mature ways of ways of understanding and ways of thinking, which are absent for many high school and college students. For example, when a class of calculus students was asked what $y = \sqrt{6x - 5}$ meant to them, many were unable to say more than what one of their classmates said, "It is a thing where what you do on the left you do on the right." For many students the act of interpreting algebraic symbols can be characterized as being free of quanti-tative meaning.

It may not be easy to get accustomed to the technical distinction between the terms "way of understanding" and "way of thinking" as is made here, partly because the two terms are often used interchangeably (and without exact def-initions). Also, the phrase "way of " seems to connote a sort of a process and, hence, a dynamic image, whereas the definition of "way of understanding" as a *product* of a mental act may connote an outcome, a static image. One reason for the use of the phrase "way of " is to accentuate subjectivity, as was the intent for the definitions of "proof " and "proof scheme." The term "understanding" usually entails a quality judgment. The term "way of understanding" intends to highlight neutrality, remove subjective judgment, and maintain a student-centered orienta-tion. Thus, in DNR, "ways of understanding" and "ways of thinking" are dis-tinguished from their values, in that one's ways of understanding and ways of thinking can be judged as correct or wrong, useful or impractical in a given context, for example.

The constructs of ways of understanding and ways of thinking are central in the DNR conceptual framework for different reasons; the one pertaining to this chapter is that these constructs give the basis for how *mathematics* is defined in this framework:

Mathematics consists of two complementary subsets: the first subset is a collection, or structure, of structures consisting of particular axioms, def-initions, theorems, proofs, problems, and solutions. This subset consists of all the institutionalized ways of understanding in mathematics throughout history—those the mathematics community at large accepts as correct and useful in solving mathematical and scientific problems. The second subset consists of all the ways of thinking, which characterize the mental acts whose products comprise the first set. (Harel, in press)

Two critical implications follow from this definition. The first implication is that the instructional objectives teachers set for their classes should correspond both to ways of understanding and to ways of thinking. Currently, the instructional focus is largely on subject matter in terms of the products of mental acts—the ways of understanding—such as particular definitions, procedures, techniques, theorems, and proofs, with little or no attention to the characteristics of the mental acts—the ways of thinking—that students possess and are to develop by learning particular subject matter. Hence, with respect to the proving act, questions of interest are whether university faculty set as an instructional goal for their students to acquire the deductive proof schemes and what instructional interventions they employ to achieve this goal.

The second, and related, implication has to do with the DNR's concept of *teacher's knowledge base*. Building on Shulman's (1986, 1987) work, and consistent with other views (Cohen & Ball, 1999, 2000), Harel (1993) suggested three critical components that define teachers' knowledge base: *knowledge of mathematics, knowledge of student epistemology,* and *knowledge of pedagogy.* Generally speaking, *knowledge of mathematics* refers to the breadth and, more importantly, the depth of the mathematics knowledge possessed by the teacher. In light of the definition of mathematics given earlier, teachers' knowledge of mathematics consists of their ways of understanding *and* ways of thinking, not just the former as is commonly held. *Knowledge of student epistemology* refers to teachers' understanding of fundamental psychological principles of learning, such as how students learn and the impact of their previous and existing knowledge on the acquisition of new knowledge. *Knowledge of pedagogy* refers to teachers' understanding of how to teach mathematics in accordance with these principles. This includes, among other things, an understanding of how to assess students' knowledge, how to utilize assessment to pose problems that stimulate students' intellectual curiosity, and how to help students solidify and retain knowledge they have acquired.

A Desirable Knowledge Base for Mathematics Instructors

In this section, we discuss certain elements in a desirable knowledge base for mathematics instructors. The goal is not to compare and contrast what we view as desirable and what appears to be instructors' current knowledge base. Rather, we point to certain knowledge base elements, the possession of which by a mathematics instructor can help her or him be a more effective teacher, particularly in relation to the teaching of proof.

An implication of the DNR definition of mathematics is that instruction should focus on both categories of knowledge: ways of understanding and ways of thinking. According to this definition, mathematics includes historical ways of understanding and ways of thinking; hence, it must also include those that might be judged by contemporary mathematicians as imperfect or even erroneous. The notion of "proof scheme" as a way of thinking is intended to capture this stance. A "proof," on the other hand, is a way of understanding—it is one's way of understanding why a particular assertion is true (or false). This subjective stance highlights the student as learner and implies that university faculty must take into account what constitutes ascertainment and persuasion for their students and then offer, accordingly, instructional activities that can help them gradually refine and modify their proof schemes into desirable ones.

The question of critical importance is this: what instructional interventions can bring students to see an intellectual need to refine and alter their current proof schemes into deductive proof schemes? An important source for addressing this question is the development of proof in the history of mathematics. For example, an investigation into the transition from Greek mathematics to modern mathematics and into the role of Aristotelian causality in the development of mathematics during the Renaissance has provided us with insights as to possible epistemological obstacles students encounter with certain deductive proofs (see Harel & Sowder, 2007) as well as to possible instructional interventions to address them.

These instructional interventions are based on the three foundational instructional principles of DNR. The duality (D) principle concerns the developmental interdependency between ways of understanding and ways of thinking: Students develop ways of thinking only by constructing ways of understanding, and the ways of understanding they construct are determined by the ways of thinking they possess. According to the necessity (N) principle, problem solving is not just a goal but also the means—the only means—for learning mathematics. Learning grows only out of problems intrinsic to the students, problems that pose an intellectual need for them. The repeated reasoning (R) principle is complementary to the duality principle and the necessity principle, in that its aim is for students to internalize what they have learned through the application of these two principles. Through repeated reasoning in solving problems—appreciated as such by the students—the application of ways of understanding and ways of thinking become autonomous and spontaneous.

The Necessity Principle

The necessity principle (the N of DNR) is as follows: "For students to learn what we intend to teach them, they must have a need for it, where by 'need' is meant intellectual need, not social or economic need" (Harel, 2007). What are the experiences that constitute intellectual need in mathematics, especially for proofs and for careful definitions? Although we are claiming neither completeness nor uniqueness, here are five categories of such experiences that seem to capture the essence of the intellectual need manifested in mathematical practice: (1) need for certainty, (2) need for causality, (3) need for computation, (4) need for communication, and (5) need for structure. These categories are not static constructs; rather, they have developed through millennia of mathematical practice and are likely to continue to develop in the future. They are inextricably connected, making it very difficult to discuss them separately from each other. In what follows, we provide but a brief illustration for each (a more extensive discussion is underway in Harel, in preparation).

NEED FOR CERTAINTY

When an individual makes an observation he or she conceives it either as a *conjecture* or as a *fact*. A person's observation ceases to be a conjecture and becomes a fact in her or his view once the person becomes certain of its truth. The *need for certainty* is one's desire to render a conjecture into a fact. Proof schemes throughout the history of mathematics have not been static but have varied from civilization to civilization, generation to generation within the same civilization, and community to community within the same generation (see Harel & Sowder, 1998; Kleiner, 1991). For example, the Babylonians merely prescribed specific solutions to specific problems, and so their proof schemes were mainly empirical. The deductive proof scheme—that is, the approach of establishing certainty by deducing facts from accepted principles—was first conceived by the Greeks and continues to dominate the mathematics discipline to these days. However, some historians (e.g., Kleiner, 1991) argue convincingly that during the 16th to 18th centuries there was a decreased attention to the standards of rigor set by the Greeks.

"So you can be certain that it will always be true" is often offered by instructors as a reason for deductive proofs, but book- or instructor-given results are naturally rarely questioned so there is no felt need for certainty. If, contrariwise, instructors plan instruction so that student-generated conjectures arise naturally, it is easier to appeal to a need for certainty.

NEED FOR CAUSALITY

Certainty is achieved when an individual determines, by whatever means he or she deems appropriate, that an observation is true in a particular system. Truth alone, however, may not be the only aim for the individual: he or she may desire to know *why* the observation is true—the *cause* that makes it true. The individual may be certain of the truth of an observation and still strive to understand what liberates her or him from doubt. "Proofs really aren't there to convince you that something is true—they're there to show you why it is true," said Gleason, one of the solvers of Hilbert's Fifth Problem (Yandell, 2002, p. 150). Two millennia before him, Aristotle, in his Posterior Analytic, asserted, "To grasp the why of a thing is to grasp its primary cause." In a particular deductive system, tracing and emphasizing which earlier result or results lead to a later result can be an aspect of causality.

NEED FOR COMPUTATION

In general, humans need to construct objects, to determine their measurements or properties, to transform them into desired forms, or to realize relations among them. Obviously, the nature of the means by which one fulfills such needs depends on particular circumstances and the characters of the objects conceived. In mathematics, one such means is computation: to *compute* is "to determine by mathematical means, especially by numerical methods" (the American Heritage Dictionary). Specifically, the *need to compute* in mathematics refers to one's desire to carry out a symbolic transformation act for the purpose of finding a missing object, constructing an object, determining a value or property of an object, transforming a given object into a desired form, or determining relations among objects. Finding a matrix that transforms one set of given vectors into another set of given vectors is an example, as is the algorithm for expressing the greatest common divisor of two integers as a linear combination of the integers. In addition, the need to compute includes one's desire to minimize or routinize computation.

NEED FOR COMMUNICATION

Students may be satisfied with their intuitive explanation of why $\lim_{n \to \infty} \frac{1}{n} = 0$, which typically is something to the effect, "because the larger n gets, the closer $1/n$ is to 0." A teacher preparing students for the epsilon–N definition of limit might proceed, upon hearing this explanation, by writing it on the board along with the graphs of $f(n)=1/n$ and $g(n)=-1$. Then the teacher may point out to the students that based on their own justification one can rightly argue $\lim_{n \to \infty} \frac{1}{n} = -1$, because, by their own words, "the larger n gets, the closer $1/n$ is to -1". Our experience suggests that this exchange usually results in a sort of a conflict for the students, and they see a need to modify their conception of limit. This is an example of a need for communication and hence for careful definitions. The need for communication is more than just to expose and clarify one's own thinking. It can lead, as this example illustrates, to the development of new ideas.

NEED FOR STRUCTURE

Euclid's geometrical edifice is an example of a structure consisting of a sequence of theorems and proofs, each depending on previous results. The two-millennium attempt to prove the parallel postulate was, in effect, an attempt to perfect that structure. Showing, for example, that any exterior angle of a triangle is greater than either non-adjacent interior angle without using the parallel postulate both highlights the importance of that postulate and suggests that the result holds in other models or contexts. With the addition of the parallel postulate, a central result such as any exterior angle of a triangle is equal to the sum of the non-adjacent interior angles shows that the postulate contributes in making a result sharper (and in other important ways, later). An important question is, when does a need for structure become a need for students? Proof-oriented courses often build structures, but the students may at some point profit from some explicit attention to the nature of such structures, for example, by imparting to the students the responsibility to establish a structure for what they have learned.

Current Knowledge Base of Some Mathematics Instructors

The Interviewees

The views we extracted from interviews of university mathematics faculty illuminate certain aspects of their knowledge base as it pertains to the proving act.[2] We take for granted that as mathematics faculty our interviewees' knowledge base with respect to the first component—mathematical knowledge—is broad and deep in both ways of understanding and ways of thinking. Thus we focus on the other two components of their knowledge base with respect to the proving act: knowledge of student epistemology and knowledge of pedagogy.

Twenty-two faculty members from five universities were interviewed. All were at the associate or full professor level with research areas, collectively, including algebra, analysis, number theory, and topology. The universities were fairly representative, including a large public research university, three large public multipurpose universities, and one midsized private university. Our contacts at the universities identified faculty members reputed to be among the more reflective and more effective teachers at those locations.

The mathematics faculty members were questioned individually. In most cases, they were given a list of seed questions before the interview to orient them to the general thrusts of the interview. Interviews were audiotaped, and noteworthy statements were later transcribed for organization. What appears here as a description of part of the faculty knowledge base is the result of our effort to "glue" many of the responses together, with proof schemes in mind. Adjacent but separate quotes are from different faculty members.

The Importance of Proof [3]

We assumed that all the faculty members would highly value the learning of proof by their students. Yet the faculty's views on the importance of proof for undergraduate students did vary. One position was that some proofs are given out of a "sense of guilt"—that is, we should "do right" by the students and at least show them proofs, or present a deductive development even though that may not be an important goal for a particular course. Another position was that teaching proofs is a "moral obligation," though they are primarily "for the A student," and "[one] should be able to get a (terminal) bachelor's in math without necessarily being able to be a *master* at proving things."

Thus, perhaps surprisingly, university faculty do not uniformly seem to see the deductive proof scheme as an objective for all students, but only for the ablest students in their mathematics classes. That is, apparently for some instructors the focus is on the students' ways of understanding (e.g., as expressed by exposure to the proofs of theorems) rather than including much attention to their ways of thinking (i.e., their proof schemes) beyond the deductive proof scheme, ignoring the likelihood that the students' proof schemes may have been less mature.

Is Proof an Obstacle?

Typical responses to the question, "Is proof an obstacle for the undergraduate major?" include the following: "The majority of the students have trouble with proof," "Their eyes roll back in their heads," "Their expectations overwhelm them, when they hear the word 'proof'," "Students expect proof [in this course], and that's what scares them," and "Many students believe they can't do it." Some faculty mentioned that they even avoid the word "proof" because of its negative connotations, using instead "explain," "argue," or "convince."

Why is proof an obstacle, in the instructors' views? Besides the need for some mathematical maturity—mentioned by several and called by one person "an initiation into a mystical brotherhood"—there were other ideas. To their students: "Proof is not important. Maybe they think it's abstract. Computation is what's important to the students . . ." "Why [I the student] have to do proofs is not clear, even though the student may see value in proofs." A few noted that they themselves had shared similar feelings when they were undergraduates: "I understood proofs okay as an undergraduate, but I did not appreciate their significance. I wanted to *solve* problems, and it was not clear to me that proof was a part of problem solving." Another recalled that in differential equations, he was impressed with the results, but was just accepting of the proofs. He remained unimpressed with the importance of proof until a particular graduate course. Another said: "I've come to believe that students don't understand what mathematics really is, to the point that they don't understand that mathematics is learning to use a set of tools, one of which is logic and reason, and not just imitating examples." We faculty may share some of the blame, according to some: college faculty are "too good at it" to appreciate the students' difficulties. Or, "Unfortunately a lot of faculty members tend to be out of contact with students. The faculty don't view [proof as being] as great an obstacle as the students do. Students may successfully mimic what faculty do but I suspect that there's not much thought going on—and often students write gibberish . . . Faculty have a lot of faith that what they do in class communicates."

Thus, proof appears to be an obstacle for many if not most students, according to our faculty interviewees. Most of the comments were descriptions of external manifestations (that students experience difficulties with proof) and of affect (that these difficulties have to do with students' expectations, anxiety, and self-confidence). Some, however, attended to a possible cognitive cause for students' difficulties with proof, namely, that the students may have no appreciation for the value of proofs in mathematics because they see little or no intellectual necessity for them. Notice the relevance of the necessity principle: how can instructors help their students to see the need for proof in mathematics?

Assumptions about Students' Proof Backgrounds

An important aspect of an instructor's knowledge of pedagogy is to have an awareness of what knowledge and expectations the students have. Most faculty made only mild assumptions about proof knowledge and abilities. Those whose students had had a lower division course in linear algebra usually assumed some familiarity with indirect proof, and those whose students had had a discrete mathematics course assumed exposure to mathematical induction as well as indirect proof, realizing that the students had not actually done much writing of proofs. "When I say I assume they are familiar, I don't say they understand it. They've heard the terms, and seen them done . . . but they haven't *done* much [proof]." A teacher of an upper division transition course felt that "In the beginning, students have no concept of 'axiom' or 'postulate.' Numbers are physical objects, and properties of numbers are *experimental* properties. Proof to them is hardly necessary, and they can't distinguish between which things do or don't require proof." A geometry teacher noted: "[I once] just assumed that they understood indirect proofs—probably naively. I would now be more skeptical about what they know, and I try to listen much more now" in class as well as during office visits.

Thus, there is doubt that any ideas about proof from high school or lower division courses have been internalized as parts of genuine deductive proof schemes. Only an occasional instructor was aware of evidence of empirical proof schemes on the part of the students. Overall, few expected their students to be familiar with axiomatic proof schemes, or even the significance of terms like axiom or postulate. Again, an awareness of students' likely proof schemes and of many students' lack of readiness for a structural approach may help an instructor in planning a course usually devoted to deductive proof schemes and a structure for the topic of the course.

Students and Their Difficulty with Definitions

Perhaps no other part of the interviews aroused such dismay as a question about students' struggles with definitions, an important element in deductive ways of thinking. "They don't learn definitions, they don't value definitions, they don't understand the significance of definitions" and "Good Lord, I wish I knew [why students do not learn definitions] . . ." are typical faculty responses. Although another asks for definitions on tests to stress their importance, in class "it's water off the duck's back." "Even though I have lots of exercises that require checking whether definitions fit, the students still confuse definitions—they'll rotely memorize them and mix up two definitions." One faculty once taught a year-long linear algebra course, giving six examinations, each containing a question asking for the definition of linear independence—and each time some students failed to give a definition!

Why don't the students learn definitions? "They don't understand the basic nature of proof. . . . In part, it's because they don't *understand* the definitions." One outlined the problem this way: "What is done in the math books is not natural. I think we do the formal definitions because we have to. I don't think there's any mathematician who just, well, actually I do know a few, but most mathematicians really don't think their absolute favorite thing is to have some obscure, detailed, page-long definition of something, except you're *forced* to do it. You just need that level of precision if you're really going to prove something."

Faculty, of course, recognizes that definitions are central in deductive proofs, but it also acknowledges that their students do not seem to have an appreciation for them. Some seem to recognize that this is a major source for the difficulty students have in using and remembering definitions. As with proof, a major difficulty that students have with definitions may reside in their lack of appreciation for the *necessity* of definitions.

More About Definitions

Undoubtedly, the role of definitions in mathematics is crucial, but the question is when and how to introduce them so that students appreciate their need. From the proof schemes viewpoint, to appreciate definitions fully one needs to possess an advanced deductive proof scheme, the axiomatic proof scheme. As the label suggests, this proof scheme includes an awareness of the roles of the building blocks of an axiomatic system, including definitions. Attending to definitions can be viewed as a way of thinking, in the sense it is a characteristic of one's act of interpreting concepts in the process of comprehending mathematical content or justifying mathematical assertions. Our faculty interviewees did not seem to be aware that this way of thinking is a cognitive precursor to utilizing and appreciating definitions. A need for a definition should precede a definition, for students with limited ways of thinking.

The distinction between definition and image can perhaps help to sharpen an instructor's understanding of where many students are, in their (lack of) appreciation for definitions. For a given concept, Vinner (1992) distinguishes between the concept's definition and the concept's image:

> The concept image is a non-verbal entity associated in our mind with the concept name. It can be a visual representation of the concept in case the concept has visual representations; it can be a collection of impressions or experiences. [These can often be translated into verbal forms.] But it is important to remember that these verbal forms were not the first thing evoked in our memory when we heard or saw the concept name. (1992, p. 197)

For example, when a student sees the word "function," all that might come to the student's mind is his/her concept image—perhaps particular graphs, $y = f(x)$, $f(x)$ = particular algebraic or trigonometric expressions, linear and quadratic functions, a vertical line test, a table of values, such terms as domain and range, and for more experienced students, terms like one to one and onto, arrows from one region to another, composition, adjectives like continuous and differentiable, and so forth.

A student's concept image is idiosyncratic, it can vary with the context, and it may be flawed, depending on what features of the student's experiences are most prominent in his or her concept image in a particular situation. Edwards and Ward (2004) noted the strong influence of concept images among even their advanced students. Apparently such students do not have a need for definitions, particularly when their concept image seems sufficient for what they are doing, as is the case for many terms in elementary mathematics (e.g., angle, polygon, kite) but also for more advanced terms, as the earlier example for "limit" suggests.

The dominance of the concept image, and its prevalence and usefulness in non-proof settings, is one reason that students have difficulty when deductive proof is a paramount concern—when "the game has changed" from computation to abstract concepts. One faculty member cited a colleague's view that in ordinary English, once you've learned a concept, you can forget the definition, but in mathematics it is very different. That is, in everyday life, concept images work well, but reasoning in proof demands explicit concept definitions.

Definitions may nonetheless play an important role in understanding a concept, for some faculty and presumably for some students. One faculty member mentioned that the definition of subspace "opened the door" for his understanding of subspaces and that learning definitions as a first step seemed reasonable to him. Another felt, by way of contrast, that definition first did not reflect the course of development in mathematics. The fact that in high school work and even in college courses there is often no careful definition for "proof" itself means that students may be guided *only* by their concept images for proof as it appears in mathematics, and possibly even after they have had a body of upper division work.

It is interesting that psychologists' thinking about concepts has also evolved in a way that parallels the concept definition/concept image distinction. At one time deciding whether something exemplified a particular concept was viewed as a checking off of properties of the concept—that is, example identification is based on a concept definition. More recent work (see Medin, 1989), however, suggested that such decisions are often based on comparisons with a prototype or known examples—that is, example identification is based on a concept image. Hence, definition-based learning may not be natural and itself requires some learning. Halmos refers to "the maladroit ineffectiveness of mathematics at matching precise definitions to our intuitive preconceptions" (1985, p. 63), suggesting again the important role of concept images and the difficulty of even creating a concept definition.

One faculty member mentioned telling students to study examples to "shore up their intuition"—that is, in Vinner's terms, to embellish their concept images. If these views of the importance of examples are true for all students, then we have the dilemma, on the one hand, of needing to bolster concept images to aid in understanding concepts and yet, on the other hand, of wishing to emphasize concept definitions for the purposes of proof. Perhaps dealing with this conflict comes with more time spent with proof; one faculty member offered, "Students don't value definitions as much as we'd like them to—it comes as they do more proofs." Or as Sfard puts it, students have "to be sufficiently mature in the mathematical culture to see the role of definitions in mathematics as binding logically and not as *descriptions of certain aspects of an object otherwise known by sense or insight*" (1992, p. 47, emphasis added to highlight the relevance of concept images and the influence of empirical proof schemes).

What Do Instructors Do to Help?

The responses to the question, "Are there any things you do in particular to 'teach' proof making? Proof appreciation?" were much less fruitful than expected, although we did not expect anything so principled as the DNR framework to appear. Some ideas—pointing out that mathematics is like a language, giving the idea of a proof before giving it, working a specific example before the proof, reviewing crucial points of logic, describing generic templates for different proof forms, including false steps in proofs for students to find—might indeed help students. Other ideas, such as using pop quizzes for definitions or excusing high scorers from the final examination, are not in the spirit of this chapter. Overall, in our terms, the instructors appeared less concerned with ways of thinking than with ways of understanding. We should acknowledge that we have no direct observational evidence of how the teachers actually did teach, in general.

Is It Just a Matter of Experience?

All the opinion and other evidence suggest that proof is troublesome for many students entering upper division work. And instructors often proceed without well-articulated, principled guidance for their pedagogy. It is natural to wonder whether with more and more proof exposure under their belts, students do nonetheless eventually become comfortable with the role and value of proof in mathematics and do become able to write proofs—"There really is a thing called mathematical maturity," or "Practice helps students get better at proof," or "Some do reach a breakthrough in proof writing." After all, by the conclusion of their degree work, students will have experienced deductive proof in several settings and might well have profited from the repeated exposure.

Some faculty members noticed quantum jumps within one course: it is "apparent in abstract algebra, where some catch on at once, and for others—after several weeks—the light turns on. Maybe if there were two weeks more, the light would have turned on for more." Some faculty members had themselves had "lights turn on." One mentioned that, although he had been getting good grades in abstract algebra, he did not really understand the material until studying for the final examination: "All this stuff [homomorphisms, isomorphisms] kind of fell into place. It hit me like a ton of bricks. It was like a veil had been lifted from my eyes!" One college geometry instructor has been fairly pleased with his students' ability with proof by the end of his courses, but he noted that at the start they tend to be guided by their visual images and to miss the point of axioms.

Some faculty members expressed opinions that many students finishing the coursework in the mathematics major would have at least a fair grasp of proof: "It's a matter of how much time you spend at something. A lot of these people are going to catch on pretty well." "A fair number can handle proofs, especially shorter ones or ones not involving a 'trick'." But this latter faculty member also felt that "there would be a large number for whom the word 'proof' still induces fright." Some were also less sure that, at least within one course, students adapt completely: "To a degree. It depends on your expectations." As the semester goes on, "They get better at the basic ideas, but the trouble is, the material gets harder." In another's opinion, some students "can't handle it. Once you get to graduate math courses, the kid gloves come off. Nobody starts kidding anybody about anything, that you don't have to know proofs." Still another says, "Perhaps not even OK by the end of the semester. They know they can get through—there's enough computation stuff—but I'm not sure that *all* finishing majors can write proofs. A few will be able to write decent proofs." Although some felt that students who had had other exposures to proof-oriented courses did perform better than those in their first such course, others were less certain, feeling that proofs in abstract algebra, for example, are so different from those in advanced calculus that proof *performance* at least might be independent of background—"Algebraists regard proofs in analysis as ad hoc, and vice versa." Case-studied students in one report (Sowder & Harel, 2003) were collectively at virtually every point in their proof schemes, from authoritative to empirical to deductive.

Blaming the messenger is not in the spirit of this chapter, but the question of the students' ability and motivation did arise. Most of us have probably encountered students who have made us give a silent "Yes!" to this faculty member's rhetorical question, "Is teaching proof [to some students] like teaching music to the tone deaf?" Or perhaps this view: "Can or can't—I'm in both camps. Given enough time I believe most can master it, but there exist people that, given two lifetimes, I'm not sure they could." Or "30 to 40% get it. If I really pounded on [an abstract topic], it would be maybe 42%." Whether students' motivations are adequate or proper also came up. "Students come in now, and the object of learning is to pass the test, and not to think of it as something that is going to be used." Learning for the long term doesn't seem to be a goal. One faculty member noted that he himself seemed to have trouble with long-term memory for areas of mathematics that were of low personal priority to him, and he saw something like this as a problem for some of his students. That some students come in with, and continue to have, a get-answers image for mathematics would also explain their difficulties in attaching importance to proof.

Blaming the students is at the lowest level of the four levels in a "spectrum of pedagogical awareness," based on repeated interviews of six mathematics tutors, with "tutor" to be taken in the European sense (Nardi, Jaworski, & Hegedus, 2005). So it is encouraging and may speak to the quality of the sample of faculty interviewed that there were more optimistic views—"I just hate writing people off." "Students make mistakes, but there are reasons behind their mistakes." "Some people catch on a lot faster than others as to what's expected. Others could, I think, if given more time." One interviewee looked at the faculty: "The problem of [student ability] has as much to do with [the instructor's] personality as with whether the students have 'it' or don't. There's a bell-shaped curve [for instructors], but there's a pretty big group in the middle that have a perhaps elitist view about mathematics. I'm not saying everyone does, but there's an elitist view that only a select few can do math. There's a temptation to want to spend time only on the best students."

Thus, there is mixed opinion about whether finishing undergraduate majors have reached a satisfactory level of proof knowledge and proof schemes, perhaps because a word like "satisfactory" can have different meanings. Although some faculty had noticed substantial improvements within one course for some students, others were naturally dubious about giving an unreserved stamp of approval for finishing majors.

Closing Statement

With Pólya's (1962) division of problems into problems-to-find and problems-to-prove, it is not surprising that giving proofs, a form of problem solving, is difficult for undergraduate mathematics majors—and for the mathematics researcher: "The problem . . . is that there's no way of proving a theorem to order. If someone comes up with a gun and says, 'Prove this theorem,' well, then, you're just dead." (John Conway, in Kolata, 1993). But it is nonetheless grating and deflating when a mathematics major says, "I like math except for the proofs."

The aphorism, "The good student is one who can fill in the gaps in instruction," is overly simplistic, but there is no loss and a possible gain if we do look for some of those gaps in our instruction. We think that an instructor's awareness of the variety of ways of thinking as exemplified by the proof schemes, some of which are not the kind that will be properly featured in an upper division mathematics course, and of the DNR principles in general, can help the instructor address some of the students' deficiencies vis-à-vis deductive proof schemes and to answer the key question: what instructional interventions can bring students to see an intellectual need to refine and alter their current proof schemes into deductive proof schemes?

Notes

1. For a more detailed discussion of DNR, see Harel, 1998, 2001, 2007, in press.
2. Collections of opinions such as the ones we note can only be suggestive; they naturally are limited to those interviewed, and there can be no claim of generalizability to the whole community of mathematics professors.
3. "Proving" is a mental act, as was defined above. Recall that while a "proof" is a particular product of the proving act—a way of understanding—a proof scheme is a characteristic of this act—a way of thinking. We are interested in the faculty's views about students' proof schemes, including their teaching actions aimed at advancing the deductive proof scheme among students. Thus, the phrase "importance of the deductive proof scheme" is more consistent with these definitions than the phrase "importance of proof" used here. The latter phrase, however, matches the language used by the interviewees.

17

Understanding Instructional Scaffolding in Classroom Discourse on Proof[1]

MARIA L. BLANTON, DESPINA A. STYLIANOU,
AND M. MANUELA DAVID[2]

Although proof is widely accepted as central to mathematical thinking, many students find proofs difficult to develop and understand. Research has significantly advanced our understanding of how students learn to read and construct proofs through studies on the learning of proof as an individual, cognitive construction (e.g., Harel & Sowder, 1998; Porteous, 1986; Senk, 1985). What is also needed is an understanding of the teaching and learning of proof *as a social process*. Taking a distributed view of mind as a collective embodiment of classroom activity, Herbst and Balacheff (Chapter 3, p. 40) maintain that "public activity in mathematics classrooms embodies an epistemology" whose understanding is essential in knowing how to craft instruction that fosters the learning of proof. This chapter aims to address the social aspect of teaching and learning proof by providing a framework for analyzing teacher and student utterances in classroom discourse[3] and how these public utterances scaffold student learning of proof.

The Social Aspect of Proof

Proof is ultimately a socially constructed object whose purpose is to communicate the validity of a statement to a community based on criteria established by that community. Even within the mathematics community proof is at the center of mathematical discussions that allow an argument to evolve in a dynamical manner (Hanna, 1990; Lakatos, 1976), although that argument is subject to formal rules agreed on by that community. Moreover, an accepted proof may be found flawed at a later point in time, or may be questioned as the criteria and premises used by the community shift. As such, proof and the criteria used to judge its validity might be construed as negotiable objects.

Despite this, teaching proof has often been limited to demonstrating completed proofs to students. Little of the kinds of discussion and communication inherent in how mathematicians develop proofs are reflected in these demonstrations. It is not surprising, then, that only a small percentage of students achieve an understanding of proof (e.g., Chazan, 1993; Senk, 1985) and that the formal teaching of proof is not effective (Schoenfeld, 1986). Davis (1985) suggests that the use of demonstrations of completed proofs obscures from students the social aspect of creating a proof. What, then, is the effect of instruction that, instead, embeds the development of proof in the social activity of the classroom, where proofs are constructed by, not demonstrated to, participants? The study reported here is a first attempt in our work to understand what that instruction might resemble and how we might interpret learning in that context.

Using a Sociocultural Perspective to Study Teaching and Learning Proof

Discourse as a Lens for Understanding Teaching and Learning Proof

Our work draws on the view that learning, an embodiment of institutional, cultural and historical experiences, cannot be understood apart form the social context in which it occurs (Vygotsky, 1962/1934). Moreover, Vygotsky maintained that "higher voluntary forms of human behavior have their roots in social interaction, in the individual's participation in social behaviors that are *mediated by speech*" (Minick, 1996, p. 33, emphasis added). On the significance of speech in development, Halliday reasoned that the "distinctive characteristic of human learning is that it is a process of making meaning—a semiotic process; and the prototypical form of human semiotic is language" (1993, p. 93). This social view of mind and the mediating role of speech in development is extended in the work of Bakhtin, who argued that "utterances are not indifferent to one another, and are not self-sufficient; they are aware of and mutually reflect one another . . . [and] every utterance must be regarded primarily as a *response* to preceding utterances" (1986, p. 91).

The notions of socially constructed knowledge, speech as a psychological tool of development, and utterance as a unit of analysis constitute the guiding principles in this study. They suggest that the ability to understand and construct proofs is shaped by the social context in which learning occurs and that the speech occupying that social context encodes a story of development that can be deconstructed through discourse analyses which treat utterance as "the *real unit* of speech communication (Bakhtin, 1986, p. 71).

Conceptualizing Teaching and Learning through Instructional Scaffolding and the ZPD

Vygotsky postulated the *zone of proximal development* (ZPD) as a way to conceptualize learning. The ZPD is defined as the space characterizing one's potential for development through the assistance of a more knowing other (Litowitz, 1993; Vygotsky, 1962/1934). As a diagnostic, the ZPD intends to assess not only those cognitive functions that one possesses, but also those that are in the process of development by virtue of the learner's interaction with more knowing others, cultural tools, and so forth (Kozulin, 1998). Since learning is viewed as a product of social interaction, it follows that one's development within the ZPD is affected by the intellectual quality and developmental appropriateness of these interactions (Diaz, Neal, & Amaya-Williams, 1999). That is, one's development within the ZPD is predicated in part on how the more knowing other organizes, or *scaffolds*, the task at hand. Thus, understanding development within the ZPD, entails understanding how instruction scaffolds learning.

Instructional scaffolding is a mechanism for observing the process by which the learner is helped to bring about his or her potential learning (Stone, 1993). Practically speaking, it refers to the "provision of guidance and support which is increased or withdrawn in response to the developing competence of the learner" (Mercer, 1995, p. 75), and it is based on the appropriation, not simple transfer, of ideas between teacher and student. However, understanding the subtleties by which this appropriation occurs is a complex process that requires sensitivity to the learner's goals as these goals emerge in the course of activity (Wells, 1999). For example, knowing how to give hints that focus and challenge students' thinking requires a deep knowledge of students' individual learning capacities with respect to the task at hand. It presupposes that the teacher is continuously attending to students' thinking in order to access their ZPD. The complexity increases for the teacher because hints are often given in large group settings (e.g., whole class discussions) that necessarily conceal individual differences and thus diminish the teacher's capacity to attend to them.

From our theoretical framework, we came to view the nature and effect of instructional scaffolding as a critical part of understanding how students learn to construct proofs. Thus, we aimed to develop a framework that would allow us to critically examine instructional scaffolding in classroom discourse on the development of proof. Such a framework should allow the researcher to study forms of instructional scaffolding, including the possible ways in which the teacher may scaffold student thinking and evidence of student development within the ZPD. In particular, we considered the following questions:

1. What are the forms of instructional scaffolding in classroom discourse on the development of mathematical proof?
2. How do teacher utterances scaffold student thinking and what counts as evidence of student development within the ZPD?

Background

Data Corpus

Our work is based on data collected from a 1-year classroom teaching experiment conducted in an undergraduate discrete mathematics course that emphasized learning to develop mathematical proofs. The course consisted of approximately 40 undergraduate mathematics and computer science majors. Classroom instruction was videotaped and small group discussions were audiotaped. Pre- and post-assessments, given to identify the progression in students' arguments, provided evidence for shifts in students' capacity for self-regulatory thinking (Blanton & Stylianou, 2002). The study reported here focuses on verbal data collected during whole-class discussions and small group episodes.

Instructional Context

The teacher for the course, one of the authors, worked to establish expectations that students explain their reasoning and make sense of and challenge each other's explanations and justifications. A typical lesson began with a problem introduced by the teacher and was followed by small group proof construction alternating with whole class discussion of students' thinking. Throughout the course, there was a concerted focus on both written and verbal expression of student thinking and alternative forms of assessment.

Data Analysis for the Development of the Framework

Transcripts of whole class and small group discussions were selected based on their focus on proof construction[4] and analyzed by the research team. For whole class discussions, each teacher utterance was coded independently by the authors to determine the instructional intent of the utterance. The distinct forms of instructional intent ultimately comprised the coding scheme. Codes were mutually negotiated in order to develop and refine a framework describing the nature of instructional scaffolding in whole class discourse. To illustrate the framework, we focus here on a detailed analysis of one classroom episode, where the instructional goal was to prove that the square root of two is irrational. The 90-minute episode occurred during week 4 of the first semester and spanned two class periods. It was selected because it occurred at a point in the semester when students were beginning to develop the rudiments of proof technique and because it highlights what we see as an important type of whole class discussion. While it was chosen to illustrate our framework, the nature of the episode in terms of teacher–student interaction was fairly representative for this class.

In keeping with Bakhtin's argument that utterances be considered in relation to other utterances in speech communication (1986; see also Wertsch, 1999), we then developed a coding scheme to describe the cognitive nature of student utterances in the context of already coded teacher utterances. We used this scheme to look for evidence of student development within the ZPD. We then looked for connections between student development and instructional scaffolding by examining how teacher and student utterances mutually interacted with respect to the two coding schemes.

Finally, from the theoretical perspective that students begin to internalize the language of proof and argumentation as they actively participate in whole class discourse, we then used the coding scheme characterizing student utterances to examine how the structure of *small group* conversation in one episode occurring near the end of the second semester resembled the *whole class* discussion analyzed here.

Description and Discussion of the Framework

Nature of Instructional Scaffolding

We based our framework for discourse analysis on the work of Kruger (1993) and Goos, Galbraith, and Renshaw (2002). In particular, we found Kruger's (1993) use of *transactive reasoning*, characterized by clarification, elaboration, justification, and critique of one's own or another's reasoning helpful in initially identifying each person's contribution to the collaborative structure of the whole class interaction. However, it became apparent that the teacher's utterances, generated by a more knowing other with instructional intent, were fundamentally different than those of students, and analysis of whole class discussion had to attend to the dynamic created by the different purposes of the speakers. Since the work of Kruger (1993) and Goos et al. (2002) focused on peers, we needed to extend their work to account for the idiosyncratic ways in which the teacher participated in discourse. Moreover, attending to differences in teacher and student utterances was not only essential to interpreting the teacher's utterance, but also to understanding how utterances (from both the teacher and students) extended students' development within the ZPD.

Our analysis suggests that the teacher's utterances could be categorized in one of four ways:

- transactive prompts
- facilitative utterances
- didactive utterances
- directive utterances.

In what follows, we discuss each of these and then examine how they functioned as instructional scaffolding.

TRANSACTIVE PROMPTS

We define *transactive prompts* to be a form of scaffolding in which the teacher's utterances are intended to promote transactive reasoning in students. Accordingly, we define transactive prompts by the teacher to consist of *requests for* critique, explanations, justifications, clarifications, elaborations, and strategies, where the teacher's intent is to prompt students' transactive reasoning.

We claim that transactive prompts are a critical part of discourse because they serve as a scaffolding tool to build practices of argumentation necessary in the development of proof. We see argumentation as a process of higher mental functioning developed first socially between people and then internalized as part of one's private speech. Thus, we see one's ability to develop private mathematical arguments as a function of the public development of argumentation. It follows that the way in which students interact with mathematical ideas in the classroom, whether passively or actively, structures their (private) thinking about mathematics (Wertsch & Toma, 1995). Thus, when a teacher gives a transactive prompt, she is potentially shifting students' cognitive stance from passive to active, from accepting to questioning, and these transactive prompts scaffold the development of structures of argumentation.

We found that an important characteristic of the teacher's transactive prompts is that they were in the form of questions that required an immediate response from students and served the teacher's overall goal to direct students towards transactive reasoning. In other words, the teacher used transactive prompts to either initiate a discussion or redirect one already in place. Because one of the teacher's main aims was to help students develop a capacity for public, and subsequently private, argumentation when reading or writing proofs, transactive prompts were used as a tool towards achieving this goal.

The following excerpt from the whole class discussion on the proof that the square root of two is irrational illustrates this form of instructional scaffolding. As the excerpt opens, students are investigating the claim that $2q^2 = p^2$:

4[5]	*Darrell*[6]	It fails for odd numbers.
5	*Teacher*	It fails for odd numbers.
6	*Teacher*	What do you mean? [*transactive prompt—request for elaboration*]
7	*Darrell*	[. . .] If p is odd, then p^2 is going to be odd.
8	*Teacher*	OK, hang on a second.
9	*Teacher*	[Speaking to the class] Do you buy that? [*transactive prompt— request for critique*]
10	*Adam*	It's true.
11	*Teacher*	Why is that true? [*transactive prompt—request for justification*]

In this protocol, the teacher offers transactive prompts directly requesting students to engage in transactive reasoning of three types: (1) to elaborate and further explain their ideas and suggestions (line 6); (2) to critique an idea or strategy presented to the class (line 9); and (3) to justify their claims (line 11). These requests are part of the teacher's long-term goals to aid students in developing a habit of argumentation when constructing a proof. In the protocol that follows, the teacher provides a transactive prompt for clarification (line 31):

28   *Degan* $2q^2$, the 2 just being a scalar [times] q^2. So if we just temporarily

forget that it's there. Temporarily. Well, we're going to see that q^2 = p^2. Well, how can something squared equal the same thing squared times a scalar?

[. . .]

31 *Teacher* So you're trying to say how can it ever be because you've got a scalar times p^2, or times q^2? [*transactive prompt—request for clarification*]

FACILITATIVE UTTERANCES

We use this characterization to describe instances when the teacher re-voices or confirms student ideas, or attempts to structure classroom discussion. The teacher might repeat or rephrase a student utterance, thereby lending authority to students to proceed along the student-suggested path. Structuring involves summarizing a discussion, pacing a conversation, or redirecting an utterance to focus students and direct their ideas or arguments.

When using facilitative utterances, the teacher shares responsibility for the proposed action through tacit approval, while transactive prompts indicate a transfer of responsibility from the teacher to the students. A second difference between the two types of utterance is with respect to the immediate impact on the conversation. When using transactive prompts, the teacher may redirect the classroom discussion by requesting students to supplement or alter their reasoning. However, as the following protocol illustrates, the teacher's role in scaffolding through facilitative utterances is less proactive:

12 *Anthony* We could prove that an odd times an odd is an odd.
13 *Teacher* Yeah, we could do something like that. [*facilitative—confirmation*]
[. . .]
15 *Degan* We already know that 2 times any integer is going to be an even number anyway.
16 *Jacob* That's what I was going to say. The left side $(2q^2)$ is always even.
17 *Teacher* OK, So here's something $(2q^2)$ that's always going to be even, so you're saying that if p is odd, [then] p^2 is odd. [*facilitative—re-voicing*]

In lines 13 and 17, the teacher's utterances are intended to support students as they take initial steps in their reasoning. Line 8 provides an example of facilitative structuring, as the teacher slows the pace of the discussion by the utterance "OK, hang on one second." Rittenhouse (1998) argues that, just as in learning a new language, novices in mathematical discourse may face difficulties in participating in a discussion that occurs in "normal" speed. However, when the discussion is slowed down via the use of repetition and rephrasing of what has been said, the likelihood that novices can understand the discussion increases. In line 17 the teacher re-voiced the students' utterance as a statement of affirmation, not as a question to which the student needed to respond.

DIDACTIVE UTTERANCES

We define didactive utterances to mean teacher utterances on the nature of (mathematical) knowledge. Students may not be obvious participants in conversation and the teacher's utterances are not ideas to be negotiated. That is, the teacher brings to the conversation big ideas such as the nature of mathematics, axioms and principles, or historically developed ideas that students are not expected to reinvent. The following excerpt illustrates this type of utterance:

33 *Teacher* Well, so in your intuition you may say it can never be [referring to claim in line 28], but mathematically you have to say why it can never be [true]. [. . .] So at some point, one of the things we really haven't addressed in here is when is a proof a proof? And at some point, somebody might accept that and say that kind of argument would be OK. But then somebody else might say, "OK, I want to push you further than that." [*didactive*]

Unlike the previous categories of teacher utterances, here the teacher appeared to "step out" (Rittenhouse, 1998)[7] of the role of scaffolding student-generated ideas and moved into a more explicit role of most knowledgeable person. In transcripts not included here, she continued by providing explanation on the nature of mathematical proof and, in particular, "proof by contradiction" as a logical technique. The teacher's utterances are (tacitly) not open to discussion or negotiation; as the most knowledgeable person on the topic, the teacher clarifies the norms of the discipline with respect to proof.

DIRECTIVE UTTERANCES

Directive utterances are utterances that provide students with either immediate corrective feedback or information towards solving a problem. The teacher follows her own line of thought rather than attempt to engage students in some form of inquiry. Unlike didactive utterances, with directive utterances students could develop the particular notion under discussion through transactive prompts, but the teacher makes the choice to *tell directly* rather than elicit information indirectly. The intent of the teacher could be to accelerate the process of solving a particular problem when she is confident that with extra help the students could do it by themselves, or perhaps more commonly, directive utterances reflect the "teacher as teller" instructional paradigm. From our data, the teacher's utterance "Since we're doing a proof by contradiction, we should find a flaw in this" was classified as a directive utterance. Because students were already familiar with "proof by contradiction," the teacher could have questioned students to elicit the next step in the process; instead, she chose to give the information directly.

Frequency and Structure of Teacher Utterances

As part of our analysis of teacher utterances, we also analyzed the data from this classroom episode further by recording the frequency with which each type of teacher utterance appeared in the entire classroom episode. Percentages were calculated as the number of instances of each utterance type with respect to the total number of teacher utterances. As is shown in Table 17.1, teacher utterances were mostly facilitative (47%) or transactive (40%). The remaining 13% of the utterances were directive or didactive. Thus, we claim that when the teacher participated in the classroom discussion, her role was primarily to direct students towards transactive reasoning about mathematical proof.

We further analyzed teacher utterances by looking for structure in the overall coding. A more detailed examination suggests a general ordered pattern with which the four types of scaffolding utterance were used in the teacher's guided development of the proof. After initially posing the problem, the teacher tended to open the discussion to the students and follow a pattern of facilitative–transactive utterances in response to student comments. In particular, the teacher opens the classroom discussion with a facilitating statement followed by a transactive prompt and from that point on shapes the discussion around the thinking of the students. Once a student offers a suggestion or shares her or his thoughts with the class, the teacher proceeds to provide a facilitative response (mostly to help structure the conversation), followed by a transactive prompt to the class to follow up on the student comment. When students respond, the teacher again contributes to the discussion with a pair of facilitative–transactive utterances. The pattern of student comment followed by teacher facilitative or transactive utterance reflected spiraling towards an increasing number of student comments. From this, we maintain that the structure identified in this episode is qualitatively different than others such as the IRE structure identified by Mehan (1979), and reflects a continuum in discourse from authoritative to internally persuasive (Bakhtin, 1981; Wertsch, 1991). The positioning of the teacher utterances enabled us to visualize classroom discourse as a continuum between authoritative and internally persuasive dimensions, where the responsibility of conversation transitioned from teacher held (authoritative) to teacher shared to student held (internally persuasive). Utterances with greater potential for initiating or

Table 17.1 Frequency of forms of instructional scaffolding (teachers)

Type of teacher utterance	Frequency	Percentage
Transactive prompt	37	40
Facilitative utterance	43	47
Directive utterance	7	8
Didactive utterance	5	5
Total	**92**	**100**

continuing an internally persuasive discourse (e.g., transactive prompts) were predominant in the episode (see Table 17.1).

Wertsch (1991) explains Bakhtin's distinction between these two categories of discourse "in terms of the degree to which one voice has the authority to come into contact with and interanimate another" (Wertsch, 1991, p.78). According to Bakhtin; "The authoritative word is located in a distance zone, organically connected with a past that is felt to be hierarchically higher. It is, so to speak, the word of the fathers. Its authority was already acknowledged in the past. It is a prior discourse" (Bakhtin, 1981, p. 342). Having said that: "The internally persuasive word is half-ours and half-someone else's. Its creativity and productiveness consist precisely in the fact that such a word awakens new and independent words, that it organizes masses of our words from within, and does not remain in an isolated and static condition" (Bakhtin, 1981, p. 345).

Our analysis raises a larger issue concerning how different frequencies and structures in teacher utterances affect student learning. What, for example, can be said about student understanding of proof when teacher utterances are largely transactive prompts or facilitative utterances (as they are here) for a period of weeks, months, or even years? Or, if teacher utterances are largely directive, how does this scaffold structures of argumentation in students' thinking? As such, the instructional scaffolding framework described here potentially gives us a way to think about how particular types of utterances support students' capacity for proof.

Student Development within the ZPD: Establishing the "Scaffolding" in Teacher Utterances

We have just described four forms of instructional scaffolding identified in one classroom episode on proof. However, while we have termed these forms as scaffolding, some analysis is required in order to establish to what extent and in what ways the utterances did, in fact, scaffold students' thinking. Thus, we turn our attention to identifying evidence of student development within the ZPD by analyzing student utterances and their connection to teacher utterances.

As a starting point for our analysis of student utterances, we considered the analysis of metacognitive and transactive functions of student utterances by Goos et al. (2002). They identified metacognitive acts as when "new ideas" were suggested or when an "assessment" of aspects of the solution was given.[8] Their notion of transactive reasoning is consonant with our use of the term here.

From our analysis and building on Goos et al. (2002), we delineated five forms of student utterance[9] that we took as evidence for development within the ZPD:

1. *Proposal of a new idea.* This occurs when a student brings to the discussion new and potentially useful information. It can be in the form of a new concept for which the student notices a connection with existing ideas, or a new representation that may potentially reveal a different aspect of the existing information, an extension of an idea, or even an elaboration of an existing idea towards a

new direction. New ideas can be correct or incorrect, as long as they are relevant to the proof being developed.

2. *Proposal of a new plan or strategy.* This classification is applied when a student suggests a potentially useful strategy or plan for developing a proof or some aspect of the proof. In essence, it represents a course of action. We differentiate it from "proposal of a new idea" because new ideas might not entail a plan or strategy.

3. *Contribution to or development of an idea.* This type of utterance adds to existing ideas and is often made by students other than those who made the initial suggestions, thus indicating that suggested ideas are embraced by others in the class.

4. *Transactive questions.* These are students' requests for clarification, elaboration, critique, justification, or explanation of their peers' utterances.

5. *Transactive responses.* Transactive responses are either direct or indirect responses to explicit or implicit transactive questions and serve to clarify, elaborate, critique, justify or explain one's thinking. They provide an assessment of the potential usefulness or execution of an idea or strategy and are made in response to a transactive question or prompt.

Note that transactive responses are not necessarily distinct from proposing a new idea or strategy or contributing to the development of new ideas. As we will see in the data that follow, it could be the case that one's transactive response comes, for instance, in the form of a proposal for a strategy or new idea. However, for purposes of analysis, it was helpful to identity whether or not utterances specifically brought new ideas or strategies—or contributions to these—into the conversation.

In addition to these categories, we identified a sixth code as *general confirmation* to characterize students' utterances of agreement (e.g., "yeah," "true," or "OK"). However, we found that these utterances did not convey sufficient evidence for determining how and if a students' ZPD had been accessed. Finally, utterances by either the teacher or student that did not directly relate to the task (e.g., "May I use a calculator?") were not coded.

Using this scheme, we coded student utterances in the entire whole class episode. We illustrate the results of the coding and analysis of teacher and student utterances from an excerpt of this episode. The analysis focused on characterizing the structure of dialogue surrounding teacher utterances in the whole class discussion in order to understand how the teacher was able to scaffold student thinking and to identify what counted as evidence for student development within the ZPD:

11	*Teacher*	Why is that true ["$2q^2 = p^2$" fails for odd numbers]? [*transactive prompt—request for justification*]
12	*Anthony*	We could prove that an odd times an odd is an odd. [*transactive response—proposal of new plan*]
13	*Teacher*	Yeah. We could do something like that. That would certainly work. [*facilitative—confirmation*]

14	*Teacher*	That would be a more general case in fact, instead of a particular case. [*direct*]
15	*Degan*	We already know that 2 times any integer is going to be an even number anyway. [*building on or developing a suggested plan*]
16	*Jarrod*	That's what I was going to say. The left side [$2q^2$] is always even. [*building on or developing a suggested plan*]
17	*Teacher*	OK, so here's something [$2q^2$] that's always going to be even. [*facilitative—confirmation*]
18	*Teacher*	So you're saying that if p is odd, [then] p^2 is odd, so you'd have an odd number equal to an even number? [*transactive prompt— clarification*]
19	*Jarrod*	Yeah. [*general confirmation*]
20	*Teacher*	True. So if p is odd, it fails. [*facilitative—confirmation; re-voicing*]

In attempting to scaffold students towards a construction of the proof, the teacher offers here four forms of utterance: (a) requesting justification—*transactive prompt*, (b) re-voicing and/or confirming a student utterance—*facilitative utterance*, (c) requesting clarification of a student's idea—*transactive prompt*, and (d) a *directive utterance*. To answer whether and how these utterances scaffolded students' thinking, we needed to look at their utterances. The episode just presented illustrates two types of student utterance: *development of a suggested plan* and *proposal for a new plan*. The first case is illustrated in lines 15 and 16 as students responded to the teacher's transactive prompts[10] (hence, the utterances can also be seen as transactive responses). In each case, students were elaborating ideas initially generated by *other* students. For example, Deagan's claim (line 15) is an extension of Anthony's claim (line 12). We take this as evidence that Deagan's own thinking is furthered by Anthony's claim and the teacher's transactive prompt. Moreover, while Deagan does not yet have a complete plan to develop the proof, he is beginning to make connections between the existing situation (the need to prove that $\sqrt{2}$ is irrational) and his existing mathematical knowledge and skills in proving. Hence, we claim that Deagan's (and similarly, Jarrod's) utterances provide evidence that the classroom discourse has accessed his ZPD. However, we suggest that Jarrod's *general confirmation* (line 19) does not necessarily provide evidence of further ZPD *advancement*, but rather an acknowledgement of his cognitive status.

The case of Anthony, who brought to the discussion a new idea (line 12), falls in a different category of utterance. While it is difficult to identify what triggered Anthony to suggest this new idea, we maintain that his utterance supports the claim that classroom discourse had accessed his ZPD. Not only did he participate in the classroom dialogue, but he also advanced the discussion by suggesting a new idea for constructing the proof.

We maintain that students' proposals of new ideas or strategies and plans and their subsequent elaboration and justification of them in a way that furthered the construction of a proof indicate their development within the ZPD. Results of a written pre-/posttest analysis administered individually suggest that prior to instruction, students were unable to construct this proof or even more basic proofs dealing with generalizations about sums of even numbers and odd numbers (Blanton & Stylianou, 2002). However, the teacher's transfer of the proof responsibility through transactive prompts and facilitative utterances supported students in making significant contributions to the proof. By re-voicing and confirming student-originated ideas (lines 13, 17, 20), we claim the teacher tacitly shared responsibility for the proposed action with students, and as the more knowing other, lent authority and confidence to students to proceed with their thinking. By prompting students to engage in transactive discussion (lines 11, 18), the teacher transferred responsibility for constructing the proof to students, thus shifting the discourse structure from authoritative to internally persuasive.

As with teacher utterances, we found it useful to analyze the frequency of student utterances in relation to the entire classroom episode. As is shown in Table 17.2, the majority of student utterances were elaborations of existing ideas (40%), while 12% of their utterances were proposals of new ideas and plans, and almost one-fifth (18%) consisted of critiques and questions. In summary, about 70% of student utterances indicated active participation in the collective construction of proof.

Tables 17.1 and 17.2 suggest that new ideas and strategies were contributed primarily by students, while the teacher's role (even when using directive utterances) was to facilitate the discussion. The question that arises then is what causes students to suggest new ideas and plans? Can we attribute these new ideas to a specific type of utterance by the teacher or even a fellow student? In an attempt to gain some insight into this question, we examined the chronological sequence of codes (including both teacher and student utterances) in the transcripts. However, the data did not suggest any identifiable pattern. New ideas and plans appeared to be generated randomly throughout the discussion and were not always connected to the specific utterance preceding it. The lack of any

Table 17.2 Frequency of forms of instructional scaffolding (students)

Type of student utterance	Evidence of ZPD access	Frequency	Percentage
Proposal of a new idea	Yes	3	4
Proposal of a new plan	Yes	6	8
Contribution to an existing idea	Yes	29	40
Transactive response	Yes	7	10
Transactive questions	Yes	6	8
General confirmations	Not necessarily	9	13
No code	No	12	17
Total		**72**	**100**
Total utterances indicating ZPD access		**51**	**70**

discernable pattern should probably not surprise us; classroom discussions are complex events where it is hard (if not impossible) for the researcher to identify to which previous utterance any new utterance is linked. In other words, when students produce a new idea, this could be linked to the immediately preceding comment, to any of the other comments that preceded it during the episode, or to discussions during previous classes. In this sense, individual and small group interviews might allow us to trace the origin of each thought and search for patterns in the development of a solution or a proof with more detail than a classroom discussion allows. However, in these settings we lose the complexity of whole class instructional scaffolding and how they support students' development of proofs.

Furthermore, as discussed earlier, Goos et al. (2002), in their study of collaborative problem solving, concluded that successful problem solving is characterized by a high incidence of transactive and metacognitive transacts. Because proofs are constructed through carefully developed "new ideas," we hypothesize that successful proof construction in a collaborative setting is also characterized by a high incidence of transactive reasoning. However, in this classroom setting we observed dense transactive reasoning and new ideas and elaborations on new ideas tightly interwoven throughout the discussion, making it difficult for us to isolate specific utterances that led to new ideas. Further research is needed to explore this issue.

Small Group Discourse: Moving Beyond the Teacher's Guidance

Concerning the notion of scaffolding, Stone (1993) notes that little attention has been paid to the mechanism by which the transfer from mentor to student is accomplished. This is particularly true of research on instructional scaffolding in undergraduate mathematics settings. As such, the intent of this study was to establish a framework for interpreting whole class and small group discourse as a way to understand forms of instructional scaffolding and how that scaffolding supported students in learning to construct proofs. We conjecture that those prompts that encouraged transactive discussion, that drew students towards presenting new ideas, or clarifying, justifying, elaborating, and so forth, were most crucial in students' construction of proof. That is, these forms of scaffolding seemed to be the primary mechanism by which students' ideas were publicly identified and negotiated. Moreover, students' capacity to engage in the conjecturing and negotiation of mathematical ideas seemed to be scaffolded mainly through the teacher's transactive prompts and facilitative utterances as opposed to directive or didactive utterances. As such, the forms of instructional scaffolding operating in the classroom studied here—transactive prompts, facilitative utterances, directive utterances, and didactive utterances—seemed to function differently in how they were able to extend student thinking.

We recognize a tension here in that, by their very nature, didactive and directive utterances are those in which students are not invited into conversation. Unlike transactive prompts, which help make students' privately held notions visible, these types of utterance can conceal learning that does occur. While this is a point for further research, our theoretical perspective suggests that it is as students engage in public speech that their privately held notions are mediated. In addition, because scaffolding must be considered in conjunction with the learning it intends to promote, we also considered as part of this study what counted as evidence that students' ZPDs had been accessed through classroom discourse. We suggest that when students propose a new idea or plan, contribute to an existing idea through elaboration, clarification or justification, or assess, critique, or evaluate an idea, this reflects the extension of their thinking about proof. Moreover, that the robustness of students' ideas is acquired over time notwithstanding, we take the collective construction that occurred in the episode analyzed above as evidence of student learning. This has particular implications for the nature of whole class discourse that occurs in advanced mathematical settings, specifically those that deal conceptually with mathematical proof. It suggests that students can internalize public argumentation in ways that facilitate private proof construction if instructional scaffolding is appropriately designed to support this.

We turn, then, to classroom evidence that students did begin to internalize public, whole class argumentation in their private[11] discussions and were beginning to engage in more sophisticated arguments.

We analyze an episode that occurred near the end of the 2nd semester of the teaching experiment, where students are trying to establish that the *center* H of group G, where H = {a|ag = ga, $\forall g \in$ G}, is a subgroup of G. Students were challenged by the meaning of H and how to interpret the definition of a subgroup for this particular type of set. (Note that this is a sophomore level course.) The excerpt opens about 20 minutes into this discussion, with students trying to establish the closure property for H. However, because the definition of H already involves two elements, $a \in$ H and $g \in$ G, students seem to have difficulty grasping the need for a third element, $b \in$ H:

Thom: If "a dot[12] g" is in there and "a dot g" equals "g dot a" then "g dot a" is in there too . . . [inaudible] [*contribution to or development of a new idea*]

Atram: [inaudible] If you know the original . . . the original operation must be in there. [*transactive response*]

Muel: OK, how does it look? [*transactive question*]

Atram: [inaudible] You say something then you say it again. [Writing on a piece of paper] You know that, you know, "a dot g" is "g dot a". If it didn't, [a] would not be in H. [*transactive response*]

Muel: Right. [*general confirmation*]

Thom: But we don't know that "g dot a" is in H. [*transactive question*]

Atram: No we don't. You're right. Not yet. [*transactive response*]

Atram: So you know that the operation "a dot g" equals "g dot a" which has to be in H [inaudible]. [*contribution to or development of a new idea*]

Thom: We know that a is in H. We don't know that g is H. [*contribution to or development of a new idea*]

Brot: Well if that's the case then g has to be in H. [*transactive response*]

Student: The way that I read it, I don't think that g actually exists in H [inaudible]. [*transactive response*]

Muel: [Points to something on the table] No, we have definitely established that they are both in there. [*transactive response*]

Thom: H? [*transactive question*]

Muel: Yeah ... [*general confirmation*]

Mac: Well, it didn't convince me. [*transactive response*]

Muel: ... because we said that the things that are in H are the things that are commutative with everything else in g. [*transactive response*]

This excerpt illustrates the framework of transactive questions and responses and proposing or developing new ideas and strategies as a way to interpret how students were beginning to internalize public forms of argumentation. It is important to note that, prior to instruction, these students viewed "proving" as typically an inductive process of convincing through examples. They had not studied techniques of mathematical proof and did not express their mathematical ideas in abstract forms. However, this last episode suggests that some students were learning to engage with much more complex mathematical ideas using more abstract, symbolic representations.

We take the amount of transactive reasoning and the influx of new ideas that mark this episode as evidence that public practices of argumentation, led by the teacher as more knowing other, were being internalized into students' private small group conversations. That is, the *structure* of the conversation as a flow of transactive reasoning and proposing and developing ideas seemed to reflect the structure of whole class conversation. Moreover, students' ideas were being negotiated (although they had not reached a final conclusion) *as they engaged in* transactive reasoning, proposing new ideas, and so forth. Thus, we claim that students' ZPDs were accessed through these practices of discourse.

Conclusion

The study reported here focused primarily on data in two episodes. While the discourse was representative of that which occurred throughout the year, a longitudinal analysis that carefully details linkages between teacher utterances and student development within the ZPD is an important next step in our work. Moreover, to understand the full story of acculturation from public to private speech, more research is needed to detail the nature of instructional scaffolding, its longitudinal affect on students' capacity for small group and individual proof construction, and how particular forms of instructional scaffolding impact diverse learners who bring different abilities to whole class instruction. The small group analysis provided here is a starting point in that work.

We hypothesize that the use of transactive utterances by students in whole class discourse is preliminary evidence that the teacher's scaffolding prompts did support the development of private, internal discourse structures that facilitated students' thinking about proof. Moreover, analyses that would establish the increasing use of transactive discussion in students' individual proof construction subsequent to whole class and small group discussion would provide a critical link between instructional acts and one's transition to mathematical proof.

This study assumes that every teacher utterance affects how students learn to think about, speak about, and argue mathematics. No utterance is neutral. It therefore places a great deal of responsibility on those of us involved in undergraduate mathematics education to carefully examine how and if our words facilitate student learning.

Notes

1. The research reported here was supported in part by the National Science Foundation under Grant #REC- 0337703. Any opinions, findings, and conclusions or recommendations expressed in this material are those of the authors and do not necessarily reflect the views of the National Science Foundation.
2. M. M. David received partial support for the preparation of this article by a grant from the Coordenação de Aperfeiçoamento de Pessoal de Nível Superior (CAPES) to develop her sabbatical program in the University of Massachusetts Dartmouth, USA. The opinions expressed herein are those of the authors and do not necessarily represent the position or policies of CAPES.
3. We take "discourse" here to mean verbal utterances.
4. Some course concepts did not focus on developing proofs. Corresponding data were not analyzed.
5. The lines numbers shown in the text correspond to the line numbers in the original classroom protocol. The sign [. . .] indicates omission of lines from the original protocol.
6. All names are pseudonyms.
7. We use here the term "step out" introduced by Rittenhouse (1998). However, we need to point to a significant difference: in her teaching, Lampert "steps out" of the classroom discussion as coded by Rittenhouse to discuss norms of discourse. The teacher here steps out to provide discussion on the nature of proof.
8. In a departure from Goos et al., we do not characterize the suggestion of a new idea as metacognitive.
9. We cannot say what non-participation (silence) indicates one's development. While other forms of data could shed light on this, we were interested specifically in learning as evidenced through verbal utterances in classroom discourse.
10. Deagan's utterance followed a facilitative–direct utterance of the teacher (lines 13–14). However, we suggest that it is a *response* to the earlier transactive request (line 11).
11. Here, we use the notion of "private" broadly to refer to conversation among students, without the teacher. There is another level of analysis of private, individual thought that is beyond the focus of this chapter.
12. "Dot" refers to the operation in H and G.

18

Building a Community of Inquiry in a Problem-based Undergraduate Number Theory Course

The Role of the Instructor

JENNIFER CHRISTIAN SMITH, STEPHANIE RYAN NICHOLS,
SERA YOO, AND KURT OEHLER

When asked to imagine a typical undergraduate "introduction to proof" or "transition" course, many readers would envision a university classroom with an instructor at the board lecturing about various proof techniques, such as induction or contradiction, giving a series of examples of each, and assigning proofs for the students to complete for homework. The students, mostly undergraduate mathematics majors, would dutifully take notes, ask a few questions, and then later attempt to replicate the proofs that were presented simply and logically in class. This may be what a "typical" transition course looks like, but a growing body of evidence suggests that many students fail to develop a solid understanding of mathematical proof by the end of their undergraduate programs (Harel & Sowder, 1998; Tall, 1992; Weber, 2002). In this chapter, we examine one alternative to a traditional lecture-based approach for teaching proof; in particular, we discuss the ways in which an experienced instructor teaching a problem-based introduction to proof course assumed and relinquished leadership of the class in order to engage students and mediate the discussion of content—in this case, formal mathematical proof.

The results reported here come from a video-based study of an upper division number theory course at a large state university. The purpose of this chapter[1] is to present an analysis of the ways in which the instructor's actions changed over the course of the semester with regard to assuming and relinquishing leadership of the class discussion, and to show that through his instruction he encouraged a view of mathematics as a human, social activity by engaging students in discourse about mathematics and proof.

Relevant Background

A growing body of research indicates that participation in a community of learners can be a vital part of students' success in mathematics. From a sociocultural perspective, learning can be regarded as the product of the reflexive relationship between communally developed and shared classroom processes and individual constructive activity (see Cobb & Yackel, 1996). In recent years, researchers have begun to examine mathematics learning that takes place in *active* classrooms, in which social negotiation of mathematical meaning is commonplace. For example, cognitively guided instruction [CGI] has been shown to be effective in helping children in elementary classrooms develop mathematical thinking, computation, and problem-solving skills (see Carpenter, Fennema, Franke, Levi, & Empson, 1999). An important feature of CGI is that students develop their own strategies for solving problems with little or no direct instruction from the teacher. Children communicate their mathematical ideas and solution strategies to each other, and the role of the teacher is to listen to and understand how the children are thinking, and then to push their thinking forward through questioning and discussion. Problems are chosen based on the students' current ways of thinking, research-based knowledge on students' learning and the analysis of the domain in elementary arithmetic rather than exclusively on a predetermined curriculum.

Similarly, at the secondary and undergraduate levels problem-based mathematics courses have shown to be effective in providing opportunities for students to develop their mathematical problem-solving and communication skills (e.g., Goos, 2004; Stephan and Rasmussen, 2002; Yackel, Rasmussen, & King, 2000, etc). Many studies of such courses focus on the class as a learning community and the instructor's role in guiding the community in the development of mathematical understanding.[2] For instance, Goos (2004) reported on a secondary mathematics class in Australia in which the teacher modeled mathematical thinking and sense making, encouraged students to clarify, validate, and elaborate on their own work, and waited to introduce mathematical conventions and symbolism when necessary. The students in the course reported that they appreciated opportunities to solve interesting problems and valued the role of explanation and proof in their mathematical work. Yackel, Rasmussen, and King (2000) studied an undergraduate differential equations course in which the instructor actively and consciously worked to influence the social and sociomathematical norms for participation in the class. They report that it was this explicit attention to the interactions between individuals that made possible the sophisticated reasoning demonstrated by the students. In a subsequent study of this course, Stephan and Rasmussen (2002) documented the development of students' collective understanding of concepts and found that classroom mathematical practices developed in non-sequential ways, with regard to both time and mathematical structure. As before, the instructor played a key role in the development of these practices by facilitating discussion and argumentation in the classroom.

The classrooms described here all allow for students to participate in what Goos (2004) describes as a *classroom community of inquiry*, that is, one in which "students learn to speak and act mathematically by participating in mathematical discussion and solving new or unfamiliar problems" (p. 259). They also are all examples of classrooms that involve what we will call inquiry-based learning (IBL).

Many researchers in mathematics education (see Wood, 1999) and science education (see Newton, 1999) have described the critical role of the teacher in facilitating discussion and argumentation at all levels. McClain and Cobb (2001) studied the role of a 1st grade teacher in developing sociomathematical norms in her classroom. Their research analyzed the processes that took place in order for these norms to emerge; in particular they focused on the role of discussions and the teacher's ways of symbolizing student reasoning. They found that through the teacher's active guidance the students developed a sense of intellectual independence as well as a stronger mathematical disposition. Wood (1999) examined the role of a teacher in a 2nd grade classroom, focusing on the ways in which the teacher helped to develop a community of inquiry where discussion and disagreement were valued. Wood found that in this classroom the teacher put a great deal of effort into establishing the norms for discussion, participation, listening, and disagreement in this classroom community, but noted that her level of participation declined throughout the course of the year as the students became more confident in their own roles as participants in the learning community. Martino and Maher (1999) note that is it important not only to ask the right questions, but teachers must also "back away strategically when communication and reasoning flourish, [to allow] students to play more active roles in their own and each other's learning" (p. 75).

The "Modified" Moore Method and Inquiry-based Learning

For the last few years, we have followed the experiences of students and instructors in undergraduate number theory courses at a large southern state university. Several sections of this course are taught every semester; at least two are typically taught using the "modified" Moore method (MMM), a teaching approach that employs mathematical discourse among students and is similar to the inquiry-based learning courses described previously. (The other sections are usually taught in a traditional lectured-based style.) While this course is often referred to as a MMM course because of the similarity of this method to that of R. L. Moore, the community norms that are established are similar to those in the IBL classrooms described in the previous section. Because of this, as well as the instructor's own preference, we will refer to these particular courses as IBL rather than MMM.

In Smith (2006), we reported that this instructional approach appears to encourage students to *make sense* of the mathematics when attempting to construct proofs. This study examined the differences in approach to proof by students enrolled in IBL and lecture-based sections of the number theory course. The most striking differences were seen in the strategies the students employed when presented with a statement to prove and in their approaches to validating proofs of others. In general, the IBL students demonstrated a tendency to make sense of the mathematical ideas, while the traditional students were more concerned with finding quick and easy ways to complete tasks. This suggests that the structure of an IBL course can provide students with an opportunity to learn mathematics in meaningful ways.[3] In the remainder of this section, we provide some background information about the structure of the course studied.

The IBL course we studied is a *problem-based* approach to teaching mathematics. In this course, students are given a carefully constructed list of problems to solve or theorems to prove on their own, with little or no direct instruction from the instructor.[4] The students then present their solutions to the problems in class, and the instructor (or student presenter) facilitates a whole group discussion of the solution. This presentation of student work and subsequent critique through classroom discourse is an important characteristic of this approach; the instructor plays a quiet yet critical role in the facilitation of the discussions about the presented proofs. In general, the instructor is not an arbiter of mathematical correctness in the classroom, but acts more as a mathematical mentor, providing guidance and direction when necessary.

The approaches to teaching and learning mathematics represented by the classrooms discussed in the previous section as well as in the particular classroom we studied have a great deal in common. All emphasize problem solving and sense making over transmission of information, and so shift the focus in the classroom from the actions of the teacher to the actions of the students. In many of these classrooms, the level and pace of the course is adjusted to students based on the teacher's perception of students' understandings and conceptions of mathematical ideas (Carpenter, Fennema, Peterson, Chiang, & Franke, 1989; Mahavier, 1997; Mahavier, 1998; Parker, 1992). They share the idea that students construct meaningful mathematical knowledge when they engage in solving carefully selected problems, based on an analysis of the domain of mathematics and the development of students' mathematical thinking. By having students solve a variety of problems and explain their thinking, classroom activity is focused on the development of students' mathematical understanding and creative thinking. (Carpenter, Fennema, & Franke, 1996; Mahavier, 1998) Finally, the roles of students and teachers are considered important and interdependent. As an active learner, the student is expected to participate in the learning process and to be responsible for his or her learning. As a facilitator, the teacher is expected to guide students' learning, pose questions and assist group work and classroom discussion (Clark, 2001; Knapp & Peterson, 1995; Mahavier, 1998).

Role of the Instructor

The role of the instructor in the development of a classroom community can be broken into two general categories of responsibility: the teacher facilitates and directs the mathematical activity of the class, and the teacher guides the development of the culture of the learning community (Hiebert et al., 1997). In this section, we describe the ways in which one IBL instructor did both of these things. In particular, we discuss the ways in which his choices to assume and relinquish the role of leader in the classroom affected students' participation in the learning community and facilitated their engagement in mathematical discussions about presented proofs.

The IBL instructor described in this chapter is a full professor in the department of mathematics. "Mark" has more than 20 years of experience teaching using the IBL, and has been recognized by the university as a distinguished teacher. He is generally regarded by students and colleagues alike as one of the best teachers in the department; however he has never received any educational training beyond his own experiences with IBL courses as a student. He does not prescribe to any particular learning theory, but chooses to teach this way because he believes it allows for students to become independent thinkers. The structure of Mark's number theory course is similar to what was described in the previous section. Every few weeks, students were given handouts called "course notes"; these contained a list of theorems to be proved, along with necessary definitions and some exploratory problems. These course notes were developed by Mark and a colleague over a period of several years and were designed to guide students through a logical sequence of mathematical propositions and concepts. The students in the course worked through these notes outside class and presented their proofs during class meetings. At the beginning of each class, Mark asked for volunteers to present the next three to six proofs in the problem sequence from the course notes. These students wrote their proofs on the board during the first 10 minutes of class, and then returned to the board to explain their proofs one at a time. After each proof was presented, Mark (or the student presenting) asked for questions or comments, with the goal of facilitating a discussion of the presented proof. Mark did not indicate whether the presented proofs were correct or incorrect, but allowed the class to discuss the proof until there were no further questions or comments.[5]

From the very first day of class, Mark did not assume the traditional role of a lecturer at the front of the room. At times, he was the focus of the students' attention and clearly the authority figure in the class, but at other times he was in the background, letting the students direct the flow of the discussion. He appeared to purposefully relinquish and assert his role as leader and mathematical authority during class sessions, both with the whole group and with individual presenters. These shifts in leadership were generally indicated by his position in the classroom, his voice (tone, volume, and silence), and his physical gestures. Over the semester we observed Mark claiming a leadership role in the classroom on multiple occasions.

The data presented in this section come from a semester-long study of this course, its instructor, and the students enrolled in it. There were 22 undergraduate students enrolled in the course, most mathematics majors, with a few computer science majors; about one-fourth of the students were preservice secondary teachers. The number theory course serves as a transition-to-proof course at the university and satisfies a program requirement for all of the students who

312 • Jennifer Christian Smith, Stephanie Ryan Nichols, Sera Yoo et al.

were enrolled. For most students, this course represented their first experience with mathematical proof at the undergraduate level. We videotaped every class meeting of the course during the semester of study, and these videos were organized into meaningful clips 30 seconds to 5 minutes in length. These clips were coded using a framework developed by the research team. In order to analyze the patterns in Mark's leadership, we examined every clip and counted the number and examined the context of instances in which we observed Mark stepping forward to lead, redirect, or assume control of a class discussion during a proof presentation. Within these instances we found many could be further described by the following three categories:

- *Motivating participation* in the developing classroom community by organizing the presentations of proofs and modeling the types of questioning and commentary necessary for understanding.
- *Facilitating whole group discussions* of presented proofs by asking students to comment on and discuss a presented proof, as well as directing students to refine their questions and comments.
- Discussing or questioning students' *strategies for proof.* For example, Mark would frequently comment on the initial strategies used to begin constructing a proof, the use of notation, and the use of prior knowledge and experiences. He would emphasize the importance of concrete examples for gaining insight into a statement to be proved, and would occasionally give advice, tips, or comments about general strategies for constructing proofs.

Although these categories emerged from our analysis of the data, we should note that they correspond with ideas discussed in the previous section. In particular, Mark's actions resonate with those of the teachers studied by Goos (2004) and Yackel, Rasmussen, and King (2000).

Before we began our analysis, we expected to find that the instructor asserted his leadership most frequently in the beginning of the semester, with incidences tapering off as the students gained more confidence and understanding of their role in the class. However, we found that there were fewer incidences of the instructor asserting his leadership at the beginning and end of the semester than there were in the middle, as can be seen in Figure 18.1.

This trend occurred not only in the overall instances of Mark asserting himself as leader, but was also present when we separated the events into the three categories described above, as can be seen in Figure 18.1. We kept track of instances where the instructor *reclaimed* leadership of the class discussion after having relinquished it; thus an increase in instances indicates that leadership was being asserted and relinquished more frequently than at other times in the semester. In order to understand why this spike occurred in the middle third of the semester, we will examine each third of the course and describe what was happening at that time.

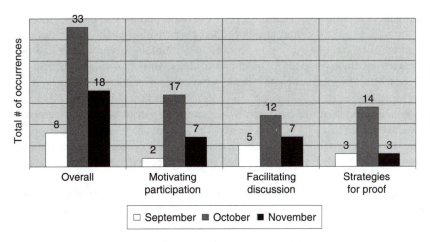

Figure 18.1 Assuming the role of leader

September: Mark Encourages Student Response to Presented Proofs

During the first part of the semester, most of the instances in which Mark asserted himself as leader occurred either during whole class discussions or when he was organizing student presentations. At this point in the course, the material was easy for most students; the goal of the early part of the course was to give them experience writing relatively simple proofs. As many of them had little previous experience with writing proofs, much of the class discussion at this time focused on whether or not a presented argument constituted a valid proof of a statement. Mark generally did not answer students' questions or present himself as a mathematical authority in the validation of proofs. Rather than concentrating on the mathematical content of the material, Mark's questions were general and directed to the whole class in an attempt to facilitate discussion of the presented proofs. Later in the semester, Mark's role as a facilitator changed to focus on the quality of the mathematical discussion by asking more content-specific questions, but early in the semester Mark motivated participation mainly by modeling the types of question he wanted students to ask of their peers. It was relatively infrequent that Mark asserted leadership to motivate participation at this point in the semester, which seems counterintuitive. Motivating student participation early in such a course would seem to be critical, since the structure of the course depends on students learning how to discuss mathematical proof. However, Mark's apparent reluctance to lead discussions himself may have actually motivated more student participation. Early discussions were often characterized by long periods of silence during which Mark did not speak or assert leadership, and students would eventually ask questions and make comments.

In order to compare the dynamics of the classroom community at this point to those later in the semester, we consider the instructor's role in the classroom as a mathematical expert or representative of the larger mathematical community. At the beginning of the semester, the instructor did not present himself as a figure of mathematical authority in the classroom. In addition, the students were uncertain of his role, so the students showed limited participation in the classroom community. They were beginning to "move" their thinking beyond what had been expected of them in past mathematics courses, but many were hesitant and unsure of the expectations for their behavior and participation early in the course.

As an example of this behavior, we present a transcript of a short segment of the class during this part of the semester. The students have just seen Alex present a proof of the existence of an integer q that serves as a quotient when two natural numbers are divided. His entire proof and the theorem statement are given in Figure 18.2.

The end of his presentation was met with silence, and many of the students were looking at Mark, as if trying to determine his reaction to the proof:

Mark: Okay, do you have any questions for Alex?
[pause]
Kevin: Just one thing. I don't think it said anywhere that n has to be less than m. Because over here on [another problem they had solved] n is 45 and m is 33.
Alex: That's right. [Corrects this part of his proof.]
[silence]
Mark: Victoria?
Victoria: How do we know it works for other numbers too?
[pause]
Barry: I would also, I would also say that's it not any integer of cause I mean you could . . . Wouldn't it be, uh, the least integer such that it didn't get the most?
[silence, and then students start to raise their hands]
[Mark moves from far side and rear of classroom, where he has been the whole time, towards the front.]
Mark: What do you think? Uh, I mean, I'm trying to get a sense of the crowd here. I mean, do you think this is a good proof or not?

Theorem: Let m and n be natural numbers. Then there exist integers q and r such that $m = nq + r$ and $0 \leq r \leq n-1$.

Alex's proof as written on the board:

1.21: Division algorithm. $m = nq + r$
We are given m and n are any integers such that $m > n$. Therefore q would be an integer of $\dfrac{m}{n}$ and r would be the remainder.
n is the divisor.
For example $27 = 3*7 + 6$.

Figure 18.2 Alex's entire proof and the theorem statement for the existence of an integer q

Kevin: The one thing I'm just hesitant about is we're proving the division algorithm and he used *m* divided by *n* in the proof.

Mark: Uh-huh.

Kevin: He used division in the proof. I don't know.

Mark: Mm-hmm. Andrus, what do you think?

Andrus: I don't know, I just, I got kind of lost. I don't know what he, how he got to his conclusion or any of that. I don't know, seems kind of . . .

Mark: Mm-hmm. Yeah, Sandra?

Sandra: I'm not sure what it means to be an integer floor of *m* divided by *n*.

[Mark moves toward back of the room.]

Alex: Well, um, hmm. Let's see. [Pause.] It'll be like computer science. If you divide one integer by the other integer and you will get something like 94.6 and we take the floor of the integer.

Ali: The least integer?

Alex: The greatest integer.

Sandra: Sorry, forget I said anything.

Alex: I have bad vocabulary.

Mark: Patrick?

Patrick: Um, I was trying to do something similar. Since you said computer science, I was using a computer science mindset too. I guess you're a CS major as well.

Alex: Yeah.

Patrick: All right. Cause, well I was thinking that if you get two integers and use the division operation then you would get a real number but if you take away everything after the decimal you'd be given the integer. That's what you were thinking as well.

Alex: Exactly.

Patrick: And then you would just get the remainder by using that newfound integer and subtracting it to get the remainder. Is that correct?

Alex: Pretty much.

Patrick: [Turning towards Mark.] Can we use that? That'd be cheating though.

Mark: Well, I don't know, what do you think?

We see that Mark did not say very much of mathematical substance during this excerpt. The first time he spoke, he moved to the front of the room and appeared to be trying to get students to respond and participate in a discussion of the presented proof. He gave no indication of his impression of the proof. During this segment he called on students to voice their opinions, but did not direct the content of the conversation. The students seemed hesitant and uncertain, as if they were not sure what was expected of them. When Patrick directly asked Mark if it was permissible to use a particular tool he was familiar with from computer science, Mark deflected the question back to the class. At this point, we can see that Mark was simply facilitating discussion by calling on students and was not directing the conversation. The students were beginning to offer feedback, but not yet in ways that were very meaningful or critical. They looked to Mark for validation and direction, but Mark did not make it clear that he would provide this for them.

October: Mark Focuses and Supports Mathematical Critique

During this section of the semester, there were many more instances in which Mark assumed the role of leader than there were in the beginning and end of the course. In addition, there was a greater variety in the types of situation during which these instances occurred. For example, Mark frequently interrupted a discussion in order to assist a student who was struggling with a difficult proof, address a particular point or step with the whole class (such as a strategy for constructing a proof), or to help the class find an error in an incorrect proof. At this point in the semester, the students had become more comfortable with sharing their ideas in class and critiquing one another's proofs, but were struggling with the mathematical content more than they had in the first part of the course.

When asserting himself as leader in order to facilitate discussion, his comments were focused on encouraging the students to become more critical of the presented proof. Sometimes, Mark asked the students to rephrase their comments in order to better articulate what they disagreed with or did not understand in a proof. As the theorems became more difficult to prove, Mark's motivation took the form of helping the students call into question the points in the proof they were challenging. For instance, he would ask students to rephrase an argument in their own words or ask a student to explain a proof line by line. In addition, there were times when the entire class had difficulty getting started with a proof. In these cases, his guidance took the form of suggested strategies to better understand the theorem statement. These included using examples, rephrasing the theorem statement, or making comments about general strategies for constructing proofs.

The students were, at this point, exploring new ways of mathematical thinking through their participation in the course. As they explored new avenues of thinking, they needed feedback about the norms and rules of participation in mathematics as it is done in the larger mathematical community. The instructor at this point began to provide more input and to refine the students' thinking. At this point, the instructor has taken a more active role in helping shape students' thinking.

In the following example from this part of the course, Kevin has just finished presenting a proof that contains some errors. The other students sense there is a problem with the proof, but cannot seem to articulate it. Mark assumes the leadership of the class in order to help the students articulate their comments:

Mark: Let me see if people are actually following this. So, let me ask somebody at random to assert whether or not you personally are following and if so, to rephrase it in your own words. So let me ask um . . . [Long pause.] Well, Michael, how about you?

Michael: Well, all right. So you've got some function f acting on a set and it just multiplies elements in that set together, and then you've got some function g, and you have g act on a number out of the set of all prime factors of that number.

Mark: So . . . let me see if I understand that, see if you understand that. Do you understand that?

Michael: I think so.

Mark: Are there any problems with that that you see?

Michael: It just kind of seems weird . . . I dunno, to let a function act on a set. I . . . I dunno, is that all right?

Mark: [Shrugs, doesn't answer.]

[Students laugh.]

Michael: I think it is.

Mark: Okay, well, let me ask this question, I mean is . . . how about if you have the set $\{2, 2, 3\}$?

Kevin: Yeah, I think that's where the flaw in my reasoning is. I don't know how to express if there's multiple prime factors that are equal to each other.

Mark: Ah.

Kevin: That was the problem with my initial thing. I wasn't exactly sure how to define that. But if I could do that, then the rest follows.

The theorem and Kevin's proof are given in Figure 18.3.

The goal of Mark's interruption of Kevin's presentation was to ensure that the students were thinking about the proof. At this point in the semester, the students knew how and when to participate. So when students did not immediately offer comments, Mark asserted himself in order to focus the attention of the class to a particular point in the proof. In the preceding excerpt, he asked one student to paraphrase the problematic part of the argument. He went on to ask a more specific question using an example in order to push the students to think about the mathematics in the argument. After this excerpt, the students appeared to have a better understanding of the problem and began to ask these questions of the presenter himself:

Terrence: What is D complement? I'm sorry, I haven't taken set theory.

[Kevin explains what the complement of a set is.]

Terrence: Okay, I got it. So what is the set you're working with then, that you're taking the complement of?

Kevin: I'm dealing with natural numbers. Yeah, I didn't make that clear. I didn't place that restriction on it, but it would be that because we're dealing with prime numbers, and nothing else.

Victoria: Do you have to restrict it to primes, or should it be the natural numbers?

Kevin: Natural numbers.

Victoria: Are you sure?

Kevin: Um, because . . . yes, you're right, because the sets . . . they're just collections of numbers and aren't just primes . . . uh . . .

Theorem: Suppose a and b are integers and that the greatest common divisor of a and b is d. Then the gcd $\left(\frac{a}{d},\frac{b}{d}\right)=1$.

Kevin's proof as written on the board:

Let $f(X)$ be a function on a set. $f(X)=$ product of all elements in X.
$f(\phi)=1$.
Let $g(y)$, y in $\mathbf{N}$, s.t. $g(y)$ is the set of all prime factors of the integer y.
Let $A=g(a)$, $B=g(b)$, $D=g(d)$, $g(1)=$ empty set.
$(a, b)=d$. By def of gcd (thrm 2.14), $(a,b)=f(A\cap B)$.

By definition of division using sets, $\dfrac{a}{d}=(A-D),\dfrac{b}{d}=(B-D)$.

So by def of gcd (2.14), $\left(\frac{a}{d},\frac{b}{d}\right)=f((A-D)\cap(B-D))$.
By def of set subtraction,
$\left(\frac{a}{d},\frac{b}{d}\right)=f((A\cap D')\cap(B\cap D'))$
By commutation of sets
$\left(\frac{a}{d},\frac{b}{d}\right)=f((A\cap B)\cap(D'\cap D'))$
$=f((A\cap B)\cap D')$. But $A\cap B=D$, so
$=f(D\cap D')=f(\phi)=1$
So $\left(\frac{a}{d},\frac{b}{d}\right)=1$

Figure 18.3 Kevin's entire proof and the theorem statement for function g

Victoria: I'm just really wondering whether it would work with natural numbers. [pause]
Kevin: It would not because of the way I defined it, the null set with one, but with natural numbers . . . one actually is a natural number, so it wouldn't correspond with the null set. So . . . did that answer your question?
Victoria: Well . . . when you're dealing with complements . . . [Points out a specific part of the proof on the board.] It just seems like somewhere in there you could get in trouble there if you have natural numbers that are not primes floating around.
Kevin: But . . . they're all primes. Restricted to primes.
Ali: I think that line that she was pointing out, you will have problems with that. You will leave out all of the factors when you take the complement. I think that is going to mess it up.
[Ali talks Kevin through an example with a complement of the set, and this helps Kevin see where the flaw in his proof is.]
Kevin: I see what you're saying, but I want to make it work so bad!
[Another student offers a suggestion.]
Mark: Well, first of all, let me just say I like this very much. They're pointing out basic issues—Victoria and Ali and others—they're pointing out that there's a

problem. [He summarizes the problem with the proof.] So people are pointing this out and that's great.

Mark did not speak for several minutes during this excerpt, but when he did speak he did not deflect the issue back to the class. He paraphrased the students' comments while commending their participation. The presenter at the board was calling on students who had their hands raised, and those students asked their questions to him directly, instead of through the instructor, as they had done earlier in the course. The discussion occurred between the presenter and the other students, while Mark was standing at the back of the room, listening but not guiding the discussion. Immediately after this excerpt, Mark went to the board to talk explicitly about the problem with the proof and make suggestions for how it might be fixed. He did this after they had successfully identified the flaw in Kevin's proof, convinced him of it, but needed help to proceed with the task of fixing the proof.

November: Mark as an Expert Participant

At the end of the semester, the number of instances of Mark assuming the role of leader decreased. In general, occurrences of the instructor asserting himself as leader arose during discussions of difficult proofs. It is important to note that even though the proofs were continuing to get more difficult the instructor needed to assume leadership far less.

The occasions on which Mark interjected during a presentation or class discussion were primarily focused on providing suggestions for making the written proof as explicit and rigorous as possible. He seemed to be offering critique as an expert, yet as an equal participant in the classroom community. His comments primarily served to help students to refine presented proofs, in the same spirit as other students were attempting to do. Motivating participation took the form of modeling how students should aid the presenter in improving their presented proof. Unlike earlier in the semester, at this point most of his participation consisted of brief comments rather than lengthy efforts to facilitate discussions and focus students' thinking.

At the end of the semester, the students have a better awareness of the norms of the mathematical community than they did in October and are comfortable with the instructor's expectations for participation and with their own mathematical thinking. The instructor and his influence as a representative of the mathematical community is less than in the middle of the semester because the norms of the mathematical community and the classroom community have been integrated and accepted by the students.

In the following excerpt, Max is reading a proof he has written on the board. Without prompting from Mark or from a student, he pauses in the middle of the proof to ask the other students if they understand what he has presented so far:

[Max reads the proof he has written on the board. After he reads the first three lines.]

Max: Is this part clear?

Ali: Can it not be limit as x goes to infinity of infinity over infinity?

Max: Cause we don't know about this number. It could be positive or negative.

Kevin: Even if you ignore the fact that we all know it would be zero, you can actually multiply by a factor that will give you zero that way, and more clearly shows what you've done. Actually, it's the reason why what you have there is true.

[Max asks him to repeat this, and Kevin leads him through it. Max works this idea into his proof on the board.]

Student: Or you could just do L'Hopital's Rule n times.

[laughter]

Kevin: Well, yeah, but calculus depends on these proofs. You should do it.

Max: [Nodding.] So is this clear enough? Any more questions? Patrick?

Patrick: Have you been able to do this without the limits? I tried to. This is a really clever way, but I . . .

[pause]

Mark: I think this is the right question, can you do this without the limits? In other words, the limit in a sense is sort of hiding something. Your goal is to prove that the denominator, ax to the n, is bigger than all the rest of the terms and in doing the limit you somehow capture that, but it would be nice to be more explicit.

Max: Okay.

Mark: So, maybe you could do that. You have the proof here; it's just a question of reformulating it in direct terms.

The theorem and Max's proof are given in Figure 18.4.

In the preceding excerpt, the student at the board was interacting with the other students in a way that suggests he expected them to understand his proof. He paused when he was not sure if his proof was clear and asked them if they understood, and then several students offered suggestions for improvement. This was typical of discussions of presented proofs at this point in the course, and the contrast between this excerpt and the one from early in the course is dramatic. Mark's role as facilitator of discussion had shifted to be an expert who could provide suggestions and guidance when needed. In the last excerpt, Mark did not speak until he sensed that Max was unsure how to proceed with the suggestions that had been made by the other students. Unlike in the previous section, Mark's interjection ended there: Max gave an overview of how he would change his proof and the class was satisfied, so they moved on to the next presentation.

Concluding Remarks

Yackel (2002) notes that it is a misconception that the instructor should remain silent during classroom discussions. On the contrary, the teacher's choices of when and how to speak are of critical importance in the development of a learning community in which the students take active roles in their own learning. We

Theorem: Suppose $f(x) = a_n x^n + a_{n-1} x^{n-1} + \Lambda + a_0$ is a polynomial of degree $n > 0$ and suppose $a_n > 0$. Then there is an integer k such that if $x > k$, then $f(x) > 0$.

Max's proof, as written on the board:

13) Suppose $f(x) = a_n x^n + a_{n-1} x^{n-1} + \Lambda + a_0$ is a poly w/degree n ($n > 0$) w/all a's in **Z**, $a_n > 0$. Then there is an integer k such that for all $x > k$, then $f(x) > 0$.

Proof: Assume $f(x) = a_n x^n + a_{n-1} x^{n-1} + \Lambda + a_0$, $n > 0$, $a_n > 0$.
Since we're going to prove $\forall x \mid x > k$ for all $x > k$, we assume x goes to ∞.
Now consider this: Since $a_n > 0$ & $n > 0$

$$\lim_{x \to \infty} \frac{a_{n-1} x^{n-1} + \Lambda + a_0}{a_n x^n} = 0$$

We conclude that as x goes to infinity, $a_n x^n$ dominates
$a_n x^n + a_{n-1} x^{n-1} + \Lambda + a_0$ (or $a_n x^n > a_n x^n + a_{n-1} x^{n-1} + \Lambda + a_0$)
Since $a_n > 0$ & $n > 0$, a_0 is positive.
Since $f(x) = a_n x^n + a_{n-1} x^{n-1} + \Lambda + a_0$ & $a_n x^n$ is positive
& $a_n x^n + a_{n-1} x^{n-1} + \Lambda + a_0$. I conclude that as
x goes to ∞. $f(x)$ is positive. So there must
be an int k where for all $x > k$ $f(x) > 0$.

Figure 18.4 Max's entire proof and the theorem statement as x goes to infinity

have shown in this chapter how the role of one instructor changed over the course of the semester as the students became more adept at participating in a classroom community of inquiry. The instructor's choices for when to speak, when to remain silent, and what to say, influenced the way students participated in the course. At the beginning of the semester, Mark's comments were primarily intended to facilitate discussion. He rarely commented on the mathematics being presented; rather, he asked questions that encouraged the students to examine the presented proofs and express their opinions about them. As the semester proceeded, the course material became more difficult, and Mark began to interrupt discussions and assume leadership of the class more frequently. His comments during this time were more focused on the mathematics being presented and pushed the students to examine and reflect on strategies for constructing proofs. Toward the end of the semester, the students were comfortable with the level of mathematics and with participating in the class, and the frequency of teacher interruptions of discourse was reduced. This is in contrast to what Wood (1999) found in a 1st grade inquiry-based classroom where the participation of the instructor declined over the course of the year.

By the end of the semester, these undergraduate students appeared not only to be comfortable discussing and presenting their mathematical ideas, but had also come to see the mathematics in the course as something that required active participation, individual effort and thought, and discourse with peers. It is our belief that the structure of the course and Mark's careful choices to assume and relinquish leadership of classroom discourse facilitated the development of this level of comfort with mathematics as a human, social activity. Such classroom environments remain rare at the university level, but research is beginning to demonstrate that they can be effective in helping students develop "mathematically mature" modes of thinking and participating in mathematics courses.

Notes

1. Although the purpose of this chapter is not to report the impact of this course on the students' understanding of proof, we should note it is our belief that the result of this instructional approach was that the students in the course developed a remarkably mature conception of mathematical proof as an essential activity of doing mathematics. For more information, see Smith, 2006.
2. Although little research has been conducted as to the efficacy of these inquiry-oriented methods, particularly at the undergraduate level, the work of Kwon, Rasmussen, and Allen (2005) is an example of a quantitative study which found that students in a problem-based undergraduate differential equations course showed comparable procedural knowledge as their counterparts in a traditional course as well as better retention of their conceptual knowledge 1 year later.
3. Although our goal in this chapter is not to show that courses taught in such a way are necessarily superior to those taught in a traditional lecture style, we should note that based on our prior and ongoing research (as well as current research on the nature of teaching and learning mathematics), we assume that an active learning environment provides more opportunities for meaningful learning in undergraduate mathematics courses than do lecture-based environments.
4. Depending on the instructor, students may or may not be allowed to use other texts or each other as resources. In the "original" method devised by Moore, students worked entirely individually and were not allowed to talk to each other at all outside of class. See Mahavier (1998) and Renz (1999) for more information.
5. In general, students would notice significant errors in presented proofs and point them out. When students did not seem to notice that a proof was incorrect, Mark used several strategies to get them to take a closer look. For example, he would state that he wasn't following part of the proof and would ask someone in the class to explain that part in their own words. In other instances, he would suggest a counterexample that would suggest there was an error. However, there were no instances when he stated precisely what was wrong with a proof without students in the class having found the error first.

19

Proof in Advanced Mathematics Classes

Semantic and Syntactic Reasoning in the Representation System of Proof

KEITH WEBER AND LARA ALCOCK

Proof is fundamental to mathematics and students should be engaged in the activity of proving throughout their mathematics education. However, the character of proof changes when students enter advanced mathematics courses—i.e., upper level, proof-oriented university courses such as real analysis and abstract algebra. In earlier mathematics courses, a proof is sometimes defined as an argument that convinces one's community that an assertion is true (e.g., Balacheff, 1987). Proofs in these courses are often presented verbally by children using their own language (NCTM, 2000, see especially p. 58). These proofs may employ a variety of types of argumentation, including the use of manipulatives and informal representations of mathematical concepts, as well as justifying general claims by looking at specific cases (e.g., Ball & Bass, 2000; Maher, Chapter 7, this volume; Schifter, Chapter 4, this volume).

In advanced mathematics courses, students are expected to produce *mathematical proofs*, or arguments that obey well-defined conventions that are agreed on by contemporary mathematicians. As Tall (1989) emphasizes, not all convincing arguments constitute mathematical proofs in this sense. For an argument to be considered a mathematical proof, it must be based on accepted axioms and definitions. Further, it is expected that proofs will be written so as to maximize clarity, meaning that logical notation will be used when appropriate and standard forms of words will indicate where a standard proof technique has been employed. Reasoning must proceed deductively, and many previously acceptable forms of argumentation—including diagrammatic reasoning and justifying general claims by reasoning from generic examples—are no longer permissible. Perhaps the sharpest shift in the character of proof is that in earlier grades, the validity of a proof is primarily dependent upon its *content*, while in advanced mathematics courses, the *form* of the proof is also paramount in determining its validity (Harel & Sowder, 2007).

A growing body of research in mathematics education demonstrates that undergraduate students find the character of mathematical proof to be deeply perplexing and that this confusion can inhibit them from successfully constructing proofs (e.g., Moore, 1994). Some researchers have compared learning to construct mathematical proofs to learning a new language (Downs & Mamona-Downs, 2005), or mastering a different genre of speech or writing (Nardi & Iannone, 2005; Selden & Selden, 2003). In keeping with these comparisons, we will argue that learning to prove in the college classrooms requires students to work within a new *representation system* (in the sense of Goldin, 1998). Our emphasis in this chapter will be on the fact that working with any new representation system is simultaneously limiting and empowering. In the case of mathematical proof, statements that may be easy to justify with natural language or less rigorous argumentation now have to be written in appropriate format, often prompting undergraduate students to complain, "I can see it, but I just can't prove it." A good example of such a statement is the Intermediate Value Theorem, which can easily be demonstrated with an informal graphical argument but whose proof relies on a relatively sophisticated argument making use of the completeness axiom. In contrast, proving other statements that may be nearly

impossible for a calculus student to justify informally (e.g., $\sum_{i=1}^{n} i^3 = \left(\frac{n(n+1)}{2}\right)^2$

for all natural numbers n) becomes a routine exercise for one who is familiar with standard proving techniques.

The purpose of this chapter is to discuss both the challenges that students face when working within the representation system of mathematical proof and the affordances it offers. First, we review Goldin's (1998) description of a representation system, including what constitutes syntactic and semantic reasoning within a representation system. We then describe what we mean by the representation system of proof and discuss syntactic and semantic reasoning within this system. Following this, we first present ways in which syntactic reasoning may facilitate undergraduate students' understanding and abilities to prove and then describe difficulties that undergraduate students have with syntactic reasoning. We illustrate each of the points raised in this section with excerpts of students' proof-related behavior from empirical studies that we have conducted; then do that for semantic reasoning. Finally, we discuss the pedagogical and research implications of our findings.

The Representation System of Proof

Representation Systems

In Goldin's (1998) theory, a representation system consists of primitive characters, configurations, and structures. *Characters* are discrete entities that are the building blocks of a representation system, and, as elementary entities, are not yet ascribed further meaning or interpretation. Characters may include elements from a well-defined set (e.g., the letters in the alphabet, English words, numerals), but can also include less well-defined entities, such as curves or physical objects. A representation system further has (sometimes ambiguous) rules for combining characters into *permitted configurations*. Words labeled as parts of speech (taken as primitive signs) can be joined to form sentences (the configurations); numbers, variables, and operation signs may form mathematical equations; and so forth. Typical representation systems have more *structure* than this. Perhaps most important is that representation systems often have (frequently ambiguous) rules for *moving* from one permitted configuration in the system to another, or from a set of configurations to another set of configurations. Representation systems may also include other kinds of structure, including relations between configurations, and (possibly ambiguous) rules for valuing configurations (e.g., the assignment of truth values of logical propositions).

One example of a well-known representation system is propositional calculus. (To avoid misinterpretation, we will *not* claim that the representation system of proof can be reduced to a formal logical system.) Propositional calculus consists of a set of characters, including letters from the English alphabet and logical symbols (e.g., $\equiv$, $\rightarrow$), and rules for combining strings of characters into permissible configurations (i.e., well-formed formulas). There are rules of inference (e.g., *modus ponens*) that permit one to move from a set of well-formed formulas to another well-formed formula. There are also procedures to evaluate configurations (e.g., to determine whether a logical proposition is true or a proof—i.e., a sequence of permissible configurations—is valid).

Syntactic and Semantic Understanding and Reasoning in a Representation System

Goldin (1998) describes two ways in which one can give meaning to and reason about characters and configurations within a representation system. One could understand a character and configuration with respect to its relations with other configurations *within* the structure of that representation system. For instance, in propositional calculus, the negation symbol $\neg$ could be understood in terms for the rules governing what constitutes permissible configurations involving this system and how configurations with this system can be treated. Such an understanding could perhaps include "$\neg\neg P \equiv P$" and "$\neg(P \rightarrow Q) \equiv P \wedge \neg Q$." This is referred to as a *syntactic understanding*. One can also view a configuration within the system as representing a configuration from a different representation system. For instance, $\neg$ can be viewed as representing the word "not" in the English language. The expression "¬P" can be thought of as "an expression that is true if and only if P is false." This is *semantic understanding*.

When one is given a mathematical task to complete, one is often asked to complete that task within a particular representation system by moving from a collection of initial configurations to a desired configuration. For instance, in solving an algebraic equation, a student is often asked to apply algebraic operations on this equation until deriving a statement of the form $x = n$, where n is a number or set of numbers. One may complete such tasks by working *within* the system, i.e., by applying the rules for that system without linking the configurations being studied to configurations in other systems. This is *syntactic reasoning*. In contrast, one can use semantic understanding of relevant configurations to choose which moves to make. In the case of solving an algebraic equation, one might determine how many solutions there will be by graphing each side of the equation to see how many times the graphs intersect. If the graphs do not intersect, we can then demonstrate the equation has no solutions. This is *semantic reasoning*.

The Representation System of Mathematical Proof

In this chapter, we do not wish to give a comprehensive description of the representation system of mathematical proof. Doing so would be tantamount to giving a definition of mathematical proof, something that is arguably impossible and certainly beyond the scope of this chapter. Rather, we wish to describe some distinguishing characteristics of the representation system of mathematical proof and use these as a starting point to discuss what challenges and affordances are offered by this system. Distinguishing features of the representation system of mathematical proof include:

Characters in this system include mathematical symbols, logical symbols, and a subset of the English language (or the natural language of the proof's intended audience). Words used in proofs, especially nouns, have a precise, unambiguous mathematical meaning.

Permissible configurations include well-formed formulas from first-order logic, grammatically correct sentences in the English language, and sentences combining English words and logical symbols. Configurations that are not sentences of this type (e.g., graphs, manipulatives, diagrams) are not permissible.

Valid proofs must employ acceptable proof frameworks (Selden & Selden, 1995), where a proof framework specifies what can be assumed in the beginning of the proof and what must be concluded by the end of the proof, based on the statement that is being proven.

Reasoning about a mathematical concept must either be based on the definition of that concept or on other established theorems deduced from that definition. Arguments cannot be based on impermissible informal representations of that concept.

Assertions in a proof must either be assumptions permitted by the proof framework, or statements that are socially agreed on as established or deduced from previous assertions in the proof. By this, we mean that there is a socially accepted general mathematical principle that can be explicitly stated that can be used to deduce the new assertion from previous assertions stated in the proof (Weber & Alcock, 2005).

The mathematical community argues that couching mathematical arguments within this representation system increases the reliability of proofs, assuring the maxim that proven statements stay proven (Jaffe & Quinn, 1993), in part because this system facilitates a skeptic's ability to determine if a purported proof is flawed (Selden & Selden, 2003). While we agree that the representation system of proof has these merits, it is not our purpose in this chapter to advocate that this system alone be used by practicing mathematicians or be featured in the collegiate mathematics classroom. Rather, we accept that this system is imposed on undergraduate students in their advanced mathematics courses and we examine what benefits this system can have for undergraduate students in advanced mathematics courses as well as what difficulties it may cause.

Syntactic and Semantic Proof Productions

When a student is asked to prove a statement, he or she can accomplish this task in (at least) two qualitatively distinct ways. First, a student may attempt to construct this proof by working within the representation system of proof. That is, the student can choose a proof framework, list his or her assumptions, derive new assertions by applying established theorems and rules of inference, and continue until the appropriate conclusion is deduced. All this can be accomplished without considering configurations in other representation systems, such as graphs, informal arguments, or prototypical examples of relevant mathematical concepts. We call attempting to prove in this manner *syntactic reasoning* and proofs successfully produced in this way are dubbed *syntactic proof productions* (Weber & Alcock, 2004; see also Alcock & Weber, in press).

Alternatively, when asked to prove a statement, a student can first attempt to understand that statement semantically by linking aspects of the statement to configurations in another representation system. The student could then try to find an explanation for why the statement is true within this alternative representation system. For instance, if a student were asked to prove that increasing functions did not have global maxima, that student could think of graphs of increasing functions and graphs of functions with global maxima. The student could then build an informal explanation for why no graph could have the property of being both increasing and possessing a global maximum. Informal explanations could then be expressed formally within the representation system of mathematical proof. We call this type of reasoning *referential* or *semantic reasoning*, and we define proofs produced in this way as *semantic proof productions* (Weber & Alcock, 2004; see also Alcock & Weber, in press).

328 • Keith Weber & Lara Alcock

Syntactic Reasoning and Proof Productions

Opportunities Afforded by Syntactic Reasoning

When constructing a proof, a student may do so entirely syntactically, without needing to refer to informal representations of mathematical concepts. To illustrate how a student can accomplish this, we present an excerpt from an exploratory study that we conducted in which we examined the proving processes of undergraduate students in a transition-to-proof course. In this study, Carla was presented with the following definitions:

A function $f: \mathbf{R} \to \mathbf{R}$ is said to be **increasing** if and only if for all $x, y \in \mathbf{R}$, $(x > y$ implies $f(x) > f(y))$.

A function $f: \mathbf{R} \to \mathbf{R}$ is said to **have a global maximum at a real number c** if and only if for all $x \in \mathbf{R}$ such that $x \neq c, f(x) < f(c)$.

Carla was then asked to think aloud while proving that an increasing function could not have a global maximum. Her response was as follows:

Carla: So . . . I'm thinking the way to prove this is using contradiction. So, I would start out by assuming . . . there exists . . . a c . . . for which . . . [*writing*] . . . f of x is less than f of c, when x is not equal to c. Okay. [Pause.] So now I'm trying to use the definition of increasing function to prove that, this cannot be. Um . . . so there exists a real number for which f of x is less than f of c, for all x . . . and there's . . . f . . . is an increasing function . . . for . . . all x . . . *writing* . . . y in R, x greater than y implies f of x greater than f of y. Mm . . . I guess what I'm trying to show is if x is in reels, and they are infinite . . . for all x . . . there will be . . . some function f of c greater than f of x. [Long pause.] So . . . there exists . . . an element . . . in R . . . [writing] . . . greater than c. Um . . . for x . . . because . . . f is an increasing function . . . f of x will be greater than f of c. Um . . . a contradiction . . . so that . . . there is no c for which f of c is greater than f of x . . . for all x. (from Alcock & Weber, in press)

In this excerpt, Carla appears to construct a proof without considering any informal representations of increasing functions of global maxima, and without forming an intuitive explanation for why this statement is true. (Carla's comments later in the interview strongly support this interpretation.) Indeed, Carla arguably could have constructed this proof without any previous experience with the concepts of increasing functions or global maxima. Some mathematics educators might find Carla's proof problematic, arguing that she has constructed a proof without understanding. (In the next subsection, we will see that such suspicions are justified to some extent.) However, there are some cases when it is appropriate and desirable for students to construct proofs in this way. First, there are some important theorems in undergraduate mathematics courses that are difficult to justify intuitively (e.g., trigonometric identities or the statement that the sum of two continuous functions is continuous). In other words, there may be times when it is easier to justify why a statement is true within the representation system of proof than within more informal representation systems. In these cases, a syntactic proof production may be the only justification that undergraduates can provide.

Second, undergraduates often possess inaccurate intuitive understandings of concepts in advanced mathematics or may lack an intuitive understanding about a concept entirely. Students may lack the ability to construct links between the formal representation system of proof and other representation systems. Again, constructing a proof syntactically may allow an undergraduate to produce a valid argument when they otherwise could not produce.

Another opportunity afforded by syntactic reasoning rests on the fact that the range of acceptable proof frameworks available to a prover is relatively small. Once one chooses a proof structure, stating the first and last line of the proof can be viewed as a procedural skill. For instance, when using a direct proof for an "if–then" statement, the antecedent of the implication is assumed and its consequent is deduced. In undergraduate mathematics there are statements such that once one chooses a proof technique and applies definitions, it becomes straightforward to prove that statement.

For some classes of theorems, there are "proof templates" that one can use to prove a class of statements, often reducing the decision-making process involved in proving to an exercise in algebra or arithmetic. To illustrate this, we present an excerpt from an exploratory study from real analysis. One student in the study, Erica, met with the first author shortly after she had completed a homework assignment in which she proved that particular sequences converged to particular limits. In our interview, the first author asked Erica to "think aloud" as she proved that the sequence $(\frac{n-1}{n})$ converged to 1. Looking at her class notes, Erica presented the following argument:

Erica: You start this proof by writing, "Let ε greater than zero be given. Let N equal." Now he [the professor] uses scratchwork over here to find the N. [The professor] says, let's see . . . [pause] OK to show this converges to 1, we . . . yeah, OK, we start with the absolute value of $(n - 1)/n$ minus 1, and we'll rewrite this as 1 minus $1/n$ minus 1 . . . which is $1/n$ and let's see . . . yeah so he writes that here as less than $1/N$ which is less than epsilon So we have N is greater than 1 over epsilon. OK, so let N equal 1 over epsilon and . . . OK, we write if the absolute value of a_n minus 1 is less than epsilon . . . (from Weber, 2005)

Erica continued to produce the proof in the manner transcribed here. She would first look at her notebook and examine a similar proof that the professor did. She would then infer the actions used by the professor and apply similar actions in the proof that she was constructing. Erica continued in this way until she had completed a valid proof.

In some cases, a proving procedure can even become *automated* so that after recognizing that the proving procedure is appropriate to use, constructing the proof becomes a routine exercise requiring little thought or decision making. Although many mathematics education researchers understandably question whether students should learn proving procedures to which they cannot attach meaning, there are two potential benefits for having students do so in some cases. First, Sfard (2003) argues that learning new mathematics often consists of reflecting on one's mathematical actions. To reflect on the processes that we perform, a certain mastery of these processes is required, perhaps to the degree of automatization (Sfard, 2003, p. 365). In Weber (2005), the first author shows that one way in which some undergraduate students develop an understanding of proof is by first learning proving processes that they do not fully understand and then reflecting on those processes. Hence, while mastering a proving procedure should not be the end goal of instruction, it can be used as a means to develop students' understanding. A second benefit is that it may be useful for low-level details and lemmas in a complicated proof to be dealt with routinely and with little cognitive effort, allowing the student to deal with more complicated and conceptually rich considerations.

Difficulties with Syntactic Reasoning

An important purpose of proving is to develop an intuitively meaningful understanding of *why* an assertion is true (e.g., Hanna, 1990). One way in which such an understanding is achieved is by building strong links between intuitive understanding of relevant mathematical concepts and the proof that he or she produces (Raman, 2003; Weber & Alcock, 2004). In syntactic proof productions, one does not develop or make use of links between the representation system of proof and other informal representations of mathematical ideas. This can leave the prover feeling somewhat unconvinced by the argument that he or she produced (e.g., Fischbein, 1982; Raman, 2003; Weber & Alcock, 2004). In the previous subsection, we illustrated Carla constructing a proof syntactically. After she constructed her proof, she lamented to the interviewer that she did not feel that she fully understood it, arguing the proof felt "flaky," since she was able to write the proof "without being able to imagine what's going on":

Carla: [The proof] seems a bit flaky, I'm not too sure still.
I: Can you explain in what way it feels flaky, in that case?
Carla: I don't know, it just doesn't make sense for me. It, it feels like, I just, it's just proved systematically, without being able to imagine what's going on. So that's why it feels flaky ... Can I look at it once more? [Pause.] I guess this part where, I say that it's an increasing function, and the whole part where I say there exists an element in R greater than c ... it seems like it would need more proof to it? (from Alcock & Weber, in press)

Another potential problem with syntactic reasoning is that undergraduates often have trouble mastering the rules for legally constructing permissible logical sentences, as well as manipulating and drawing deductions from logical assertions. One consequence is that many of the proofs that undergraduates produce contain logical errors. Moreover, even when students are proficient with the rules and structure of the representation system of proof, this does not ensure that they will be capable of proving a wide range of novel theorems. When proving a theorem, there may many valid inferences that one can draw, only a small number of which will be useful. Undergraduates often fail to construct proofs not because they lack the procedural, logical skills or the factual knowledge to do so (e.g., they fail to negate an assertion properly or are not aware of an important theorem), but because they cannot choose the correct actions to take from so many legal alternatives. To illustrate, we present an excerpt taken from an exploratory study on group theory. In this study, students were asked to think aloud while attempting to prove challenging propositions about group homomorphisms. One student, James, was asked to prove the following proposition:

> Let G and H be groups and f be a surjective homomorphism from G to H. Let G be a group of order pq, where p and q are prime. Show that H is abelian or H is isomorphic to G.

James' response was as follows:

James: So let's see . . . I'm not quite sure what to do here. Well . . . injective . . . well if G and H have the same cardinality, then we are done. Because f is injective then. And f is surjective. G is isomorphic to H. With the isomorphism being f. OK, so let's suppose their cardinalities are not equal. So we suppose f is not injective. Show H is abelian . . . OK f is not injective so we can find distinct x and y so that f(x) is equal to f(y). (Weber, 2001, p. 109)

James proceeded to draw a series of valid inferences, concluding with the group element xy^{-1} was in the kernel of f, before abandoning the proof attempt as he did not know how to proceed. On a subsequent paper-and-pencil test, James demonstrated that he was aware of the facts and theorems needed to prove this theorem and when he was specifically told which facts to use, he was able to construct the proof. This excerpt illustrates a fundamental difficulty with syntactic reasoning within the representation system of proof. The approach that James took in proving this theorem and the inferences that he drew were all permissible in the representation system of proof, but he wasn't able to choose a productive approach or draw useful inferences from among his many permissible options. Weber (2001) observed that many students had difficulties similar to James— they could state the facts and apply the logical rules, but they failed to construct proofs because they were not able to make good decisions about which theorems and rules to apply.

Semantic Reasoning

In the last section, we described benefits of working syntactically within the representation system of proof—primarily that this representation system allows students to justify some propositions that would otherwise be inaccessible to them, either because they lack an adequate conceptual understanding or because the statements are intrinsically difficult to justify in alternative, informal representation systems. However, we also highlighted difficulties with this type of reasoning. We believe that each difficulty stems from the fact that the students involved did not link aspects of the representation system of proof to other systems: if the rules for manipulation in such a system seem completely arbitrary, one will have trouble learning them and applying them correctly. If one views a proof *only* as a sequence of words and symbols obeying arbitrary conventions, proofs will not seem convincing. If one views the process of proving *only* as searching for deductions to get to a desired "goal" statement, proving will feel like blindly working through a maze—a maze that becomes prohibitively large as the mathematical domains being studied become more complex. In this section, we first show how semantic reasoning can alleviate these difficulties, then illustrate new difficulties that semantic reasoning introduces.

Opportunities Afforded by Semantic Reasoning

We have observed elsewhere that those who are competent in a given mathematical domain are likely to have access to multiple informal representations of specific and generic examples within that domain (Weber & Alcock, 2004). Often, both undergraduate students and mathematicians can represent examples of concepts informally and use these representations to draw quick and convincing inferences about the concept, inferences that may not be obvious within the representation system of proof. For instance, in Alcock & Simpson (2004), undergraduate students were observed reasoning about limits throughout their real analysis course. In the excerpt that follows, Cary was asked whether convergent sequences must necessarily be bounded. He sketched a graph of a prototypical convergent sequence, described it, and decided that the statement is true:

Cary: I've drawn . . . convergent sequence, such that . . . I don't know, we have . . . curves . . . approaching a limit but never quite reaching it, from above and below, and oscillating either side. [Pause.] I was trying to think if there's a sequence . . . which converges yet is unbounded on both sides. But there isn't one. Because then it wouldn't converge. So I'll say [the statement] is true.

Cary then considered several possible properties of convergent sequences, assessing whether these could be used as the basis for a proof by evaluating them against other types of example:

Cary: If it converges . . . that has to be . . . well, I don't suppose you can say bounded. It doesn't have to be monotonic [. . .] I'm trying to think if there's like . . . if you can say the first term is like the highest or lowest bound but it's not. Because then you could just make a sequence which happens to go . . . to do a loop up, or something like that.

Cary's informal reasoning provided him with a strong conviction that convergent sequences are indeed bounded, and this conviction was obtained without the laborious process of constructing a formal proof. This can be contrasted with Carla's lack of comfort in her syntactic proof production described in the previous section. Using Fischbein's (1982) terminology, we might say that Carla's proof provided her with *formal extrinsic conviction* that increasing functions do not have global maxima. She applied a technique that she understood would yield a valid proof and was convinced the technique was applied properly. However, she did not attain the *internal intuitive conviction* that Cary gained by exploring representations of sequences that were internally meaningfully to him. The fact that many undergraduate students can gain stronger psychological conviction from informal reasoning than from rigorous argumentation is a potential obstacle for having undergraduate students appreciate the value of the representation system of proof.

A second benefit of semantic reasoning is that it can aid in the construction of formal proofs by helping the prover decide what line of reasoning to pursue. As Raman (2003) notes, inspection of non-formal representations of mathematical concepts can provide the "key idea" for how a formal proof about that concept should proceed. We illustrate this through Ellen's work in the transition course study. In this study, Ellen was asked to describe the processes that she used to construct proofs on her mid-term examination. Ellen had successfully produced a correct answer to a question that according to her professor (and to other participants from her class) was difficult for most to even begin. In this question, a relation was defined on $\mathbf{R} \times \mathbf{R}$ such that (a, b) is related to (c, d) if both $a \le c$ and $b \le d$. The question asked the student both to prove that any two-element subset of $\mathbf{R} \times \mathbf{R}$ had an upper bound with respect to this relation and to find the least upper bound of $\{(-1, 2), (3, -4)\}$. In explaining her thinking in producing her answer, Ellen said:

Ellen: For the second part [. . .] like I sort of skipped it and went to the third part where he gave the actual numbers. [. . .] And I just realized that I could draw a graph of it [see Figure 19.1 for Ellen's graph.] So I did over here and I realized

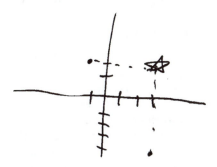

Figure 19.1 Ellen's diagram of constructing proof

the upper bound would be the corner of the sort of rectangle that would make. [. . .] So I just thought that any upper bound would sort of just be to the right and up of your two points. So for this one it was obvious that it was the corner of the square. The least upper bound would be sort of on the lines and then um, so I went back to the second part then and [inaudible] um since it's real numbers for um well it's um, I guess pairs of real numbers so either way you can go an infinite amount. So for any um two pairs that he takes, you can always go um larger I guess. So that's why there's an upper bound (Weber, Alcock, & Radu, 2005).

The importance of the graphical representation in proving this statement was further stressed when Ellen was asked whether she had had any difficulties with this question:

Ellen: Um, I couldn't really conceptually visualize it, I guess. Um until I got to the third part. And then it sort of hit me, what he meant by it. Because um the things that we went through usually is not two ordered pairs. It was like one object and another object. Like x and y. That's why it was a little difficult to me. To see what he means. (from Weber, Alcock, & Radu, 2005)

Here we can see an affective response associated with semantic reasoning. The sense of having grasped the meaning seems important to Ellen, as it did to many of those students who regularly used semantic reasoning in both of the afore-mentioned studies (see, in particular, the case of Brad in Alcock & Weber, in press). Indeed, both mathematicians and students who engage in this type of reasoning may consider it to be an essential part of mathematical activity. For example, Adam, a student of real analysis who regularly invoked both specific examples and generic images in his reasoning, talked about the necessity of this search for meaning in his interactions with the $\varepsilon - N$ definition of convergence.

Adam: You can just write down the definition without, kind of remembering what it means, I find. So, you have to like . . . sort of write it, and think like . . . what does that mean, and what have I got, and how can I put it into that form?

Difficulties with Semantic Reasoning

One potential problem with semantic reasoning, as many researchers have observed, is that an example or image may incorporate properties that are not universally true, and therefore mislead the reasoner into trying to prove untrue general claims (e.g., Tall, 1995). A second is that an individual may be able to develop a mathematically correct understanding of why a proposition is true, but be unable to frame their argument within the representation system of proof. For example, in the transition course study, Hannah was asked to think aloud as she tried to prove an increasing real-valued function could not attain a global max-imum—the same task that Carla completed in the previous section. Having read the provided definitions of strictly increasing and having a global maximum at *c*, she appeared to be comfortable with the assertion that follows:

Hannah: . . . well, *c* would be saying that the function is like, finite in the sense that it's . . . it has a maximum. And . . . this one, if it's increasing, then . . . yes. If it's increasing, then each . . . as *x* gets higher, um like the *x*, *y* coordinate plane . . . *y* is increasing, so like it wouldn't have a maximum.

She also drew some diagrams (see Figure 19.2) to back up her assertions.

She verbally re-expressed and refined this basic idea at several times during the next few minutes, and made several attempts to introduce notation and begin a proof. However, she was unable to express her ideas formally, even though the required definitions were in front of her on the page (Brad's attempts, detailed in Alcock & Weber, in press, were very similar). The interviewer's questions about her attempts to begin led to the following exchange:

I: One other thing that I noticed you said at the beginning [. . .] "Well it's intuitively . . . ," and then you kind of stopped. Can you explain what was going on there?

Hannah: Just, it's—it's so like . . . it's so . . . o . . .

I: You can say that if you want.

Hannah: It's so obvious, it's so like . . . so like a 2nd grader could tell you this, sort of. Maybe 6th grade, I don't know [. . .] but that doesn't mean much because you still have to figure it . . .

Hannah's tone during this exchange was one of comedic frustration, and frustration is commonly expressed by students in this situation, who feel that they

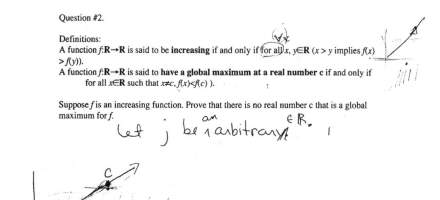

Figure 19.2 Hannah's diagrams to back her assertions

understand the relationships between the concepts they are considering but are unable to express this understanding in an acceptable way. Overall, they may come to feel that the proofs they see in class are quite unrelated to their own understanding.

Discussion

The preceding discussion implies two goals for advanced mathematics courses at the collegiate level. First, undergraduate students should develop strong links between their informal reasoning and the representation system of proof. These links will provide them with the skills to prove complicated theorems and to obtain conviction and meaning from the proofs that they are shown and that they produce. Second, undergraduate students should also develop the skills to reason syntactically within the representation system of proof, if only to prove statements that are difficult to justify intuitively or to prove statements about concepts that they are still learning about. The important pedagogical question then becomes, how can these goals be achieved?

One interesting feature of Goldin's (1998) theory of representations is that the links between representation systems do not have directionality. Within Goldin's theory, configurations in each representation system are seen as simultaneously signifiers and signified. This suggests two theoretical trajectories that an undergraduate student might take toward understanding the representation system of proof. The student could first use his or her existing understanding of informal proofs to give meaning to proofs within the formal representation system. After formal rules have been extracted, students could then learn how to reason syntactically, perhaps by examining and comparing the syntactic structure of multiple related arguments. Alternatively, the student could first learn how to work within this system syntactically, treating proving as an activity (initially) somewhat independent from their other ways of reasoning. After becoming comfortable working within the representation system of proof, the student could then relate the rules and structures of proof to more intuitive ways of reasoning, perhaps by examining the ways in which syntactic steps relate to specific examples of the concepts involved.

Initially, like many mathematics educators, we believed that students would be most likely to succeed by following the first of these trajectories—viewing the representation system of proof as a formalization of their prior ways of forming convincing arguments. We argued that students who tried to produce proofs in this way would be more successful than those who tried to produce proofs syntactically (see Weber & Alcock, 2004). However, in recent years, a growing body of research suggests that our initial assumption was too simplistic—that the best way for students to learn the representation system of proof *depends on the student.* For example, Pinto and Tall (1999) analyzed the behavior and learning strategies of undergraduate students in an introductory real analysis course in the UK. This course was the first exposure that most students had to the representation system of proof. Some students attempted to learn to work within this system by *giving meaning* to the symbols within the system. They would use graphs, prototypical examples, and other informal types of reasoning to recall, make sense of, and reason from the definitions provided for them. Others would *extract meaning* from the definitions—initially memorizing a definition as a sequence of symbols and performing logical deductions on it, and only later attempting to link the definitions to other representations. Two important findings from Pinto and Tall's analysis emerged. First, individual undergraduate students were consistent in the reasoning strategies that they used, suggesting that they had qualitatively distinct *reasoning styles*. Second, neither strategy appeared to be superior to the other strategy; there were successful and unsuccessful students who used each strategy. Several empirical studies that we have conducted (Alcock & Simpson, 2004, 2005; Alcock & Weber, in press; Weber, Alcock, & Radu, 2005) have found similar results. Different undergraduate students used different reasoning styles to make sense of the material in their advanced mathematics courses and we did not find a strong correlation between reasoning style and performance.

Another suggestion is that many mathematicians have strong reasoning preferences too. This is consistent with claims made by and about famous mathematicians (Pinto & Tall, 1999) and systematic interviews with many practicing mathematicians (Burton, 2004). Further, some mathematicians seemed unaware that there are other forms of productive mathematical reasoning than the ones that they employ (Burton, 2004). We believe that this is an extremely important point because having one's own preferences can lead one to (perhaps unintentionally) undervalue the efforts of those who reason differently. For instance, some professors hold the belief that undergraduate students must have a strong syntactic understanding of logic before learning about complex mathematical concepts (see Marty, 1986; Weber, 2004). When a student presents an informal argument to support an assertion but cannot prove it, it is natural for these teachers to play down the informal argument and direct the student to focus on syntactic reasoning (proof frameworks, unpacking definitions, etc.) Similarly, such teachers may attempt to help a student correct a proof by highlighting the logical "rule" that was violated when, in fact, an error could be due to a poor conceptual understanding of a situation.

In our observations, many students who are inclined towards semantic reasoning find advice of this type discouraging, and feel isolated because the teacher does not value their mathematical thinking. These students might be better helped via tasks that will help them refine their image of a concept or build links between a concept and its formal representation in the system of proof. In contrast, when students who reason syntactically make a logical error, it is natural for some teachers to draw attention to this by showing that the given argument does not hold for a particular counterexample. To a student who has not yet formed links between symbolic definitions and meaningful mathematical objects, the teachers' explanation may not be helpful because it does not provide any direction regarding how to correct the argument. It might be more beneficial to this student if the syntactic flaw in their logic is explicitly stated. What makes this an especially troublesome issue is that the same logical error can be the result of a flaw in either syntactic or semantic reasoning (Alcock & Weber, in press). This suggests to us that where possible, more individual attention must be given to student errors. Ideally, a teacher should determine how a student prefers to reason and whether their difficulty was due to flawed syntactic or semantic reasoning before providing instruction to remedy the student's difficulties.

We conclude this chapter by raising a question central to the nature of mathematical proof and discussing its relevance to collegiate mathematics education research. Is a mathematical proof a string of symbols that obeys the rules and conventions of contemporary mathematicians or is it an argument used to convince oneself and one's community that an assertion is true? Many mathematicians espouse the former viewpoint (e.g., Griffiths, 2000) while most mathematics educators favor the latter (e.g., Balacheff, 1987). Our response to this question is that both views are correct. Rather than viewing these positions as in opposition, we see them as complementary, and representing two sides of the same coin.

Our fear is that syntactic reasoning in proof production is sometimes undervalued by the mathematics education community. For instance, Harel (2001) refers to students who prove by "thinking of symbols as if they possess a life of their own without reference to their functional or quantitative meaning" (p. 193) as possessing a symbolic quantitative proof scheme. This is counted as an external proof scheme, which students should be led to revise or abandon in favor of a deductive proof scheme. One interpretation of Harel's passage (although not necessarily shared by Harel) is that syntactic reasoning is not desired in the mathematics classroom. However, such an interpretation would run counter to the practice of some mathematicians, such as Andre Weyl, who in describing the essence of mathematical thinking, asserts:

> We forget about what the symbols stand for. The mathematician is concerned with the catalogue alone; he is like the man in the catalogue room who does not care what books or pieces of an intuitively given manifold the symbols of his catalogue denote. He need not be idle; there are many operations which he may carry out with these symbols, without ever having to look at the things they stand for. (Weyl, 1940)

To be clear, we agree that if undergraduates emerge from their advanced mathematics classes with only a syntactic understanding of proof, this represents a grave failure in their mathematical education. However, we stress that syntactic reasoning is legitimate mathematical reasoning and is used by some students to *develop* a semantic understanding of proof. Understanding how to balance both syntactic and semantic reasoning in advanced mathematics courses remains a significant challenge for teachers of mathematics and a research topic for mathematics educators.

20

Teaching Proving by Coordinating Aspects of Proofs with Students' Abilities

JOHN SELDEN AND ANNIE SELDEN

In this chapter, we introduce concepts for analyzing proofs, and for analyzing undergraduate and beginning graduate mathematics students' proving abilities. We discuss how coordination of these two analyses can be used to improve students' ability to construct proofs.

For this purpose, in order to keep track of students' progress we need a framework richer than the everyday one used by mathematicians. We need to know more than that a particular student can, or cannot, prove theorems by induction or contradiction or can, or cannot, prove certain theorems in beginning set theory or analysis. We more usefully describe a student's work in terms of a finer grained framework, including various smaller abilities that contribute to proving and that can be learned in differing ways, at differing times of a student's development.

Developing a fine grained framework for analyzing students' abilities is not an especially novel idea. In working with higher primary and secondary students, Gutiérrez and Jaime (1998) developed a fine grained framework of reasoning processes in order to more accurately and easily assess student van Hiele levels.

For proof construction, there are already a number of abilities suitable for keeping track of students' progress. For example, in comparing undergraduates who had completed a course in abstract algebra with doctoral students in abstract algebra, Weber (2001) found doctoral students more able to use strategic knowledge. For example, when asked whether two specific groups, such as **Q** and **Z**, are isomorphic, undergraduates first looked to see whether the groups had the same cardinality, after which they attempted unsuccessfully to construct an isomorphism between them. The doctoral students considered properties preserved by isomorphism, strategically a better starting point (Weber & Alcock, 2004).

The ability to use strategic information is about students' proving processes. But it is also useful to consider the product of such processes—the proofs themselves—because differing kinds of, and aspects of, proof can correspond to various abilities needed to produce them. Thus, an analysis of kinds and aspects of proof should facilitate teaching by coordinating assigned theorems with student abilities. It should also facilitate the assessment of student abilities.

In this first section, we introduce three different global aspects of proof that we call *structures*, and illustrate them using a real analysis proof that advanced undergraduate or beginning graduate students might construct. In the next section, we discuss the coordination of aspects of proofs with student abilities. After that, we make three informal observations. We conclude by summarizing our main ideas.

Three Structures of Proofs

Mathematics education has often emphasized student learning and activity, rather than the properties of the associated objects, such as the global aspects of proofs that we are calling structures. There are two previous discussions of similar structures. Leron (1983) suggested presenting proofs (e.g., in a lecture) in a "top down" way, first describing the main ideas before filling in the details. The resulting structure is reminiscent of what we call the hierarchical structure of a proof, except that Leron was not concerned with characterizing the proof itself, but rather with providing a kind of advance organizer for its presentation.

Konior (1993) was also concerned with a hierarchical structure of proofs, but he focused on segmentation, that is, "separation of the whole [mathematical] text into a few parts ... in order to reflect the logical structure of the proof and facilitate the reconstruction of the whole proof." He discussed delimiters, such as words and paragraph breaks that indicate beginnings and endings of proof segments (e.g., subproofs). However, an investigation of delimiters is beyond the scope of this chapter.

We introduce three proof structures, that is, aspects of a proof that refer to it as a whole, how it is organized, and how it could have been written. These emerged from our attempts to understand and characterize our students' proof writing difficulties. For example, such difficulties might arise from a tendency not to focus on producing a subproof, supposing that a proof should be written linearly from the top down as if it were being read, or concentrating on the big picture to the exclusion of the "nuts and bolts" of proving. By attending to these structures a teacher might be able to arrange for the associated student difficulties to arise in a way that allowed them to be overcome.

We discuss (1) a *hierarchical* structure in which we attend to subproofs and subconstructions (such as finding the δ in an $\varepsilon - \delta$ real analysis proof); (2) a *construction path*, that is, a linear path describing one ordering of the steps[1] through which a proof could have been constructed by an idealized prover, one who never erred or followed false leads; and (3) a division of proofs into what we call the *formal–rhetorical* part and the *problem-centered* part. For any given proof these three structures, considered together, can in a qualitative way suggest the proof's complexity. That complexity, in turn, can provide one indication of the proof's difficulty, and knowledge of a proof's difficulty can be an important tool in teaching.

Of course, a particular theorem can often be proved in several ways. However, there are not likely to be very many such ways for the theorems most advanced undergraduate and beginning graduate students are asked to prove.

The Hierarchy of a Proof and a Possible Construction Path

We analyze a proof of the theorem: $f + g$ *is continuous at a point provided f and g are.* We build the proof up hierarchically, attending to subproofs and subconstructions. Simultaneously, we build a construction path giving an order in which its sentences, or subsentences, could have been written by an idealized prover. For reference, and to facilitate the subsequent analysis, individual sentences have been numbered in bold brackets (e.g., [1], [2]) in their order of appearance in the proof. For example, the sentence labeled [12, 13, 14, 15] is divided into four smaller units, one for each statement of equality or inequality.

Proof

[1] Let a be a number and f and g be functions continuous at a. [2] Let ε be a number > 0. [3] Note that $\varepsilon/2 > 0$. [4] Now because f is continuous at a, there is a $\delta_1 > 0$ such that for any x_1, if $|x_1 - a| < \delta_1$, then $|f(x_1) - f(a)| < \varepsilon/2$. [5] Also because g is continuous at a, there is a $\delta_2 > 0$ such that for any x_2, if $|x_2 - a| < \delta_2$, then $|g(x_2) - g(a)| < \varepsilon/2$. [6] Let $\delta = \min(\delta_1, \delta_2)$. [7] Note that $\delta < 0$.

[8] Let x be a number. [9] Suppose that $|x - a| < \delta$. [10] Then, $|x - a| < \delta_1$, so $|f(x) - f(a)| < \varepsilon/2$. [11] Also $|x - a| < \delta_2$, so $|g(x) - g(a)| < \varepsilon/2$. [12, 13, 14, 15] Now $|f(x) + g(x) - (f(a) + g(a))| = |(f(x) - f(a)) + (g(x) - g(a))| \leq |(f(x) - f(a))| + |(g(x) - g(a))| < \varepsilon/2 + \varepsilon/2 = \varepsilon$. [16] Thus $|f(x) + g(x) - (f(a) + g(a))| < \varepsilon$. [17] Therefore $f + g$ is continuous at a. QED.

The first step of the construction path appears as [H1], for "hidden." It expresses information needed to construct the proof, but that does not appear in the proof. It is done by rewriting the statement of the theorem in a more formal way, explicitly mentioning variables and quantifiers and using standard logical connectives. This "clarification" yields the statement, [H1]: *For all real-valued functions f, all real-valued functions g, and all real numbers a, if f is continuous at a and g is continuous at a, then f + g is continuous at a.*

This version of the statement of the theorem exposes its logical structure. It is essential to understand the logical structure in order to be certain that one's proof proves this theorem, as opposed to some other theorem. The logical structure, independent of the meaning of "function," "+," and "continuous," yields the first and last sentences of the proof. Together they form what we have called a *proof framework*[2] (Selden & Selden, 1995). The resulting partial construction path of our idealized prover is now [H1], [1], [17], ... and the resulting partial hierarchical structure is shown in Figure 20.1.

Next we clarify line [17] by applying a definition of continuous to $f + g$ at the number a. This clarification yields a new statement, [H2], that is needed to construct the proof, but does not appear in it: For every number $\varepsilon > 0$, there is a $\delta > 0$, so that for every number x, if $|x - a| < \delta$ then $|f(x) + g(x) - (f(a) + g(a))| < \varepsilon$.

342 · John Selden & Annie Selden

Proof: [1] Let a be a number and f and g be functions continuous at a

$$\bullet\ \bullet\ \bullet$$

[17] Therefore $f + g$ is continuous at a . QED

Figure 20.1 Top level of the hierarchical structure

Proof: [1] Let a be a number and f and g be functions continuous at a

[2] Let ε be a number > 0. Let $\delta = \ \bullet\ \bullet\ \bullet$
[7] Note that $\delta > 0$. [8] Let x be a number. [9] Suppose that $|x - a| < \delta$

$$\bullet\ \bullet\ \bullet$$

[16] Thus $|f(x) + g(x) - (f(a) + g(a))| < \varepsilon$

[17] Therefore $f + g$ is continuous at a . QED

Figure 20.2 Adding the second level of the hierarchical structure

This statement requires a subproof and writing a proof framework for it yields the partial construction path [H1], [1], [17], [H2], [2], [7], [8], [9], [16], . . . and the resulting partial hierarchical structure is shown in Figure 20.2.

Next we add in the construction of δ and finish the proof, yielding the complete hierarchical structure in Figure 20.3.

In Figure 20.3, we first added lines [3]–[6] of the construction path and then added lines [10]–[15] of the construction path. The first addition was treated as a new level in the hierarchical structure, but the second was not. This is because lines [3]–[6] can stand alone as the construction of δ, but lines [10]–[15] cannot stand alone as a subproof. Rather, they form a part of [2]–[16], the proof of [H2].

The final construction path of our idealized prover is now: [H1], [1], [17], [H2], [2], [7], [8], [9], [16], [3], [4], [5], [6], [10], [11], [12], [13], [14], [15].

Although the construction path and the hierarchical structure illustrated earlier can contribute to observing student abilities, one might also ask: do undergraduate or graduate students just beginning to construct proofs even tacitly understand such structures? Does this matter? Perhaps a hint at an answer can be obtained by considering the following incident.

During a chance discussion of this theorem with a mathematics education graduate student, she expressed surprise that constructing proofs might involve the kind of structures illustrated here, rather than progressing linearly from top down. It turned out that she had studied only a little advanced mathematics and much or all of that had been presented "top down" in the traditional definition–theorem–proof style.

Our experience suggests—as this incident does—that at least a tacit

Proof: [1] Let a be a number and f and g be functions continuous at a.

[2] Let ε be a number > 0

[3] Note that $\varepsilon/2 > 0$. [4] Now because f is continuous at a, there is a $\delta_1 > 0$ such that for any x_1, if $|x_1 - a| < \delta_1$, then $|f(x_1) - f(a)| < \varepsilon/2$. [5] Also because g is continuous at a, there is a $\delta_2 > 0$ such that for any x_2, if $|x_2 - a| < \delta_2$, then $|g(x_2) - g(a)| < \varepsilon/2$. [6] Let $\delta = \min(\delta_1, \delta_2)$

[7] Note that $\delta > 0$. [8] Let x be a number. [9] Suppose that $|x - a| < \delta$ [10] Then $|x - a| < \delta_1$, so $|f(x) - f(a)| < \varepsilon/2$. [11] Also $|x - a| < \delta_2$, so $|g(x) - g(a)| < \varepsilon/2$. [12, 13, 14, 15] Now $|f(x) + g(x) - (f(a) + g(a))| = |(f(x) - f(a)) + (g(x) - g(a))| \leq |(f(x) - f(a))| + |(g(x) - g(a))| < \varepsilon/2 + \varepsilon/2 = \varepsilon$. [16] Thus $|f(x) + g(x) - (f(a) + g(a))| < \varepsilon$

[17] Therefore $f + g$ is continuous at a. QED.

Figure 20.3 Adding the third level of the hierarchical structure and finishing the proof

understanding of the proof structures examined earlier matters and that, early in their exposure to advanced mathematics, some students do not grasp much about them. Such students are likely to have a view of proof construction that we believe will prevent them from succeeding.

The Formal–Rhetorical and Problem-Centered Parts of a Proof

Notice that writing steps [H1], [1], [17], [H2], [2], [7], [8], [9], [16] in the construction path of our idealized prover (Figure 20.2) calls on an understanding of the logical structures of the statement of the theorem and of the definition of continuous. This is made clear in the statements [H1] and [H2]. In addition, one needs to know, and to act on, how parts of the statement of a theorem relate to parts of its proof. We call knowledge of this kind *behavioral knowledge*[3] because a tendency to behave or act is a part of it. It is closely related to "knowing-to-act in the moment" (Mason & Spence, 1999), as well as to the ideas of "concepts-in-action" and "theorems-in-action" (Vergnaud, 1982). For example, if a formal version of a theorem started, "For all real numbers x," then in a direct proof, one might start by "introducing" the variable x, with a statement like "Let x be a real number." For some authors of proofs, this statement might be left implicit instead of being made explicit, especially when x appears in the statement of the theorem. Such an x is usually said to be "arbitrary, but fixed" and does not "vary."

Quite a lot of such behavioral knowledge is required in constructing proofs.

While it is not important that a student be able to articulate such behavioral knowledge, it is important that he/she act on it. Although behavioral knowledge neither implies nor is implied by the corresponding procedural (knowing how, without necessarily acting) or conceptual knowledge (knowing why), these can also be useful.

For a student with a reasonable repertoire of proof-related behavioral knowledge, writing the steps in Figure 20.2 can be quite straightforward. Such writing can be viewed as invoking a schema. It does not depend on a deep understanding of, or intuition about, the concepts involved or on genuine problem solving in the sense of Schoenfeld (1985, p. 74). We call this part of a proof the *formal–rhetorical* part and contrast it with the remainder of the construction path of our idealized prover, [3], [4], [5], [6], and [10], [11], [12], [13], [14], [15] (Figure 20.3), that we call the *problem-centered* part. This part of a proof *does* involve problem solving. The steps in the problem-centered part may call on conceptual knowledge, mathematical intuition, and the ability to bring to mind the "right" resources at the "right time."

Constructing the formal–rhetorical parts of proofs seems to call on a different kind of knowledge than the problem-centered parts do. Helping students acquire these different kinds of knowledge probably involves different kinds, or aspects, of teaching, even though these might be blended in a single course and taught by a single teacher. Such a blending of teaching seems especially appropriate for learning to construct proofs because the two parts of a proof interact. Indeed, there are theorems for which constructing the formal–rhetorical part of a proof can be very helpful in revealing the "real problem" to be solved. Our sample theorem is such a theorem. Since it is about continuous functions, one might expect some kind of visual–spatial intuition about functions might be useful for constructing a proof. However, the formal–rhetorical part (Figure 20.2) reveals that the problem is to find a δ that will yield the inequality $|f(x) + g(x) - (f(a) + g(a))| < \varepsilon$ in step [16]. The solution does not arise in any obvious way from intuition. Instead, it involves use of $\varepsilon/2$ in the definition of continuity for both f and g, a choice of δ as the smaller of the resulting two δs, application of the triangle inequality, and some algebraic rewriting.

In the construction path of this proof, as in many short proofs, the entire formal–rhetorical part of the proof comes first, followed by the problem-centered part. However, in longer proofs with several subproofs, each subproof may have its own formal–rhetorical part and subsequent problem-centered part, and the order of the subproofs within a construction path may vary.

Coordinating Aspects of Proofs with Students' Abilities

Next we discuss some proving abilities that a student might have or that a proof might call for. It is the coordination of these two—what a student might be able to do, or not do, and what a proof might call for—that we suggest may facilitate teaching proving, mainly through students' construction of proofs, rather than in some more teacher-centered way such as lecturing.

Kinds of Proof

Transition-to-proof textbooks typically distinguish direct proofs, proofs by contradiction, proofs by mathematical induction, proofs by cases, and existence and uniqueness proofs. However, for guiding students' progress, such distinctions are not fine enough. For example, a student might be able to construct a number of direct proofs, but not the kind discussed in the first section. This might be due to its complex structure or the number of quantifiers involved. Knowing this would allow a teacher to guide a student's current work and to assign later theorems whose proofs are similarly complex.

Formal–Rhetorical Reasoning Versus Problem-Centered Reasoning

In the first section, we distinguished between the formal–rhetorical and problem-centered parts of a proof and suggested that writing these two parts calls on different kinds of knowledge, and hence, on different kinds of teaching. We call the abilities to write these two parts, *formal–rhetorical reasoning* and *problem-centered reasoning*, and apply these ideas to the kinds of exploration preceding and surrounding proofs, as well as to writing proofs.

Here is an example of the kind of problem-centered reasoning that can precede a proof. Two students with fairly strong upper division undergraduate mathematics backgrounds were jointly attempting to prove: *if the number of elements in a set is n, then the number of its subsets is 2^n*. They had been advised: *don't forget to count the empty set, $\varnothing$, and the whole set as subsets. For example the subsets of $\{a, b\}$ are $\varnothing, \{a\}, \{b\},$ and $\{a, b\}$*. They first considered $\{a, b, c\}$ and wrote $\{c\}, \{a, c\},$ $\{b, c\},$ and $\{a, b, c\}$. When asked about this, they indicated that they had "added" c to each of the subsets of $\{a, b\}$ to generate the four additional subsets of $\{a, b, c\}$. Next they considered $\{a, b, c, d\}$ and then seemed puzzled as to how to continue. The teacher intervened, "You can get from 2 to 3, and you can get from 3 to 4. What does that make you think of?" They replied "induction," something they had not thought of. These students could benefit from an opportunity to think independently of using induction on the proof of a subsequent theorem.

Comparing the Difficulty of Proofs

For an individual, a major determiner of the difficulty of constructing proofs seems to be the nature of his/her knowledge base and habits of mind. These are discussed later in connection with problem-centered reasoning. However, in the context of a course in which students prove all, or most, of the theorems, one can often see that one proof is more difficult than another by observing that only a few students can construct it. This suggests that there are characteristics intrinsic to proofs that make some more difficult than others, at least with respect to rough judgments of difficulty. Certainly, we have found a need to make such judgments especially in teaching Moore Method courses,[4] because we prefer to ask a student to prove a theorem that will be challenging, but not so challenging that he/she fails to do so.

To illustrate how rough judgments of the relative difficulties of proofs can sometimes be made independently of specific individuals, we compare our sample proof (Figure 20.3) with the proof (Figure 20.4) of the following theorem: *if f and g are functions from A to A and f og is one to one, then g is one to one.*

The earlier proof is longer, and there are more quantifiers to contend with, but there are other less obvious reasons that the earlier proof might be more difficult. The hierarchical structure of the second proof has only two levels (Figure 20.4) while the earlier one has three levels (Figure 20.3). Also, while the problem-centered part of the second proof consists of just three consecutive steps [5], [6], [7], the problem-centered part of the earlier proof consists of two separate sections, [3], [4], [5], [6] and [10], [11], [12], [13], [14], [15], each of which depends on the other. Finally, one might think the earlier proof could be developed from some kind of visual–spatial intuition about continuous functions, but this is unlikely. Hence, the formal rhetorical part of the earlier proof plays a large role, and a kind of technical–algebraic intuition is called far.

Sets and Functions

The language of sets and functions occurs widely in proofs, and undergraduate students are often introduced to it in an abstract way. More important than being able to recall abstract definitions, students need to use them in proofs, that is, be able to carry out appropriate actions effortlessly in order to leave maximum cognitive resources for the problem-centered parts of a proof. In doing this, students need what we are calling behavioral knowledge, and we suspect this is learned more from practice at constructing proofs than from reproducing abstract definitions.

For example, the definition of set equality is usually given as $A = B$ if and only if $A \subseteq B$ and $B \subseteq A$, and students are told that this means A and B have the same elements. But in constructing a proof that two sets are equal, it should come to mind easily that this involves showing that an arbitrary element of each set is an element of the other set. That is, normally two subproofs need to be constructed: one beginning *Suppose $x \in A$*... and ending ... *Then $x \in B$*; the other beginning *Suppose $x \in B$* ... and ending ... *Then $x \in A$*. Logically the definition involving two inclusions and the above "element-chasing" view of set equality are equivalent, but psychologically they appear to be very different. In our experience, naive students often attempt to deduce $A = B$ directly, without considering elements.

Proof: [1] Let f and g be functions from A to A. [2] Suppose $f \circ g$ is one-to-one
[3] Let x and y be in A. [4] Suppose $g(x) = g(y)$ [5] Then $f(g(x)) = f(g(y))$, [6] i.e., $(f \circ g)(x) = (f \circ g)(y)$ [7] But $f \circ g$ is one-to-one. [8] So $x = y$
[9] Therefore g is one-to-one. QED

Figure 20.4 Hierarchical structure of the second sample proof

Students can have a reasonably good intuitive grasp of the meaning of one to one (1–1), but not know how to prove a function f is 1–1. They may even realize that the definition involves an implication, namely, that $f(a) = f(b)$ implies $a = b$, for all a and b in the domain. However, that does not mean they know where to begin a proof (Moore, 1994). It may not be clear to them that they should normally begin the proof by almost automatically writing, *Suppose $f(a) = f(b)$ ...* or arbitrary elements a and b in the domain, and only then attempt to use the hypotheses to arrive at $a = b$.

Logic

Logic does not occur within proofs as often as one might expect. Furthermore, the logic used in proofs is mainly propositional calculus, and there is a tendency, whenever possible, to avoid variables and quantifiers. That is, typically proofs do not contain arguments about all elements x, but instead are about an arbitrary, but fixed, element x. This cannot always be accomplished and our proof about the sum of continuous functions is an exception in that x_1 and x_2 are universally quantified.

Where logic does occur within proofs, it plays an important role. For example, students should have the ability to effortlessly convert *not (p and q)* into *not p or not q*; *if p then q* into *(not p) or q*; and *if p then q* into *if not q then not p*. In addition, they should be able to draw inferences such as *q*, given both *p* and *if p then q* (*modus ponens*). It seems to us that it would be useful to coordinate a student's abilities to do these various logical activities with proofs that might call on them.

In addition, there are logical and logic-like activities that connect what happens within a proof to the external context. For example, before starting to write a proof of a theorem, a student should be able to unpack its logical structure, making variables and quantifiers explicit, and converting to standard logical connectives, such as if–then. This can be difficult for many students (Selden & Selden, 1995). After that, a student needs a suitable proof framework. For direct proofs, we have illustrated this with our two sample proofs (Figures 20.3 and 20.4). In order to handle proofs by contradiction, students also need to formulate negations of quantified statements. They should be able to negate "for all x, $P(x)$" almost automatically to get "there is an x such that not $P(x)$."

Finally, a student must be able to connect a previous theorem, or definition, with a proof in progress. This requires unpacking the previous theorem's, or definition's, logical structure, taking an instance thereof, adjusting the symbols to fit those used in the current proof, seeing that the premises are satisfied, and writing the corresponding conclusion into the current proof.

Problem-Centered Reasoning

Abilities in problem-centered reasoning are more difficult to separate out and observe than previously discussed abilities. However, problem-centered reasoning plays a very large role—ultimately a dominant role—in constructing proofs, so it cannot be omitted. As background, Schoenfeld's (1985) analysis of problem solving should be very helpful, although the time available to solve his problems was less than that usually needed for constructing proofs. This may make a considerable difference. One of the points Schoenfeld makes in regard to control is that students often fail to monitor their work, continuing too long in an unpromising direction. Surely this also happens in student proof construction. However, our experience suggests that another kind of control, persistence, can play a very positive role in proof construction. Strategic knowledge can also be very useful (Weber, 2001; Weber & Alcock, 2004).

A major factor in proof construction is the mathematics a student knows, what Schoenfeld (1985) includes in *resources*. However, bringing such resources to mind might also be regarded as an ability, because many students cannot do so (Selden, Selden, Hauk, & Mason, 2000). For example, earlier in this section, we discussed two students who appeared to have some knowledge of proof by induction, but neither was able to bring it to mind until the teacher intervened. Bringing to mind appropriate knowledge depends on both the situation, for example, the comments of others, and the interconnected nature of a student's own knowledge.

Finally, intuition has a role in problem-centered reasoning and we suggest it would be useful to consider at least two kinds of intuition that we call *visual–spatial* and *technical–algebraic*. By *visual–spatial* we mean intuition based on pictures or diagrams that can be sketched or visualized. These might be realistic, such a graph, or visually metaphorical, such as a "blob" for an open set in a topological space. Figure 20.5 is an example of a somewhat realistic sketch of the composition of two functions, although one that does not include their Cartesian graphs. Figure 20.5 might also be regarded as a visual metaphor for a similar situation in higher dimensions. In contrast, *technical–algebraic intuition* depends on one's familiarity with the interrelations among definitions and theorems. Earlier in this section, we suggested that part of the difficulty of our proof that the sum of continuous functions is continuous was due to the probable expectation

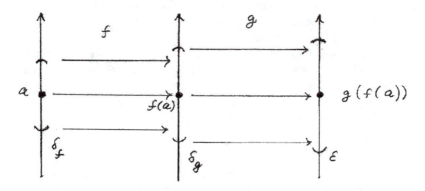

Figure 20.5 A representational sketch of g o*f*

that visual–spatial intuition might be useful, but actually technical–algebraic intuition about manipulating inequalities is called for.

Informal Observations

The Genre of Proof

When students first start constructing their own proofs, they may inquire: what is a proof? They are occasionally told just to write a convincing argument. However, proofs are not just convincing, deductive arguments, they are also texts composed in a special genre.[5] If beginning students are unaware of the need to write in this genre when asked to write "easy" proofs, they may suffer from what might be called the "obviousness obstacle."

For example, Moore (1994, pp. 258–259) reports that when one student was asked on a test to prove: *if A and B are sets satisfying $A \cap B = A$, then $A \cup B = B$,* she drew a Venn diagram with one circle, labeled A, contained in a larger circle, labeled B, and gave an intuitive argument based on her "understanding of set equality, subset, intersection, and union" using informal language, rather than a "proof based only on definitions, axioms, previously proved results, and rules of inference." According to her professor, she had not learned "the language and culture of how we write these things down." When asked what was wrong with her proof, she said "I didn't explain it well enough." Without understanding that there is a genre of proof, such obvious theorems may be very difficult for students.

However, our experience suggests that students who see themselves as learning to write in a special genre will have something positive to do, and hence, be more successful. This seems to be so, even when no detailed description of the genre of proof exists, and students must learn by trial and error or as apprentices.

Convince Yourself

Students are sometimes encouraged to prove theorems by first convincing themselves intuitively, then making their argument more and more precise, eventually arriving at a proof. We suspect that most students would interpret convincing themselves intuitively to mean they should visually or spatially manipulate the objects that occur in the statement of the theorem. This is sometimes helpful, but we suggest that for most students there are some theorems that cannot be proved this way, and students who insist on basing their work on refinements of this kind of intuition cannot prove such theorems. Our sample theorem about the sum of continuous functions is such a theorem. For contrast, we now discuss the proof of another theorem that can be obtained by a refinement of visual–spatial intuition.

The theorem is: for all real-valued functions of a real variable, if f is continuous at a, and g is continuous at $f(a)$, then g of f is continuous at a. At first glance, this theorem seems to be rather like the theorem about the sum of continuous functions. However, one can develop some visual–spatial intuition about it by examining a sketch (see Figure 20.5).

Drawing a representation like that in Figure 20.5 might well be within the grasp of an advanced undergraduate student who started with the intuitive view that continuous at a means "points close to a, map close to $f(a)$." This representation not only suggests that the theorem is true; extending the idea of closeness to an $\varepsilon - \delta$ argument also leads to a proof.

Logic

In the United States, most undergraduates who receive any significant explicit instruction in how to write proofs are provided it in one mid-level transition-to-proof course. Typically, such courses include logic and a little about topics such as sets, equivalence relations, and functions, as well as some especially accessible mathematics to provide theorems to prove. The logic is often an abstract, symbolic, decontextualized treatment of propositional and predicate calculus, including truth tables and the validity of arguments. It is often taught early in the course, presumably because logic is considered prerequisite to understanding proofs.

Although some parts of logic, such as understanding how to negate a statement, are certainly important for constructing some proofs, we doubt that formal logic often plays a large role in proofs that beginning students typically construct. This somewhat counterintuitive view is supported by an analysis of student-generated proofs from a transition-to-proof course at a large southwestern university. Although the course was not intended to be the basis of a study, all of the 62 correct student-generated proofs were preserved and later analyzed for a master's thesis (Baker, 2001). These proofs consisted of 926 lines of text whose analysis included noting uses of logic mentioned in the teacher's lecture notes or in the textbook (Velleman, 1994). Logic was used in just 29 lines. The remaining lines included advance organizers, assertions of hypotheses or conclusions, statements of whether the proof was direct, by contradiction, or by cases, applications of definitions, and use of subject matter knowledge.

We also examined our sample proof from the first section for uses of the kinds of logic often taught in transition-to-proof courses. In writing this proof, we doubt that a student would call on much formal logic. However, we can find three places where a valid logical argument, together with information from outside the proof, could have been used. These occur in deducing step [3] from [2]; step [7] from parts of [4], [5], and [6]; and the premise of [10] from [9] (and similarly, the premise of [11] from [9]).

We illustrate how deducing step [7] from parts of [4], [5], and [6] might be viewed as the result of the valid logical argument: $(P \wedge Q) \to R, P, Q, \therefore R$ First take $(P \wedge Q) \to R$ to be an instance of a statement from outside the proof, namely, *For all real numbers a and b, if $a > 0$ and $b > 0$, then $\min(a,b) > 0$.* If one then interprets P as $\delta_1 > 0$ (from [4]) and Q as $\delta_2 > 0$ (from [5]), this statement becomes: *if $\delta_1 > 0$ and $\delta_2 > 0$, then $\min(\delta_1, \delta_2) > 0$*, where R is interpreted as $\min(\delta_1, \delta_2) > 0$. Finally, invoking the valid logical argument $(P \wedge Q) \to R, P, Q$, $\therefore R$ and identifying δ with $\min(\delta_1, \delta_2)$ (from [6]), one gets $\delta > 0$, which is [7].

We believe that few, if any, students would consciously employ such cumbersome arguments. Thus, our sample proof does not call on formal logic. Because many direct proofs call on only a little formal logic, it seems that logic is not prerequisite to understanding proof. Therefore, it should be possible and helpful to teach logic in the context of proofs. Where logic is required in student proofs it plays a central role, so we are not suggesting its teaching be omitted, but rather that the teaching of logic grow out of students' own work with proofs, and thus, take a more practical than formal form.

Instead of [7] being the result of a valid logical argument, we see it as the result of an inference-generating schema that accepts both conscious and unconscious inputs, such as parts of [4], [5], and [6] and additional information, that is invoked outside of consciousness. Such schemas are not under conscious control so they might best be developed through experiences and reflections on experiences, rather than on abstractions.[6]

Teaching

Teachers of upper division and graduate mathematics courses, such as real and complex analysis, often ask students to produce proofs as a major part of assessments, presumably because well-written proofs reflect a good understanding of course content. Thus, a student with only modest proving ability is at considerable disadvantage in demonstrating understanding.

How do undergraduate students currently learn to construct proofs—a topic that is mostly part of the implicit curriculum? The only widely taught courses in the US devoted explicitly to teaching undergraduate students how to prove theorems are mid-level transition-to-proof courses (e.g., see Velleman, 1994). Given the discussion we have conducted in these pages, it seems that students should be able to start proving theorems without prior formal instruction in logic or practice with sets and functions, and develop that knowledge through experience and mentoring.

Students may also learn something of proof construction from lectures on mathematical content, such as real analysis or abstract algebra. However, for teaching proof construction, even well-presented lectures may be ineffectual, because a teacher cannot know what features of proofs students are focusing on in class and probably does not know what kinds of proofs students have, or have not, already learned to construct. Consider for example, Dr. T's teaching methods in real analysis, as described by Weber (2004). Dr. T, known for very good teaching, first discussed the formal–rhetorical aspects of proving[7] when presenting theorems about sets and functions. Somewhat later when considering sequential limits, Dr. T concentrated on demonstrating how to manipulate absolute value expressions (to find N), apparently mistakenly assuming that the students had learned and could supply the required formal–rhetorical parts of such proofs, but they could not.

Homework and tests can also provide opportunities to learn proof construction. However, these too are likely to be ineffectual because often they do not focus directly on teaching how to construct proofs or do not include mentoring students' work. In summary, none of the current ways of teaching proof construction—transition-to-proof courses and homework and tests in content courses—seems adequate. Indeed, it has been our experience that many beginning graduate students at US universities could benefit from a course designed to improve their ability to construct proofs.

What kind of course is likely to be reasonably effective in helping students improve their proof constructing abilities? We suggest that a good way to teach such a course is from a set of notes containing definitions and statements of theorems to prove, with little or no additional explanation. Proofs, as well as examples and nonexamples of definitions, can be provided by the students themselves. At first, students' proof construction might best be done in class, so the teacher can provide adequate mentoring. The teacher should not provide heavy-handed hints, but only enough intervention for students to succeed reasonably often with considerable effort. Every intervention, in a sense, deprives a student of the opportunity to succeed without it. Thus it is probably best if mentoring were not available during all, or even most, of a student's proving of a particular theorem.

Occasionally student proving is likely to require general information, such as how to negate a universally quantified statement or how to prove a function is one-to-one. When this occurs, a teacher can add such information to a developing set of supplementary notes that students could refer to as needed. The information in such supplementary notes should be especially pertinent as it would have first arisen in context in the form of behavioral knowledge, or the lack of it, and only subsequently developed into conceptual knowledge, rather than the other way round.

In the kind of course we are describing, students seem to do well in small groups. It may be that early on small group discussions alleviate concern over working in an unfamiliar problem-oriented situation. Also, the need to convince one's colleagues of one's ideas may enhance problem-solving control.

In addition, where understanding previous mathematical content (e.g., the definition of uniformly continuous) or bringing it to mind are called for, surely several students have an advantage over one. All of this suggests that working in a small group raises the probability of a student's successfully proving a theorem. We see raising the probability of success as important because our experience strongly suggests that success breeds success.

There is another way to raise the probability of students successfully proving a theorem that we call "long-range priming." We illustrate this by referring to our proof that the sum of continuous functions is continuous. It seems clear this is a difficult proof and depends on students thinking of using minimum to find δ and the triangle inequality to complete the problem-centered part of the proof. The proof might be rendered somewhat less difficult by inserting two earlier theorems in the notes, one requiring minimum and the other requiring the triangle inequality.

Our illustration here suggests that course notes should be written as the course progresses, so that the teacher knows which features of proofs students are already familiar with and which kinds of proof they have successfully constructed thus far. The idea is to provide course notes having "just-in-time" challenges and information.

There is an additional constraint: the course notes should not be too narrow. Before writing the notes, one might wish to establish priorities on which abilities to include. This might be done in a way that would ensure that the supplemental notes include much of the background material in a transition-to-proof course. All of this suggests the notes may require considerable time to write and raises a question of practicality for most teachers.

Is such a course practical? Can the notes be written in a reasonable length of time and can a reasonable number of students be mentored in class? Our experience from transition-to-proof courses suggests that mentoring would become difficult as class size nears 36, even with groups of four. However, a properly trained student assistant could no doubt be a great help with mentoring. For more advanced undergraduate or graduate students, the writing of notes appears to be manageable for a very small class with three groups of two students each or of three groups of three to four students each.

However, a tool could surely be built that would make writing such notes practical for a larger number of students. It is much easier to select a theorem that requires particular proving abilities than it is to write one "on demand." Thus, what is needed is a reference book, or database, containing branching sequences of definitions, theorems, and proofs, together with an analysis of which previous definitions and theorems, and which abilities, are used in each proof. A teacher could then select a theorem that would "stretch" a particular ability, and working backwards, include any necessary definitions and theorems needed to join the desired theorem to the current notes.

Finally, instead of needing to teach the kind of course we have described, it might be better to integrate the teaching of proving throughout the undergraduate program, and in schools—as is called for by the *NCTM Standards* (2000). This would require setting aside adequate time for such teaching and providing a good deal of help to school teachers.

Conclusion

This chapter has concerned features of proofs, student abilities called on by such features, and how coordination of such features and abilities can be used in teaching students to construct proofs. First we introduced three proof structures: a *hierarchical* structure of subproofs and subconstructions; a linear *construction path* giving the order in which an idealized prover could write a proof; and a division of proofs into *formal–rhetorical* and *problem-centered* parts.

We then discussed how several specific features of proofs could be seen as calling on specific student abilities. For example, a proof by induction not only calls on a student's knowledge of induction, but also on the student's ability to bring that knowledge to mind. Attending to the three structures mentioned here can also provide some idea of a proof's complexity. Also, the ability to use the language of sets and functions differs from merely reproducing formal definitions. Next we discussed three informal ideas:

1. Only arguments written in a particular genre are accepted as proofs. Students who do not understand this seem to have difficulty constructing proofs of statements they see as obvious.
2. Some theorems do not have proofs that are refinements of arguments based on visual–spatial intuition. Such theorems seem to be very difficult for students who habitually depend on such intuition.
3. Logic does not occur so often in student proofs that it must be taught prior to, rather than with, proving. Furthermore, the logic used in student proofs often seems to depend more on a practical, intuitive approach than on an application of formal logic.

Finally, we described a method of teaching using mentoring and group work in which instruction is integrated into students' construction of proofs. In such teaching, kinds and features of proof are more important than specific topics. The key idea is that when students require mentoring on some ability, the teacher tries to assign a later theorem having a proof calling for that ability. We ended with a discussion of the practicality of such teaching.

Notes

1. We speak of steps, or sentences, rather than statements, because proofs can contain sentences that are neither true nor false, but instead give instructions to the reader, such as "Let x be a number."
2. Proof frameworks can be independent of the meanings of certain content words. Selden and Selden (1995, p. 130) illustrated this with a proof framework for a theorem about semigroups that can also serve as a proof framework for a theorem about real intervals. This was accomplished by replacing the words "semigroup" with "set of numbers," "group" with "interval," and "a subgroup of" with "a subinterval of" and retaining the usual meaning of the other words.
3. Behavioral knowledge should not be confused with behaviorism, the idea that physically observable, and perhaps measurable, behavior should be the only basis for a scientific treatment of psychology. Taking a behaviorist view would prevent most of the study of the mind, and hence, much current research in mathematics education.
4. In such courses, students are typically given notes containing definitions and statements of theorems, or conjectures, and asked to prove them or to provide counterexamples. The teacher provides the structuring of the notes and critiques the students' efforts. For more information, see Jones (1977) or Mahavier (1999).
5. We are not only suggesting that someone might view proofs in this way, but that it is an empirical fact. That is, with rare exceptions, proofs possess a number of stylistic commonalties not found in other forms of deductive argument. For example, definitions available outside of a proof tend not to be repeated inside it.
6. The development (i.e., construction) of knowledge through experience and reflections on experience is consistent with a constructivist viewpoint.
7. Weber (2004) refers to Dr. T's teaching at this point in the course as being in a logico-structural style.

21

Current Contributions Toward Comprehensive Perspectives on the Learning and Teaching of Proof

GUERSHON HAREL AND EVAN FULLER

In a paper with a similar title, Harel and Sowder (2007) called for comprehensive perspectives on the learning and teaching of proof. They define such a perspective as one that incorporates a broad range of factors: mathematical, historical–epistemological, cognitive, sociological, and instructional. Toward this end, they identified 13 questions to be addressed within these factors:

- *Mathematical and historical–epistemological factors*:
    1. What is proof and what are its functions?
    2. How are proofs constructed, verified, and accepted in the mathematics community?
    3. What are some of the critical phases in the development of proof in the history of mathematics?
- *Cognitive factors*:
    4. What are students' current conceptions of proof?
    5. What are students' difficulties with proof?
    6. What accounts for these difficulties?
- *Instructional–sociocultural factors*:
    7. Why teach proof?
    8. How should proof be taught?
    9. How are proofs constructed, verified, and accepted in the classroom?
    10. What are the critical phases in the development of proof with the individual student and within the classroom as a community of learners?
    11. What classroom environment is conducive to the development of the concept of proof with students?
    12. What form of interactions among the students and between the students and the teacher can foster students' conception of proof?

13. What mathematical activities—possibly with the use of technology—can enhance students' conceptions of proof?

As we read and analyzed the contributions in this volume, three additional major questions emerged: (14) What theoretical tools seem suitable for investigating and advancing students' conceptions of proof? (15) How is proof currently being taught? (16) What do teachers need to know in order to teach proof effectively?

We center our reflections around these questions, listing major questions addressed by this volume's chapters at the beginning of each section. For convenience, we also include a table summarizing the foci of each chapter. In all, our aim is to point to current contributions toward the formation of a comprehensive perspective on the learning and teaching of proof.

The historically weak performance of students at the secondary and undergraduate levels on proof can be attributed to a number of factors. Whether the cause lies in the curriculum, the textbooks, the instruction, the teachers' background, or the students themselves, it is clear that the status quo needs, and has needed, improvement. One obstacle to improvement has been the disjointedness of reform efforts, some of which focus on one aspect of proof while ignoring others. A single factor usually is not sufficient to account for students' behaviors with respect to proof. Instead, comprehensive perspectives on proof that examine all major factors are needed in order to understand students' difficulties, the roots of these difficulties, and the type of instructional interventions needed to advance students' conceptions of and attitudes toward proof. It is also worth noting that the nature of students' difficulties with proof may evolve as students are provided with opportunities to learn to prove beginning at the early grade levels and continuing throughout their education (as illustrated in several chapters of this volume).

The questions listed earlier define such a comprehensive perspective. By analyzing the contributions in this volume, we will point to which questions have been addressed and what new questions have emerged (see Figure 21.1). As was indicated in Harel and Sowder (2007), the list of questions is not exhaustive. As research on the learning and teaching of proof advances, other fundamental questions are likely to emerge, as did questions 14–16. Two features of the list should be highlighted. First, none of the questions in this list should be viewed as a stand-alone question; rather, they are all interrelated, constituting a cohort of questions that define the comprehensive perspective on proof that emerged from the work reported in this volume. The second feature of this perspective is that of *subjectivity*. While the term "proof" often connotes the relatively precise argumentation given by mathematicians, in the perspective offered here, "proof" is interpreted subjectively: a proof is the particular argument one produces to ascertain for oneself or persuade others about the truth of an assertion.

Overall, our aim here is to synthesize the contributions made by this volume toward each comprehensive perspective question, and to report on directions for further research that are indicated.

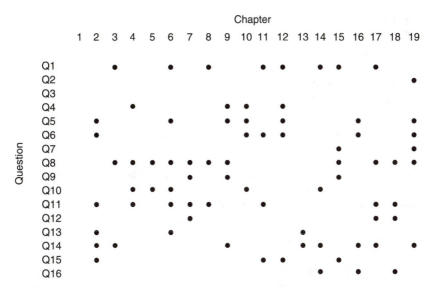

Figure 21.1 Questions addressed in each chapter

Section I: Theoretical Considerations on the Teaching and Learning of Proof

What I would Like My Students to Already Know About Proof

Hersh centers this personal essay around the question "What would we like to see happen in high school, and even earlier?" with respect to the teaching and learning of proof. While Hersh here uses his own experience as a college teacher rather than empirical data to explore this question, his reflections touch on many of the questions that the framework asks and resonate, in an anecdotal fashion, with many of the findings of past literature as well as findings reported in this volume.

Exploring Relationships Between Disciplinary Knowledge and School Mathematics: Implications for Understanding the Place of Reasoning and Proof in School Mathematics

Chazan and Lueke present several studies demonstrating that the views of students (and teachers) on proof are often in conflict with institutional ideas. They suggest that students are operating in different frames, not always choosing the appropriate one for the task. To help students with this issue, teachers must set up classroom circumstances that support the desired frames for particular contexts. We note that this seemingly innocuous suggestion is significant in several ways. First of all, it suggests that teaching the desired frames directly will not be effective, a view consistent with Davis's (e.g., 1984) work on frames. Direct instruction about how to reason can fall into the trap of metacognitive shift (Brousseau, 1997), in which instruction focuses on what students should be doing rather than on the mathematics itself. Providing necessity for the desired frames is a desirable alternative to this approach (Harel, 2007; Harel & Sowder, 2007). A classroom environment that supports the desired frames will be conducive to development of proof. In addition, the nature of current algebra teaching, with its focus on "methods," accounts for some of students' difficulties with proof because students are used to "justifying" the reason for, rather than the correctness of, steps in a solution.

Proving and Knowing in Public: The Nature of Proof in a Classroom

Herbst and Balacheff ask what kinds of representation of proof can be constructed, and how these can be described by an observer. They focus on theoretical tools that can help investigate students' understanding of proof. To this end, they propose a way to distinguish different types of justification without resorting to symbolic formalism. In addressing what proof is from this perspective, they propose three types of proof, which operate on three different timescales. The authors suggest that observers, including the teacher of a class, should examine each of these types of proof in a classroom. In this way, the extent to which the teacher has made room for proof can be described without imposing a single formal criterion. Using classroom constraints and these types of proof as theoretical tools can help the mathematics education community address issues of how students learn to prove and how proof should be taught.

Section II: Teaching and Learning of Proof in the Elementary Grades

Representation-based Proof in the Elementary Grades

Schifter suggests that generalizing and justifying emerge quite naturally when students notice a regularity in the number system and consider whether or why it will always hold. Thus, early experience with the number system may be a critical phase in the development of proof within individual students. Some elementary school students will justify claims using reasoning based on a representation, which the author considers a desirable form for proofs to take in elementary school. The author points out that because reasoning about computational claims helps students gain number sense and perform operations more fluently, it furthers local goals for student understanding as well as provides a basis for later conceptions of proof. A classroom environment that involves debating observations regarding regularities in the number system is conducive to the development of the concept of proof with students, even those at a lower mathematical level.

Representations that Enable Children to Engage in Deductive Argument

Morris identifies a problem with students' conception of proof: even in higher grades, many students do not understand that they can use general relationships. We note that this is related to Vinner's (1983) finding that, when high school students were asked to give their preference for proving a particular case of a previously proved statement, they preferred using a particularization of the deductive statement rather than citing the general result. Thus, the general proof was viewed as a method of examining particular cases rather than as a general relationship that has been shown to hold. A "schematic" demonstrates one way to scaffold student thinking with respect to development of proof: students begin with the concrete and have a fixed representation that helps them move to thinking more generally. The author argues that not only can elementary schoolchildren engage in deductive reasoning, but also that appropriate instructional support at that level can avoid later misconceptions, such as the documented pattern of inductive reasoning among older students and adults. Thus, elementary school experiences constitute a critical phase in the development of ideas related to proof.

Young Mathematicians at Work: The Role of Contexts and Models in the Emergence of Proof

Fosnot and Jacob "assume proof is based on establishing the validity of mathematical statements which are assembled in an order where accepted rules of inference allow one to infer each statement from previous statements or hypotheses" (Chapter 6, p. 102). From their work, we can infer one characteristic of a classroom environment that is conducive to students developing deductive rules: having students make arguments about equivalence in a concrete context. The mathematical activity of making general arguments about abstract situations—where the value of one or more variables was not known—appeared to further enhance students' conceptions of proof. Also, proof should be taught so that the representations used support the desired reasoning and allow students to make sense of their activity. The authors' work identifies a critical phase in the development of proof: when students are learning to re-examine their reasoning, use representations to reformulate it, and simplify the result into a compact argument.

Children's Reasoning: Discovering the Idea of Mathematical Proof

Maher claims that proofs can be constructed by students naturally as they work on problems, and that these proofs are verified and accepted through probing and discussion. She provides evidence that justification based on a detailed and thorough analysis of the problem situation can arise even when it is not set as a learning goal by an instructor, which suggests that creating a learning environment in which children are free to explore without pressure to complete tasks prematurely can facilitate "natural" emergence of proof. In addition, a classroom environment conducive to proof has students engage with problems that have both concrete and abstract aspects, build a variety of representations and share these with each other, and develop justifications for their ideas. The examples presented in the chapter involved problems that use manipulatives, which raises the question: *what cognitive issues related to proof are associated with young children's use of physical manipulatives to solve problems?* For instance, using manipulatives might encourage students to use empirical reasoning, or it could provide a concrete anchor to aid abstraction.

Aspects of Teaching Proving in Upper Elementary School

Reid and Zack infer important aspects of teaching proof by looking at two teaching experiments. First, problems should be nontrivial but accessible, and knowledge used to solve one problem can help in reasoning about another, which provides a basis for deductive proof. An emphasis on communication in different contexts is quite important to proof, as proof in mathematics is fundamentally a means of communication. Developing community standard for proving is also important, and students should expect to understand how arguments actually explain a result. Moreover, it is beneficial for students to experience a progression from problem solving to conjecturing and then to proving. We note that other research (Usiskin, 1980) has shown that students often do not appreciate the statements that they are proving and sometimes see no need to prove them. The authors point out that having students generate their own conjectures makes it more likely that they will engage in proving to explain rather than as a didactical requirement, and it may lead students to prove even when no teacher is present.

Section III: Teaching and Learning of Proof in Middle Grades and High School

Middle School Students' Production of Mathematical Justifications

In Chapter 9, Knuth, Choppin, and Bieda adapt a categorization scheme from Waring (2000) to create proof production levels, which can help researchers map the development of student conceptions related to proof. An understanding of generality appears to be critical for students to develop their conception of proof. The results from this study confirm previous findings that students do not have an adequate understanding of proof's generality (consistent with Morris's claim in this volume). The authors suggest several further questions:

1. To what extent do students recognize that a proof treats the general case?
2. To what extent do students think that examples suffice as proof?
3. What role do examples play in students' understandings of proof?
4. What is the relationship between students' proof production competencies and their proof comprehension competencies?

Although the authors do not answer these questions, they suggest that teachers' conceptions of proof should be improved during their math education and state that "curricular materials not only must provide opportunities for students to engage in proving activities, but must also support teachers' efforts to facilitate such engagement" (Chapter 9, p. 169).

From Empirical to Structural Reasoning in Mathematics: Tracking Changes Over Time

Küchemann and Hoyles found that English high school students rarely use "structural reasoning," partly because "the simplicity of number pattern responses may have a stronger appeal than the insight that might be gained from taking a structural approach" (Chapter 10, p. 183). A particularly interesting finding is that some students understood the features and power of the "divisibility principle" (the argument that if you multiply by a certain number the result is divisible by that number, which is key to finding a solution), but they felt insecure in this reasoning because of concern that other features of the situation might render the principle invalid. This observation concerns an issue for which results have been mixed: Healy and Hoyles (2000) found that most high school students understand that a proven statement covers all cases within its domain of validity, while other studies, such as that of Fischbein and Kedem (1982), have found that further numerical checks increased students' confidence in a proof they had already agreed was correct. The authors' findings provides a possible explanation for this apparent discrepancy: students may understand that, in theory, a proof covers all cases, but they are concerned that the proof does not apply or has some not-yet-spotted mistake, so they are checking that it is valid by computing particular cases.

Overall, although the students made some progress in the use of structural reasoning over the three years studied, this progress was not linear and not always retained. For these reasons, the authors suggest that single snapshots of student understanding can be misleading. This perspective implies that both standardized testing and research methodology must not give too much weight to single items. We infer that forming an awareness of different possible strategies and comparing these is a critical phase in students' development of the concept of proof. To aid students in reaching this phase, teachers should have students use different representations for ideas, making connections between these and justifying their reasoning.

Developing Argumentation and Proof Competencies in the Mathematics Classroom

Heinze and Reiss note that various standards have set the goal that students should learn mathematical reasoning as a habit of mind and gain the ability to appreciate proofs by the end of secondary school, but that students' lack of "prerequisites" (such as knowing heuristics, control strategies, and the nature of proof) makes it difficult to achieve this goal. The typical teaching style of German and American teachers—explicit guidance (Mullis et al., 2000)—provides minimal help in improving students' mathematical reasoning. This suggests that "classroom climates," which probably include emotional as well as cognitive factors, are induced by the nature of instruction in a class. As little is known about the mechanisms behind these effects, the authors call for more work to be done on how motivation and interest affect learning and how they can be stimulated effectively. Alternative environments have had some success in teaching proof more effectively; for instance, fostering metacognitive activities (Mevarech & Kramarski, 1997) or focusing on heuristics (Reiss & Renkl, 2002) improve students' problem-solving competencies. The authors note that research has also shown that positive emotions foster learning in general, but more specific work should be done on the effect of emotions relating to specific aspects of a math classroom.

Formal Proof in High School Geometry: Student Perceptions of Structure, Validity, and Purpose

McCrone and Martin presume that most teachers understand the values of proof and make earnest efforts to teach these to students. They then try to answer the question of why, this effort notwithstanding, so few high school students attain a desirable understanding of proof. They found that most students make good attempts to follow the "meta-discursive" rules of the classroom, but students perceive contradictions between their intuitive ways of making sense and the formal rules of the geometry classroom. The authors suggest possible reasons for why students' everyday notions continued to interfere with their understanding principles of proof; these can provide insight into some student difficulties. The fact that teachers give partial credit for arguments in which the statements are not in logical order may contribute to students' belief that such arguments are acceptable as proofs. Also, because students' experience in a traditional geometry class is limited to particular patterns and models demonstrated by the teacher, they learn how to write proofs without feeling a need for mathematical justification.

When is an Argument Just an Argument? The Refinement of Mathematical Argumentation

McClain examines the evolution of social and sociomathematical norms, in particular "what counts as an acceptable mathematical argument and the norms for justification for these arguments" (Chapter 13, p. 223). Because the teacher plays a major role in setting norms, the author implicitly focuses on the nature of the teacher's struggle as she attempts to understand and judge student contributions to the classroom discourse, while maintaining her mathematical agenda for both specific content and mathematical argumentation norms. The author's learning as a teacher facilitated communication that was critical for having productive mathematical discussions. However, she notes the complexity of providing students with the right amount of support. By having students compare inscriptions they had created to represent the data, the author seemed to be successful in shifting students to reflect more on the arguments than on the conclusions. Comparing different representations of arguments regarding a concrete situation that students had already analyzed appeared to enhance students' conceptions of proof.

Reasoning-and-Proving in School Mathematics: The Case of Pattern Identification

Stylianides and Silver focus on instructional issues related to pattern identification, which is "an essential component of the reasoning-and-proving activity." They include several activities as part of reasoning-and-proving, rather than taking as proof only the culminating stage of mathematical activity. For patterns, they distinguish whether a pattern is *definite*, meaning it is "it is possible mathematically for a solver (given the information in a task) to provide conclusive evidence for the selection of a specific pattern" (Chapter 14, p. 239), or merely *plausible*. A pattern is termed *transparent* if a solver recognizes whether it is definite or plausible; otherwise, it is *non-transparent*. To teach patterns in a desirable way, teachers must know more than the simplest expression for any given pattern; they must either understand that the pattern could be generated in different ways, or they must understand why the nature of the problem leads to a single pattern. Because of its prevalence in school, working with patterns is a critical phase in the development of students' conceptions of proof. The authors call for explicit treatment of distinctions between kinds of pattern in professional development so as to improve teachers' knowledge. They also argue that lack of progress in this area is partly due to researchers' lack of general frameworks and analytic tools, which has made it hard to connect to and build on existing findings in order to promote coherent objectives.

"Doing Proofs" in Geometry Classrooms

The authors focus on a specific aspect of proof that is quite important in the field of education, namely, the role of proof in high school geometry courses. Teachers who were interviewed all valued proof in geometry for its opportunity to teach logical thinking, which they considered important for many aspects of students' lives. They did not appear to value the geometry content itself or proof as a tool to figure out mathematical solutions. The authors suggest that a more desirable reason to teach proof is that proof is central to doing mathematics. From norms identified by the authors, we can infer that, typically, proofs are constructed through a sharing of labor in which the teacher sets things up, students provide ideas and reasons that are placed into a two-column proof, and the teacher verifies that the argument is adequate. Overall, the identified norms are consistent with the views expressed by these teachers, which emphasize proof as an exercise in logical thinking. To teach geometry in a more desirable way, these norms must be changed: classes must involve a broader share of labor for students, more flexible timing of actions, and more open-ended exchange.

Section IV: Teaching and Learning of Proof in College

College Instructors' Views of Students Vis-à-Vis Proof

Harel and Sowder suggest that DNR-based instruction, a theoretical framework for the learning and teaching of mathematics (see Harel, 2007), can be used to help investigate or change students' understanding of proof. They found that university faculty see the deductive proof scheme as an objective only for more able students, and they often concentrate on specific content more than proof schemes. Faculty perceive understanding proof as an obstacle for students, partly because students have little appreciation for its importance. In particular, students struggle with understanding and using definitions. Some faculty comments (reinterpreted by the authors) suggest that this difficulty arises partly because students attend mostly to concept images (Vinner, 1992) rather than concept definitions, and they do not feel intellectual need for the latter. University faculty typically have deep knowledge of mathematics but may benefit from additional knowledge of pedagogy. The authors suggest that an understanding of DNR principles and an awareness of the variety of proof schemes students may have will help instructors address some of the deficiencies in their students' proof schemes.

Understanding Instructional Scaffolding in Classroom Discourse on Proof

The authors address what proof is in terms of its social aspects: "Proof is ultimately a socially constructed, object whose purpose is to communicate the validity of a statement to a community based on criteria established by that community" (Blanton, Stylianou & David, Chapter 17, p. 290). They criticize the traditional practice of demonstrating completed proofs, and they instead argue that instruction should embed development of proof into the social aspect of the classroom and have students construct proofs. The coding scheme the authors created to examine both teacher and student utterances can be valuable for examining social interactions related to proofs in many contexts, thus providing a theoretical tool to examine the development of student conceptions related to proof. In the classroom examined, it appears that the teacher focused on student ideas and making students responsible for the proofs that are constructed, rather than directly telling information to the students, as is a common practice. This shifted the classroom discourse from more "authoritative" to more "internally persuasive." The authors suggest that better conceptions of proof will develop in a classroom environment in which students are responsible for constructing proofs and the teacher interacts with students primarily through "transactive prompts" and "facilitative utterances." The authors recommend further extensions of their work to provide an even stronger link between instructional acts and students' transition to mathematical proof.

Building a Community of Inquiry in a Problem-based Undergraduate Number Theory Course: The Role of the Instructor

Based on previous research, the authors assume that it is conducive to the development of proof for students to actively participate in a classroom community of inquiry, in which "students learn to speak and act mathematically by participating in mathematical discussions and solving new or unfamiliar problems" (Goos, 2004). Previous studies found that the teacher played a prominent role early in reform courses, while norms were being established, and then "backed away" later on. The instructor examined by the authors took the lead more frequently in the middle of the course, but less so during the beginning and end. We note that this finding may indicate a difference in "ideal" classroom environments for younger students and undergraduates, an issue that could be worthwhile for further study. Both the amount and nature of instructor intervention appear to be important factors in what students learn about proof: instructors must try to provide students with some direction, clarification, and content help while at the same time inducing students to take responsibility for producing and evaluating proofs. This highlights the importance of instructors attending to their students' current ways of understanding and thinking, rather than merely presenting correct content in a logical sequence.

Proof in Advanced Mathematics Classes: Semantic and Syntactic Reasoning in the Representation System of Proof

Weber and Alcock argue that one major reason for students' difficulty with proof might be the difficulty with the special form proofs take on because a new representation system must be learned (in the sense of Goldin, 1998) that makes some proofs easier and some proofs harder. They suggest that two major goals with regard to proof in advanced courses should be to establish links between intuition and representation systems, and to build syntactic skills. The authors present data showing that for some students it may be fruitful to establish semantic and then the syntactic skills associated with proof, while others may perform better by first learning and applying rules, then later extracting meaning from the rules. The authors conclude from these studies that teachers should attend to each particular student's preferred method of reasoning and answer student questions or provide help in solving problems differently according to that preferred method. Based on other studies, the authors suggest that mathematicians and educators should adopt a more balanced perspective, focusing on both syntactic and semantic reasoning. In particular, teachers should be aware of the two types of reasoning and use them as a theoretical tool to examine student difficulties with proof.

Teaching Proving by Coordinating Aspects of Proofs with Students' Abilities

Selden and Selden argue that, to identify students' difficulties with proof, one needs a fine-grained framework for analyzing proofs. They describe three aspects of proof that can be part of such a framework. The three aspects can be used to roughly classify the inherent difficulty in a given proof, depending on what level of each aspect is required for the proof, and to classify student understandings of proof, depending on a student's level of ability with respect to each. Thus, they can provide a framework for undergraduate professors to both identify student difficulties and select proofs that might help them overcome these difficulties. The authors use this framework to suggest that students should learn proof primarily by constructing proofs, rather than listening to a lecture. Also, it is more important for students to be able to learn to use the language of sets and functions in actual proofs than to understand the abstract definitions, so teachers should not assume the latter is sufficient. Finally, the authors argue that formal logic, though important at times, is rarely used in proofs, and that students could better learn it when it arises rather than through a preliminary unit on logic.

Conclusions

The following list attempts to synthesize the contributions made in this volume with respect to each of the questions we have identified. Although there is considerable overlap between questions and the sheer number of contributions for a specific question is not a reliable indicator of overall progress made in this area, this list provides some overall sense of what work reported in this volume has accomplished:

1. What is proof and what are its functions?
   - Proof is establishing the validity of mathematical statements using accepted rules of inference.
   - Proof consists of both actual arguments and the concept that mathematical statements should be supported.
   - Proving includes looking for relationships, making conjectures, providing deductive arguments, and providing other evidence.
   - Proof in classrooms at different levels may not resemble proof by mathematicians.
   - Proof is socially constructed and determined.
   - Proof can take on different, age-appropriate meanings.
   - Proofs are what students learn will be required of them as an acceptable answer.
   - Proof involves mathematical reasoning as a habit of mind.
   - Proofs communicate validity according to communal criteria.
   - Proofs reduce key ideas into a compact form.
   - Proofs can be used to justify statements, to discover new relationships, and to explain.
   - Proof is central to mathematics.
2. How are proofs constructed, verified, and accepted in the mathematics community?
   - Mathematicians construct proofs by setting goals and subgoals, using conceptual knowledge and intuition to achieve these goals, and linearly ordering their steps.
   - Proofs are constructed using a specialized representation system that makes proofs easier to verify.
3. What are students' current conceptions of proof?
   - While some elementary school students think that there is no way to prove a claim for all (whole) numbers because they cannot all be tested, others are able to engage in deductive reasoning and build general arguments using both abstract and number-based representations.
   - Students can appreciate proving even when they have already accepted the statement in question.
   - Young students can construct logical, deductive arguments when supported with age-appropriate representations and contexts.
   - Many students rely on empirical evidence or authority.

- Many students prefer numerical calculations as justification.
- Students understand some basic principles of proof, but also view other (non-deductive) arguments as proofs.
- Students tend to perceive only the local purpose of a proof and not its place in a structure of knowledge.

4. What are students' difficulties with proof?
   - Students have difficultly solving algebraic equations using deductive rules.
   - Students have difficulty proving statements they consider self-evident or accepted.
   - Students have difficulty using structural reasoning: they prefer to look for number patterns.
   - Some students see geometry proofs as applying only to a particular diagram.
   - Students rely on everyday ideas rather than deductive rules.
   - Students have difficulty with the special form of mathematical proofs.
   - Students have difficulty organizing their steps.
   - Students have difficulty setting appropriate subgoals.
   - Students have difficultly understanding and using definitions.

5. What accounts for these difficulties?
   - Lack of emphasis on structural reasoning in school.
   - Student concern that a theorem does not apply or has some mistake.
   - The teaching style of explicit guidance.
   - Negative emotions, including fear of failure.
   - Students operate in different, and often inappropriate, frames.
   - Focus on learning "methods" in school.
   - Students perceive contradictions between their own ideas and principles of proof.
   - The need to learn a new representation system.
   - Students rarely feel a need to justify statements in class—they already accept them as true.

6. Why teach proof?
   - To develop logical reasoning.
   - Because reasoning and proof are central to mathematics.
   - To lead students to develop meaningful understandings of why mathematical assertions are true.

7. How should proof be taught?
   - Proofs should be taught indirectly, leading students to propose conjectures and gradually formulate convincing arguments.
   - Proof should be taught using representations that support the desired reasoning.
   - Proofs should be made an explicit object of consideration.
   - Proof should be taught using algebraic reasoning.
   - Students should be asked to prove facts that arise through performing calculations.
   - Teaching of proof should include large amounts of time for solving related

problems with deep mathematical structure, and for having students establish communication standards.

- Proof should be taught using curricular materials that support teachers' efforts to facilitate student engagement in proving activities.
- Proof should be taught in different ways, both establishing a general expectation that propositions will be justified and creating a control structure for verifying specific solutions.
- Classes on proof should involve students—not only the teacher—deciding when it is appropriate to prove and setting up a proof.
- Students should be made responsible for constructing proofs.
- Teachers should provide instructional scaffolding to help students internalize the language of proof and argumentation and develop the skill of questioning their reasoning and proof construction steps.
- Teachers should take a leading role only at carefully chosen points in time.
- Teachers should focus on both semantic and syntactic aspects of proving.
- Student should learn necessary aspects of logic and set theory through constructing actual proofs.

8. How are proofs constructed, verified, and accepted in the classroom?
   - Proofs can be constructed naturally as students work on problems.
   - Proofs are verified and accepted through classroom discussion.
   - Typically, the teacher sets things up, students provide ideas and reasons that are placed into a two-column proof, and the teacher verifies that the argument is adequate.

9. What are the critical phases in the development of proof with the individual student and within the classroom as a community of learners?
   - Re-examining, reformulating, and simplifying an argument.
   - Elementary school experience with generalization.
   - Seeing ways to prove claims for an infinite class.
   - Working with patterns.
   - Forming an awareness of possible strategies and comparing these.

10. What classroom environment is conducive to the development of the concept of proof with students? Some features identified include:
    - Students are allowed to explore problems without premature pressure to complete tasks.
    - Students engage with problems that have both abstract and concrete aspects.
    - Students make arguments about equivalence in a concrete context.
    - Students debate their observations regarding regularities in the number system.
    - Students progress from problem solving to conjecturing to proving.
    - Emotions are mostly positive.
    - An environment supportive of the frames we want students to operate in.

11. What form of interactions among the students and between the students and the teacher can foster students' conception of proof?
    - Building and sharing representations, and justifying their ideas.

- The teacher interacts mostly by asking for student input and structuring discussion.
- Students question and build on each others' ideas.
- The teacher shifts responsibility to the students.
- The teacher acts as a representative of the mathematical community, but an equal participant in the classroom community.

12. What mathematical activities—possibly with the use of technology—can enhance students' conceptions of proof?
- Making general arguments about abstract situations.
- Comparing and reasoning with representations of statistical data.
- Working and reasoning graphically, making conjectures, and developing mathematical methods.

13. What theoretical tools seem suitable for investigating and advancing students' conceptions of proof?
- A categorization of proof production levels.
- Distinguishing patterns based on definiteness and transparency.
- Distinguishing types of proof in the classroom: general custom, regulatory structure, and relationships between conceptions.
- Instructional scaffolding to encourage development in the ZPD of proof learning.
- A coding scheme for instructor and student utterances.
- Distinguishing syntactic and semantic aspects of proof.
- Analyzing proof difficulty based on hierarchical structure, step ordering, and relative reliance on definitions or intuition.
- DNR-based instruction.
- Toulmin's argumentation scheme.
- Looking at classroom interactions in terms of symbolic exchange.

14. How is proof currently being taught?
- Through explicit guidance and demonstration.
- Students are taught to solve classes of problems using methods, which typically contain no justification.
- Mostly through a high school geometry course.
- As an exercise in logical thinking.
- Often without students feeling a need for justification.

15. What do teachers need to know in order to teach proof effectively?
- How specific patterns arise: are they forced by some structure in the task?
- How much and what types of help students need at different stages of a class.
- Pedagogical knowledge, perhaps including DNR principles and proof schemes.

We can make several observations from this list. In spite of the historical evidence regarding students' difficulties with proof, this volume highlights students' capacity for developing deductive arguments (even in the elementary grades) when instructional settings support this agenda. It also points to how proof can be taught to achieve this, with suggestions for improving different aspects of the classroom. However, further work seems to be required to better clarify (and come to some agreement regarding) the purpose of teaching proof, how proofs are or should be constructed and verified in the classroom (compared with how they are constructed by mathematicians), and what activities and interactions can foster desirable conceptions of proof.

Several authors have developed theoretical tools that can help study or change students' understanding of proof, and they point out that these tools should be developed further. In particular, we find calls for additional research to answer the following questions:

1. What representations can support students' reasoning at particular levels?
2. What curricular materials can both provide younger students with opportunities to engage in proving activities and support teachers' efforts to facilitate these activities?
3. What domains are most conducive to development of the concept of proof at different levels?
4. How do emotions and affect influence student learning of particular mathematical ideas related to proof?
5. How can classrooms promote deductive reasoning rather than only strategic reasoning?
6. What are the precise links between instructional acts and students' transition to mathematical proof?

Overall, while—as this volume attests—work regarding the teaching and learning of proof is progressing on a number of fronts, there remains much to be done in order to complete the K–16 "proof story." As several authors mention, a stronger coordination between research efforts can advance the field of proof education. Additionally, the field would benefit from research that validates and extends existing data, as well as longitudinal studies to better assess progress in students' learning across the grades.

References

Alcock, L., & Simpson, A. (2004). Convergence of sequences and series: Interactions between visual reasoning and the learner's beliefs about their own role. *Educational Studies in Mathematics, 57,* 1–32.

Alcock, L., & Simpson, A. (2005). Convergence of sequences and series 2: Interactions between non-visual reasoning and the learner's beliefs about their own role. *Educational Studies in Mathematics, 58,* 77–110.

Alcock, L., & Weber, K. (in press). Referential and syntactic approaches to proving: Case studies from a transition-to-proof course. *Research in Collegiate Mathematics Education.*

Alibert, D. (1988). Towards new customs in the classroom. *For the Learning of Mathematics, 8,* 31–35.

Alibert, D., & Thomas, M. (1991). Research on mathematical proof. In D. Tall (Ed.), *Advanced mathematical thinking.* Dordrecht, The Netherlands: Kluwer.

Anderson, R. C., Chinn, C., Chang, J., Waggoner, M., & Yi, H. (1997). On the logical integrity of children's arguments. *Cognition and Instruction, 15,* 135–167.

Arsac, G. (2007). Origin of mathematical proof: History and epistemology. In P. Boero (Ed.), *Theorems in schools: From history, epistemology and cognition to classroom practice.* Rotterdam: Sense Publishers.

Atiyah, M. F. (1984/1986). Mathematics and the computer revolution. In A. G. Howson & J. P. Kahane (Eds.), *The influence of computers and informatics on mathematics and its teaching.* Cambridge: Cambridge University Press.

Baker, S. W. (2001). *Proofs and logic: An examination of mathematics bridge course proofs.* Unpublished master's thesis, Tennessee Technological University, Cookeville, TN.

Bakhtin, M. M. (1981). *The dialogic imagination* (M. Holquist, Ed.; C. Emerson and M. Holquist, Trans.). Austin, TX: University of Texas Press.

Bakhtin, M. M. (1986). *Speech genres and other late essays* (C. Emerson & M. Holquist, Eds.; V. W. McGee, Trans.). Austin, TX: University of Texas Press.

Balacheff, N. (1987). Processus de prevue et situations de validation [Processes of proof and situations of validation]. *Educational Studies in Mathematics, 18,* 147–176.

Balacheff, N. (1988). Aspects of proof in pupils' practice of school mathematics. In D. Pimm (Ed.), *Mathematics, teachers and children.* London: Hodder & Stoughton.

Balacheff, N. (1991). The benefits and limits of social interaction: The case of mathematical proof. In A. Bishop, S. Mellin-Olsen, & J. van Dormolen (Eds.), *Mathematical knowledge: Its growth through teaching.* Dordrecht, The Netherlands: Kluwer Academic.

Balacheff, N. (1995). Conception, connaissance et concept. In D. Grenier (Ed.), *Séminaire didactique et technologies cognitives en mathématiques.* Grenoble: Université Joseph Fourier.

Balacheff, N. (1999). Contract and custom: Two registers of didactical interactions. The *Mathematics Educator, 9,* 23–29.

Balacheff, N. (2002). The researcher epistemology: A deadlock from educational research on proof. In Fou-Lai Lin (Ed.), *International conference on mathematics: Understanding proving and proving to understand.* Taipei: NSC and NTNU.

Balacheff, N., & Gaudin, N. (in press). Modeling students' conceptions: The case of function. *Research in Collegiate Mathematics Education.*

Ball, D. L. (1993). With an eye on the mathematical horizon: Dilemmas of teaching elementary school mathematics. *Elementary School Journal, 93,* 373–397.

Ball, D. L., & Bass, H. (2000). Making believe: The collective construction of public mathematical knowledge in the elementary classroom. In D. C. Phillips (Ed.), *Constructivism in education:*

Opinions and second opinions on controversial issues, Ninety-ninth Yearbook of the National Society for the Study of Education. Chicago, IL: University of Chicago Press.

Ball, D. L., & Bass, H. (2003). Making mathematics reasonable in school. In J. Kilpatrick, W. G. Martin, & D. Schifter (Eds.), *A research companion to principles and standards for school mathematics.* Reston, VA: NCTM.

Ball, D. L., Hoyles, C., Jahnke, H. N., & Movshovitz-Hadar, N. (2002). The teaching of proof. In L. I. Tatsien (Ed.), *Proceedings of the International Congress of Mathematicians, III.* Beijing: Higher Education Press.

Bastable, V., & Schifter, D. (2007). Classroom stories: Examples of elementary students engaged in early algebra. In J. Kaput, M. Blanton, & D. Carraher (Eds.), *Algebra in the early grades.* Mahwah, NJ: Lawrence Erlbaum Associates, Inc.

Bauersfeld, H. (1994). Theoretical perspectives on interaction in the mathematics classroom. In R. Biehler, R. W. Scholz, R. Strasser, & Winkelmann (Eds.), *The didactics of mathematics as a scientific discipline.* Dordrecht, The Netherlands: Kluwer Academic.

Baumert, J., & Lehmann, R. (Eds.) (1997). *TIMSS: Mathematisch Naturwissenschaftlicher Unterricht im internationalen Vergleich. Deskriptive Befunde.* Opladen: Leske + Budrich.

Bell, A. W. (1976). A study of pupils' proof-explanations in mathematical situations. *Educational Studies in Mathematics, 7,* 23–40.

Bills, L., & Rowland, T. (1999). Examples, generalisation and proof. In L. Brown (Ed.), *Making meanings in mathematics: Refereed Proceedings of the British Society for Research into Learning Mathematics.* York: QED Publications.

Bittinger, M. (1969). *The effect of a unit mathematical proof on the performance of college mathematics majors in future mathematics courses.* Dissertation Abstracts, 29, 3906A.

Black, M. (1962). *Models and metaphors: Studies in language and philosophy.* New York: Cornell University Press.

Blanton, M. L., & Stylianou, D. A. (2002). Exploring sociocultural aspects of undergraduate students' transition to mathematical proof. In D. Mewborn, P. Sztajn, D. Y. White, H. G. Weigel, R. L. Bryant, & K. Nooney (Eds.), *Proceedings of the 24th Annual Meeting for Psychology of Mathematics Education-North American Chapter,* Vol. 4. Athens, GA.

Boero, P. (1999). Argumentation and mathematical proof: A complex, productive, unavoidable relationship in mathematics and mathematics education. *International Newsletter on the Teaching and Learning of Mathematical Proof,* 7/8.

Boero P., Garuti R., Lemut E., & Mariotti, M. A. (1996). Challenging the traditional school approach to theorems: a hypothesis about the cognitive unity of theorems. In L. Puig & A. Gutiérrez (Eds.), *Proceedings of the 20th Annual Conference of the International Group for the Psychology of Mathematics Education,* Vol. 2. Valencia, Spain: Program Committee.

Boero P., Garuti R., & Mariotti M A. (1996). Some dynamic mental processes underlying producing and proving conjectures. In L. Puig & A. Gutiérrez (Eds.), *Proceedings of the 20th Annual Conference of the International Group for the Psychology of Mathematics Education,* Vol. 2. Valencia, Spain: Program Committee.

Borko, H., & Putnam, R. (1996). Learning to teach. In R. Calfee & D. Berliner (Eds.), *Handbook of educational psychology.* New York: Macmillan.

Bourdieu, P. (1990). *The logic of practice.* Stanford, CA: Stanford University Press.

Boyd, C., Burrill, G., Cummins, J., Kanold, T., & Malloy, C. (1998). *Geometry: Integration, applications, connections.* New York: Glencoe-McGraw Hill.

Braine, M. D. S., & Rumain, B. (1983). Logical reasoning. In J. H. Flavell & E. M. Markman (Eds.), *Handbook of child psychology: Vol. 3, Cognitive development.* New York: Wiley.

Brousseau, G. (1997). *Theory of didactical situations in mathematics.* Boston: Kluwer Academic.

Brown, A., Thomas, K., & Tolias, G. (2002). Conceptions of divisibility: Success and understanding. In S. R. Campbell & R. Zazkis (Eds.), *Learning and teaching number theory.* Stamford, CT: Ablex.

Brown, L., Reid, D. A., & Zack, V. (1998). On doing the "same problem." *Mathematics Teaching, 163,* 50–55.

Brown, V. (1996). Third graders explore multiplication. In D. Schifter (Ed.), *What's happening in math class? Vol. 1, Envisioning new practices through teacher narratives.* New York: Teachers College Press.

Bruner, J. (1960). *The process of education.* Cambridge, MA: Harvard University Press.

Bruner, J. S., Oliver, R. R., & Greenfield, P. M. (1966). *Studies in cognitive growth.* New York: Wiley.

Burton, L. (2004). *Mathematicians as enquirers: Learning about learning mathematics.* Dordrecht, The Netherlands: Kluwer Academic.

Carpenter, T., Fennema, E., & Franke, M. (1996). Cognitively guided instruction: A knowledge base for reform in primary mathematics instruction. *Elementary School Journal, 97*, 3–20.

Carpenter, T., Fennema, E., Franke, M. L., Levi, L., & Empson, S. (1999). *Children's mathematics: Cognitively guided instruction.* Portsmouth, NH: Heinemann.

Carpenter, T., Fennema, E., Peterson, P. L., Chiang, C., & Franke, M. (1989). Using knowledge of children's mathematics thinking in classroom teaching: An experimental study. *American Educational Research Journal, 26*, 499–531.

Carpenter, T. P., Franke, M. L., & Levi, L. (2003). *Thinking mathematically: Integrating arithmetic and algebra in the elementary school.* Portsmouth, NH: Heinemann.

Charles, R., & Lester, F. (1982). *Teaching problem solving: What, why and how.* Palo Alto, CA: Dale Seymour.

Charles, R., Lester, F., & O'Daffer, P. (1987). *How to evaluate progress in problem solving.* Reston, VA: National Council of Teachers of Mathematics.

Chazan, D. (1990). Quasi-empirical views of mathematics and mathematics teaching. *Interchange, 21*, 14–23.

Chazan, D. (1993). High school geometry students' justification for their views of empirical evidence and mathematical proof. *Educational Studies in Mathematics, 24*, 359–387.

Chazan, D., & Yerushalmy, M. (2003). On appreciating the cognitive complexity of school algebra: Research on algebra learning and directions of curricular change. In J. Kilpatrick, D. Schifter, & G. Martin (Eds.), A *research companion to the principles and standards for school mathematics.* Reston, VA: NCTM.

Chen, C., & Herbst, P. (2005). *The descriptive mode of interaction with diagrams in proving triangles congruent.* Paper presented at the Third East Asia Regional Conference on Mathematics Education. Shanghai, China.

Clark, D. M. (2001). *R. L. Moore and the learning curve.* Unpublished manuscript.

Cobb, P., & Bauersfeld, H. (1995). *The emergence of mathematical meaning.* Hillsdale, NJ: Lawrence Erlbaum Associates, Inc.

Cobb, P., Boufi, A., McClain, K., & Whitenack, J. (1997). Reflective discourse and collective reflection. *Journal for Research in Mathematics Education, 28*, 258–277.

Cobb, P., Wood, T., & Yackel, E. (1993). Discourse, mathematical thinking, and classroom practice. In N. Minick, E. Forman, & A. Stone (Eds.), *Education and mind: Institutional, social, and developmental processes.* New York: Oxford University Press.

Cobb, P., Wood, T., Yackel, E., & McNeal, B. (1992). Characteristics of classroom mathematics traditions: An interactional analysis. *American Educational Research Journal, 29*, 573–604.

Cobb, P., & Yackel, E. (1996). Constructivist, emergent, and sociocultural perspectives in the context of developmental research. *Educational Psychologist, 31*, 175–190.

Coe, R., & Ruthven, K. (1994). Proof practices and constructs of advanced mathematics students. *British Educational Research Journal, 20*, 41–53.

Cohen, D. K., & Ball, D. L. (1999). *Instruction, capacity, and improvement* (CPRE Research Report No. RR-043). Philadelphia, PA: University of Pennsylvania, Consortium for Policy Research in Education.

Cohen, D. K., & Ball, D. L. (2000). *Instructional innovation: Reconsidering the story.* Paper presented at the Annual Meeting of the American Educational Research Association, New Orleans.

Committee on the Undergraduate Program in Mathematics. (2004). *Undergraduate Programs and Courses in the Mathematical Sciences: CUPM Curriculum Guide.* Washington, DC: Mathematical Association of America.

Courant, R., & Robbins, H. (1996). *What is mathematics?* New York: Oxford University Press.

Cuban, L. (1993). *How teachers taught: Constancy and change in American classrooms, 1890–1990.* New York: Teachers College Press.

Cuoco, A., Goldenberg, E. P., & Mark, J. (1996). Habits of mind: An organizing principle for mathematics curricula. *Journal of Mathematical Behavior, 15*, 375–402.

Davis, P. J. (1972). Fidelity in mathematical discourse: Is one and one really two? *American Mathematical Monthly, 79*, 252–263.

Davis, P. J., & Hersh, R. (1981). *The mathematical experience.* Boston, MA: Houghton-Mifflin.

Davis, R. B. (1985). A study of the process of making proofs. *Journal of Mathematical Behavior, 4*, 37–43.

Davis, R. B. (1984). *Learning mathematics: A cognitive science approach.* Norwood, MA: Ablex.

Davis, R. B., & Maher, C. A. (1997). How students think: The role of representations. In L. English (Ed.), *Mathematical reasoning: Analogies, metaphors, and images.* Hillsdale, NJ: Lawrence Erlbaum Associates, Inc.

Davis, R. B., Maher, C. A., & Martino. A. M. (1992). Using videotapes to study the construction of mathematical knowledge of individual children working in groups. *Journal of Science, Education, and Technology, 1*, 177–189.

Davydov, V. V. (1975). The psychological characteristics of the "prenumerical" period of mathematics instruction. *Soviet Studies in the Psychology of Learning and Teaching Mathematics, VII*, 109–206.

Davydov, V. V., Gorbov, S. F., Mikulina, G. G., & Savel'eva, O. V. (1994). *Mathematics: Textbook for first grade*. Moscow: MIROS.

de Villiers, M. (1990). The role and function of proof in mathematics. *Pythagoras, 24*, 17–24.

de Villiers, M. (1999). *Rethinking proof with the geometer's sketchpad*. Emeryville, CA: Key Curriculum Press.

DfEE (Department for Education and Employment). (2001). *Key stage 3 national strategy: Framework for teaching mathematics: Years 7, 8 and 9*. London: DfEE.

Diaz, R. M., Neal, C. J., & Amaya-Williams, M. (1999). The social origins of self-regulation. In L. Moll (Ed.), *Vygotsky and education: Instructional implications and applications of sociohistorical psychology*. Cambridge: Cambridge University Press.

Dolk, M., & Fosnot, C. T. (2004). *Working with the number line: Mathematical models*. [CDROM]. Portsmouth, NH: Heinemann.

Dougherty, B. J., & Slovin, H. (2004). Generalized diagrams as a tool for young children's problem solving. *Proceedings of the 28th Conference of the International Group for the Psychology of Mathematics Education*. Bergen, Norway.

Downs, M. L. N., & Mamona-Downs, J. (2005). The proof language as a regulator of rigor in proof, and its effect on student behavior. In *Proceedings of the Fourth Conference on European Research in Mathematics Education*. Sant Feliu de Guixols, Spain.

Dreyfus, T. (1999). Why Johnny can't prove. *Educational Studies in Mathematics, 38*, 85–109.

Dreyfus, T., & Hadas, N. (1987). Euclid may stay—and even be taught. In M. M. Lindquist & A. P. Shulte (Eds.), *Learning and teaching geometry, K–12*. Reston, VA: National Council of Teachers of Mathematics.

Dubinsky, E., & Tall, D. (1991). Advanced mathematical thinking and the computer. In D. Tall (Ed.), *Advanced mathematical thinking*. Dordrecht, The Netherlands: Kluwer Academic.

Duval, R. (1991). Structure du raisonnement déductif et apprentissage de la démonstration. *Educational Studies in Mathematics, 22*, 233–261.

Edwards, B. S., & Ward, M. B. (2004). Surprises from mathematics research: Student (mis)use of mathematical definitions. *American Mathematical Monthly, 111*, 411–424.

Eliot, C., Harris, W. T., Angell, J. B., Tetlow, J., Taylor, J. M., Robinson O. D., et al. (1969). Report of the Committee of Ten to the National Education Association. In National Education Association (Ed.), *Report of the Committee on secondary school studies* (pp. 3–5). New York: Arno Press (Original work published in 1893).

Erickson, F. (2004). *Talk and social theory: Ecologies of speaking and listening in everyday life*. Cambridge: Polity Press.

Fischbein, E. (1982). Intuition and proof. *For the Learning of Mathematics, 3*, 9–18, 24.

Fischbein, E., & Kedem, I. (1982). Proof and certitude in the development of mathematical thinking. In A. Vermandel (Ed.), *Proceedings of the 6th International Conference for the Psychology of Mathematical Education*. Antwerp, Belgium.

Fosnot, C. T. (2007). *Measuring for the art show: Contexts for learning mathematics*. Portsmouth, NH: Heinemann.

Fosnot, C. T., & Dolk, M. (2001a). *Young mathematicians at work: Constructing number sense, addition and subtraction*. Portsmouth, NH: Heinemann.

Fosnot, C. T., & Dolk, M. (2001b). *Young mathematicians at work: Constructing multiplication and division*. Portsmouth, NH: Heinemann.

Fosnot, C. T., & Dolk, M. (2002). *Young mathematicians at work: Constructing fractions and decimals*. Portsmouth, NH: Heinemann.

Fowler, D. (1999). *The mathematics of Plato's academy*. New York: Oxford University Press.

Freudenthal H. (1983). *Didactical phenomenology of mathematical structures*. Dordrecht, The Netherlands: D. Reidel.

Fuys, D., Geddes, D., & Tischler, R. (1988). The van Hiele model of thinking in geometry among adolescents. *Journal for Research in Mathematics Education. Monograph 3*. Reston, VA: National Council of Teachers of Mathematics.

Galbraith, P.L. (1981). Aspects of proving: A clinical investigation of process. *Educational Studies in Mathematics, 12*, 1–28.

Galotti, K. M., Komatsu, L. K., & Voelz, S. (1997). Children's differential performance on deductive and inductive syllogisms. *Developmental Psychology, 33*, 70–78.

Gindikin, S. (1988). *Tales of physicists and mathematicians*. Boston, MA: Birkhauser.

Goetting, M. (1995). *The college students' understanding of mathematical proof*. Unpublished doctoral dissertation, University of Maryland, College Park.

Goldin, G. A. (1998). Representational systems, learning, and problem solving in mathematics. *Journal of Mathematical Behavior, 17*, 137–165.

Goffman E. (1974). *Frame analysis: An essay on the organization of experience*. Boston, MA: Northeastern University Press.

Goos, M. (2004). Learning mathematics in a classroom community of inquiry. *Journal for Research in Mathematics Education, 35*, 258–291.

Goos M., Galbraith, P., & Renshaw, P. (2002). Socially mediated metacognition: Creating collaborative zones of proximal development in small group problem solving. *Educational Studies in Mathematics, 49*, 193–223.

Götz, T., Pekrun, R., Perry, R. P., & Hladkyi, S. (2001). *Academic emotions questionnaire: Codebook for English-language scale versions* (Technical report). University of Munich: Department of Psychology.

Gravemeijer, K. P. E. (1994). *Developing realistic mathematics education*. Utrecht, The Netherlands: Freudenthal Instituut.

Gravemeijer, K. P. E. (1999). How emergent models may foster the constitution of formal mathematics. *Mathematical Thinking and Learning, 1*, 155–177.

Greeno, J. (1994). Comments on Susanna Epp's chapter. In A. Schoenfeld (Ed.), *Mathematical thinking and problem solving*. Hillsdale, NJ: Lawrence Erlbaum Associates Associates, Inc.

Gregg, J. (1995). The tensions and contradictions of the school mathematics tradition. *Journal for Research in Mathematics Education, 26*, 442–466.

Griffiths, P. A. (2000). Mathematics at the turn of the millennium. *American Mathematical Monthly, 107*, 1–14.

Gutiérrez, A., & Jaime, A. (1998). On the assessment of the van Hiele levels of reasoning. *Focus on Learning Problems in Mathematics, 20*, 2/3, 27–46.

Haimo, D. T. (1995). Experimentation and conjecture are not enough. *American Mathematical Monthly, 102*, 102–112.

Halliday, M. A. K. (1993). Towards a language-based theory of learning. *Linguistics and Education, 5*, 93–116.

Halmos, P. (1980). The heart of mathematics. *American Mathematical Monthly, 87*, 519–524.

Halmos, P. (1985). *I want to be a mathematician*. New York: Springer-Verlag.

Hamlin, M. (2006). *Lessons in educational equity: Opportunities for learning in an informal geometry class*. Unpublished doctoral dissertation. University of Michigan, Ann Arbor.

Hanna, G. (1983). *Rigorous proof in mathematics education*. Toronto: Ontario Institute for Studies in Education.

Hanna, G. (1990). Some pedagogical aspects of proof. *Interchange, 21*, 6–13.

Hanna, G. (1995). Challenges to the importance of proof. *For the Learning of Mathematics, 15*, 42–49.

Hanna, G. (2000). Proof, explanation and exploration: An overview. *Educational Studies in Mathematics, 44*, 5–23.

Hanna, G., & Barbeau, E. (2008). Proofs as bearers of mathematical knowledge. *Zentralblatt für Didaktik der Mathematik, 40*(30), 345–353.

Harel, G. (1993). On teacher education programs in mathematics. *International Journal for Mathematics Education in Science and Technology, 25*, 113–119.

Harel, G. (1998). Two dual assertions: The first on learning and the second on teaching (or vice versa). *American Mathematical Monthly, 105*, 497–507.

Harel, G. (2001). The development of mathematical induction as a proof scheme: A model for DNR-based instruction. In S. Campbell & R. Zazkis (Eds.), *Learning and teaching number theory*. Dordrecht, The Netherlands: Kluwer.

Harel, G. (2007). The DNR system as a conceptual framework for curriculum development and instruction. In R. Lesh, E. Hamilton, & J. Kaput (Eds.), *Foundations for the future in mathematics education*. Mahwah, NJ: Lawrence Erlbaum Associates, Inc.

Harel, G. (in press). What is mathematics? A pedagogical answer to a philosophical question. In R. B. Gold & R. Simons (Eds.), *Current issues in the philosophy of mathematics from the perspective of mathematicians*. Washington, DC: Mathematical Association of America.

Harel, G. (in preparation). *DNR-Based Instruction in Mathematics*.

Harel, G., & Sowder, L. (1998). Students' proof schemes: Results from exploratory studies. In

A. H. Schoenfeld, J. Kaput, & E. Dubinsky (Eds.), *Research in collegiate mathematics education III.* Providence, RI: American Mathematical Society.

Harel, G., & Sowder, L. (2007). Toward a comprehensive perspective on proof. In F. Lester (Ed.), *Second handbook of research on mathematics teaching and learning.* Greenwich, CT: Information Age Publishing.

Hart, E. W. (1994). A conceptual analysis of the proof-writing performance of expert and novice students in elementary group theory. In J. J. Kaput & E. Dubinsky (Eds.), *Research issues in undergraduate mathematics learning: Preliminary analyses and results (MAA notes 33).* Washington, DC: Mathematical Association of America.

Hawkins, J., Pea, R. D., Glick, J., & Scribner, S. (1984). "Merds that laugh don't like mushrooms": Evidence for deductive reasoning by preschoolers. *Developmental Psychology, 20,* 584–594.

Healy, L., & Hoyles, C. (2000). A study of proof conceptions in algebra. *Journal for Research in Mathematics Education, 31,* 396–428.

Heaton, R. (1992). Who is minding the mathematics content? A case study of a fifth-grade teacher. *Elementary School Journal, 93,* 153–162.

Heinze, A., Cheng, Y. H., & Yang, K. L. (2004). Students' performance in reasoning and proof in Taiwan and Germany: Results, paradoxes, and open questions. *Zentralblatt für Didaktik der Mathematik, 36*(5), 162–171.

Heinze, A., & Reiss, K. (2007). Reasoning and proof in the mathematics classroom. *Analysis, 27,* 333–357.

Heinze, A., Reiss, K., & Groß, C. (2006). Learning to prove with heuristic worked-out examples. In J. Novotna, H. Moraova, M. Kratka, & N. Stehlikova (Eds.), *Proceedings of the 30th Conference of the International Group for the Psychology of Mathematics Education,* Vol. 3. Prague: Program Committee.

Heinze, A., Reiss, K., & Rudolph, F. (2005). Mathematics achievement and interest in mathematics from a differential perspective. *Zentralblatt für Didaktik der Mathematik, 37,* 212–220.

Herbst, P. (1998). *What works as proof in the mathematics class.* Unpublished doctoral dissertation, University of Georgia, Athens.

Herbst, P. (2002a). Engaging students in proving: A double bind on the teacher. *Journal for Research in Mathematics Education, 33,* 176–203.

Herbst, P. (2002b). Establishing a custom of proving in American school geometry: Evolution of the two-column proof in the early twentieth century. *Educational Studies in Mathematics, 49,* 283–312.

Herbst, P. (2003). Using novel tasks in teaching mathematics: Three tensions affecting the work of the teacher. *American Educational Research Journal, 40,* 197–238.

Herbst, P. (2004). Interaction with diagrams and the making of reasoned conjectures in geometry. *Zentralblatt für Didaktik der Mathematik, 36,* 129–139.

Herbst, P. (2005). Knowing "equal area" while proving a claim about equal areas. *Recherches en Didactique des Mathématiques, 25,* 11–56.

Herbst, P. (2006). Teaching geometry with problems: Negotiating instructional situations and mathematical tasks. *Journal for Research in Mathematics Education, 37,* 313–347.

Herbst, P., & Brach, C. (2006). Proving and "doing proofs" in high school geometry classes: What is "it" that is going on for students and how do they make sense of it? *Cognition and Instruction, 24,* 73–122.

Hersh, R. (1993). Proving is convincing and explaining. *Educational Studies in Mathematics, 24,* 389–399.

Hiebert, J., Carpenter, T., Fennema, E., Fuson, K., Wearne, D., Murray, H., et al. (1997). *Making sense: Teaching and learning mathematics with understanding,* Portsmouth, NH: Heinemann.

Hill, F. (1895). The educational value of mathematics. *Educational Review, 9,* 349–358.

Hoffer, A., Koss, R., Beckmann, J., Duren, P., Hernandez, J., Schlesinger, B., et al. (1998). *Focus on geometry: An integrated approach.* Menlo Park, CA: Addison-Wesley.

Hoyles, C. (1997). The curricular shaping of students' approaches to proof. *For the Learning of Mathematics, 17,* 7–15.

Hoyles, C., Küchemann, D., Healy, L., & Yang, M. (2005). Students' developing knowledge in a subject discipline: Insights from combining quantitative and qualitative methods. *International Journal of Social Research Methodology, 8,* 225–238.

Jaffe, A., & Quinn, F. (1993). "Theoretical mathematics:" Towards a cultural synthesis of mathematics and theoretical physics. *Bulletin of the American Mathematical Society, 29,* 1–13.

Jahnke, H. N. (2005). *A genetic approach to proof.* Paper presented at the Fourth Congress of the European Society for Research in Mathematics Education. Sant Feliu de Guixols, Spain.

Jones, F. B. (1977). The Moore Method. *American Mathematical Monthly, 84,* 273–278.

Jones, K. (1997). Student-teachers' conceptions of mathematical proof. *Mathematics Education Review, 9,* 21–32.

Kaput, J. (2008). What is algebra? What is algebraic reasoning? In J. J. Kaput, D. W. Carraher, & M. L. Blanton (Eds.), *Algebra in the early grades.* Mahwah, NJ: Lawrence Erlbaum Associates, Inc.

Kaput, J., Carraher, D., & Blanton, M. (Eds.) (2008). *Algebra in the early grades.* Mahwah, NJ: Lawrence Erlbaum Associates, Inc.

Kieran, C. (1992). The learning and teaching of school algebra. In D. Grouws (Ed.), *Handbook of research on mathematics teaching and learning.* Reston, VA: National Council of Teachers of Mathematics.

Kilpatrick, J., Swafford, J., & Findell, B. (Eds.) (2001). *Adding it up: Helping children learn mathematics.* Washington, DC: National Academy Press.

King, I. L. (1973). A formative development of an elementary school unit on proof. *Journal for Research in Mathematics Education, 4,* 57–63.

Kleiner, I. (1991). Rigor and proof in mathematics: A historical perspective. *Mathematics Magazine, 64,* 291–314.

Klieme, E., Reiss, K., & Heinze, A. (2003). Geometrical competence and understanding of proof. A study based on TIMSS items. In F. L. Lin & J. Guo (Eds.), *Proceedings of the International Conference on Science and Mathematics Learning 2003.* Taipei, Taiwan: National Taiwan Normal University.

Knapp, N., & Peterson, P. (1995). Teachers' interpretations of "CGI" after four years: Meanings and practices. *Journal for Research in Mathematics Education, 26,* 40–65.

Knuth, E. (2002a). Secondary school mathematics teachers' conceptions of proof. *Journal for Research in Mathematics Education, 33,* 379–405.

Knuth, E. (2002b). Teachers' conceptions of proof in the context of secondary school mathematics. *Journal of Mathematics Teacher Education, 5,* 61–88.

Knuth, E. J., Choppin, J., Slaughter, M., & Sutherland, J. (2002). Mapping the conceptual terrain of middle school students' competencies in justifying and proving. In D. S. Mewborn, P. Sztajn, D. Y. White, H. G. Weigel, R. L. Bryant, & K. Nooney (Eds.), *Proceedings of the 24th Annual Meeting of the North American Chapter of the International Group for the Psychology of Mathematics Education,* Vol. 4. Athens, GA: Clearinghouse for Science, Mathematics, and Environmental Education.

Knuth, E., & Sutherland, J. (2004). Student understanding of generality. In D. McDougall (Ed.), *Proceedings of the 26th Annual Meeting of the North American Chapter of the International Group for the Psychology of Mathematics Education.* Toronto: University of Toronto/OISE.

Koedinger, K. R., & Anderson, J. R. (1990). Abstract planning and perceptual chunks: Elements of expertise in geometry. *Cognitive Science, 14,* 511–550.

Kolata, G. (1993). At home with the elusive world of mathematics. In *Themes of the Times* issue on mathematics. *New York Times,* October 12, pp. 1, 3.

Konior, J. (1993). Research into the construction of mathematical texts. *Educational Studies in Mathematics, 24,* 251–256.

Kozulin, A. (1998). *Psychological tools: A sociocultural approach to education.* Cambridge, MA: Harvard University Press.

Kruger, A. C. (1993). Peer collaboration: Conflict, cooperation or both? *Social Development, 2,* 165–182.

Küchemann, D. (2008). *Looking for structure: A report of the Proof Materials Project.* London: Dexter Graphics. (Available at http://www.ioe.ac.uk/proof/PMPintro.html.)

Küchemann, D., & Hoyles, C. (2003). *Technical report for Longitudinal Proof Project: Year 10 survey 2002,* Vol. 1. London: Institute of Education.

Kwon, O. N., Rasmussen, C., & Allen, K. (2005). Students' retention of mathematical knowledge and skills in differential equations. *School Science and Mathematics, 105,* 227–239.

Laborde, C. (1990). Language and mathematics. In P. Nesher & J. Kilpatrick (Eds.), *Mathematics and cognition: A research synthesis by the International Group for the Psychology of Mathematics Education.* Cambridge: MA: Cambridge University Press.

Lakatos, I. (1976). *Proof and refutations: The logic of mathematical discovery.* Cambridge, UK: Cambridge University Press.

Lakatos, I. (1978). What does a mathematical proof prove? In J. Worrall & G. Currie (Eds.), *Mathematics, science, and epistemology.* Cambridge: Cambridge University Press.

Lampert, M. (1990). When the problem is not the question and the solution is not the answer: Mathematical knowing and teaching. *American Educational Research Journal, 27,* 29–63.

Lampert, M. (1991). Connecting mathematical teaching and learning. In E. Fennema, T. P. Carpenter,

& S. J. Lamon (Eds.), *Integrating research on teaching and learning mathematics*. Albany, NY: SUNY Press.

Lampert, M. (1992). Practices and problems in teaching authentic mathematics in school. In F. Oser, A. Dick, & J.-L. Patry (Eds.), *Effective and responsible teaching: The new synthesis*. New York: Jossey-Bass.

Lampert, M. (1993). When the problem is not the question and solution is not the answer: Mathematical knowing and teaching. *American Educational Research Journal, 27*(1), 29–63.

Lampert, M. (1998a). Studying teaching as a thinking practice. In J. Greeno & S. Goldman (Eds.), *Thinking practices*. Mahwah, NJ: Lawrence Erlbaum Associates, Inc.

Lampert, M. (1998b). Where did these ideas come from? Lampert's story. In M. Lampert & D. Ball (Eds.), *Teaching, multimedia and mathematics: Investigations of real practice*. New York: Springer-Verlag.

Lappan, G., Fey, J. T., Fitzgerald, W. M., Friel, S. N., & Philips, E. D. (1998/2004). *Connected Mathematics Project*. Menlo Park, CA: Dale Seymour.

Lappan, G., Fey, J., Fitzgerald, W., Friel, S., & Phillips, E. (2002). *Getting to know Connected Mathematics: An implementation guide*. Glenview, IL: Prentice-Hall.

Laudien, R. C. (1998). *Mathematical reasoning in middle school curriculum materials*. Unpublished doctoral dissertation. University of Delaware.

Lawson, M. J., & Chinnapan, M. (2000). Knowledge connectedness in geometry problem solving. *Journal for Research in Mathematics Education, 31*, 26–43.

Lee, J. (2002). *An analysis of difficulties encountered in teaching Davydov's mathematics curriculum to students in a U.S. setting and measures found to be effective in addressing them*. Unpublished doctoral dissertation, Binghamton University, State University of New York.

Lee, L., & Wheeler, D. (1987). *Algebraic thinking in high school students: Their conceptions of generalization and justification*. Research Report. Montreal: Concordia University.

Lemke, J. (1990). *Talking science: Language, learning, and values*. Norwood, NJ: Ablex.

Lemke, J. (2000). Across the scales of time: Artifacts, activities, and meanings in ecosocial systems. *Mind, Culture, and Activity, 7*, 273–290.

Leron, U. (1983). Structuring mathematical proofs. *American Mathematical Monthly, 90*, 174–184.

Lewalter, D., Krapp, A., Schreyer, I., & Wild, K.-P. (1998). Die Bedeutsamkeit des Erlebens von Kompetenz, Autonomie und sozialer Eingebundenheit für die Entwicklung berufs-spezifischer Interessen. *Zeitschrift für Berufs und Wirtschaftspädagogik, 14*, 143–168.

Lin, F. L. (2000). An approach for developing well-tested, validated research of mathematics learning and teaching. In T. Nakahara & M. Koyama (Eds.), *Proceedings of the 24th Conference of the International Group for the Psychology of Mathematics Education*, Vol. 1. Hiroshima: Program Committee.

Litowitz, B. (1993). Deconstruction in the zone of proximal development. In E. Forman, N. Minick, & C. Stone (Eds.), *Contexts for learning: Sociocultural dynamics in children's development*. New York: Oxford University Press.

Livingston, E. (1999). Cultures of proving. *Social Studies of Science, 29*, 867–888.

Lyons, M., & Lyons, R. (1991). *Défi mathématique*. Montreal: Mondia éditeurs.

Lyons, M., & Lyons, R. (1996). *Challenging mathematics (Grade 1–6): Teaching and activity guide* (V. Tétrault., Trans.) Montreal: Mondia éditeurs.

Mahavier, W. S. (1998). What is the Moore Method? *PRIMUS, 9*, 339–354.

Mahavier, W. T. (1997). A gentle discovery method. *College Teaching, 45*, 132–135.

Maher, C. A. (2005). How students structure their investigations and learn mathematics: Insights from a long-term study. *Journal of Mathematical Behavior, 24*, 1–14.

Maher, C. A., & Davis, R. B. (1990). Building representations of children's meanings. In R. B. Davis, C. A. Maher, & N. Noddings (Eds.), *Constructivist views on the teaching and learning of mathematics: Journal for Research in Mathematics Education Monograph No. 4*. Reston, VA: National Council of Teachers of Mathematics.

Maher, C. A., & Martino, A. M. (1996a). The development of the idea of mathematical proof: A 5-year case study. *Journal for Research in Mathematics Education, 27*, 194–214.

Maher, C. A., & Martino, A. M. (1996b). Young children invent methods of proof: The gang of four. In P. Nesher, L. P. Steffe, P. Cobb, B. Greer, & G. Goldin (Eds.), *Theories of mathematical learning*. Mahwah, NJ: Lawrence Erlbaum Associates, Inc.

Maher, C. A., & Speiser, R. (1997). How far can you go with block towers? Stephanie's intellectual development. *Journal of Mathematical Behavior, 16*, 125–132.

Manaster, A. (1998). Some characteristics of eighth grade mathematics classes in the TIMSS videotape study. *American Mathematical Monthly, 105*, 793–805.

Martin, G., & Harel, G. (1989). Proof frames of preservice elementary teachers. *Journal for Research in Mathematics Education, 12*, 41–51.

Martin, T. S., & McCrone, S. S. (2003). Classroom factors related to geometric proof construction ability. *The Mathematics Educator, 7*, 18–31.

Martin, T. S., McCrone, S. M. S., Bower, M. L., & Dindyal, J. (2005). The interplay of teacher and student actions in the teaching and learning of geometric proof. *Educational Studies in Mathematics, 60*, 95–124.

Martin, W. G., & Harel, G. (1989). Proof frames of preservice elementary teachers. *Journal for Research in Mathematics Education, 20*, 41–51.

Martino, A. M., & Maher, C. A. (1999). Teacher questioning to promote justification and generalization in mathematics: What research practice has taught us. *Journal of Mathematical Behavior, 18*, 53–78.

Marty, R. H. (1986). *Teaching proof techniques. Mathematics in College.* City University of New York (Spring–Summer), 46–53.

Mason, J. (2008). Making use of children's powers to produce algebraic thinking. In J. Kaput, D. Carraher, & M. Blanton (Eds.), *Algebra in the early grades.* Mahwah, NJ: Lawrence Erlbaum, Inc./Taylor & Francis Group.

Mason, J., Burton, L., & Stacey, K. (1982). *Thinking mathematically.* New York: Addison-Wesley.

Mason, J., & Spence, M. (1999). Beyond mere knowledge of mathematics: The importance of knowing-to-act in the moment. *Educational Studies in Mathematics, 28*, 135–161.

Maykut, P., & Morehouse, R. (1994). *Beginning qualitative research.* Philadelphia, PA: Falmer Press.

McClain, K. (2002). Teacher's and students' understanding: The role of tools and inscriptions in supporting effective communication. *Journal for the Learning Sciences, 11*, 217–249.

McClain, K. (2004). *Form versus function: An articulated framework for action for teacher development.* Paper presented at the annual meeting of the American Education Research Association, San Diego, CA.

McClain, K., & Cobb, P. (2001). An analysis of development of sociomathematical norms in one first-grade classroom. *Journal for Research in Mathematics Education, 32*, 236–266.

McCrone, S. S., & Martin, T. S. (2004). Assessing high school students' understanding of geometric proof. *Canadian Journal of Science, Mathematics, and Technology Education, 4*, 223–242.

Medin, D. (1989). Concepts and conceptual structure. *American Psychologist, 44*, 1469–1481.

Mehan, H. (1979). *Learning lessons.* Cambridge, MA: Harvard University Press.

Mehan, H., and Wood, H. (1975). *The reality of ethnomethodology.* Malabar, FL: Krieger.

Mercer, N. (1995). *The guided construction of knowledge.* Clevedon: Multilingual Matters.

Mevarech, Z. R., & Kramarski, B. (1997). IMPROVE: A multidimensional method for teaching mathematics in heterogeneous classrooms. *American Educational Research Journal, 34*, 365–394.

Meyer, C., & Sallee, T. (1983). *Make it simpler.* Menlo Park, CA: Addison-Wesley.

Mikulina, G. G. (1991). The psychological features of solving problems with letter data. In L. Steffe (Ed.), *Soviet studies in mathematics education, Vol. 6, Psychological abilities of primary schoolchildren in learning mathematics.* Reston, VA: National Council of Teachers of Mathematics.

Miles, M. B., & Huberman, A. M. (1994). *Qualitative data analysis.* Thousand Oaks, CA: Sage.

Minick, N. (1996). The development of Vygotsky's thought. In H. Daniels, (Ed.), *An introduction to Vygotsky.* London: Routledge.

Miyakawa, T. (2002). Relation between proof and conception: The case of proof for the sum of two even numbers. In A. Cockburn & E. Nardi (Eds.), *Proceedings of the 26th Annual Conference of the International Group for the Psychology of Mathematics Education.* Vol. 3. Norwich: University of East Anglia.

Monk, S. (2008). The world of arithmetic from different points of view. In D. Schifter, V. Bastable & S.J. Russell (Eds.), *Developing mathematics ideas casebook for reasoning algebraically about operations.* Parsippany, NJ: Pearson Learning Group.

Moore, R. C. (1994). Making the transition to formal proof. *Educational Studies in Mathematics, 27*, 249–266.

Moreno-Armella, L. (2007). *A perspective on proof and situated proof.* Invited plenary address at the International Conference for Research Paradigms on the Teaching and Learning of Proof, Providence, RI.

Moretti, G., Stephens, M., Goodnow, J., & Hoogeboom, S. (1987). *The problem solver 5.* Sunnydale, CA: Creative publications.

Morgan, C. (1997). The institutionalization of open-ended investigation: Some lessons from the UK experience. In E. Pehkonen (Ed.), *Use of open-ended problems in mathematics classroom,* Vol. 176. Helsinki: Department of Teacher Education.

Morris, A. K. (1999). Developing concepts of mathematical structure: Pre-arithmetic reasoning versus extended arithmetic reasoning. *Focus on Learning Problems in Mathematics, 21,* 44–72.

Morris, A. K. (2000a). Development of logical reasoning: Children's ability to verbally explain the nature of the distinction between logical and nonlogical forms of argument. *Developmental Psychology, 36,* 741–758.

Morris, A. K. (2000b). A teaching experiment: Introducing fourth graders to fractions from the viewpoint of measuring quantities using Davydov's mathematics curriculum. *Focus on Learning Problems in Mathematics, 22,* 33–84.

Morris, A. K. (2002). Mathematical reasoning: Adults' ability to make the inductive-deductive distinction. *Cognition and Instruction, 20,* 79–118.

Morris, A. K. (2007). Factors affecting pre-service teachers' evaluations of the validity of students' mathematical arguments in classroom contexts. *Cognition and Instruction, 25,* 479–522.

Morris, A. K., & Sloutsky, V. (1998). Understanding of logical necessity: Developmental antecedents and cognitive consequences. *Child Development, 69,* 721–741.

Moshman, D. (2005). *Adolescent psychological development: Rationality, morality, and identity.* Mahwah, NJ: Lawrence Erlbaum Associates, Inc.

Moshman, D., & Franks, B. A. (1986). Development of the concept of inferential validity. *Child Development, 57,* 153–165.

Mullis, I. V. S., Martin, M. O., Gonzalez, E. J., Gregory, K. D., Garden, R. A., O'Connor, K. M., et al. (2000). *TIMSS 1999 international mathematics report. Findings from IEA's Report of the Third International Mathematics and Science Study at the Eighth Grade.* Chestnut Hill, MA: International Study Centre, Lynch School of Education, Boston College.

Nardi, E., & Iannone, P. (2005). To appear and to be: Acquiring the "genre speech" of university mathematics. In *Proceedings of the 4th Conference on European Research in Mathematics Education.* Sant Feliu de Guixols, Spain.

Nardi, E., Jaworski, B., & Hegedus, S. (2005). A spectrum of pedagogical awareness for undergraduate mathematics: From "tricks" to "techniques." *Journal for Research in Mathematics Education, 36,* 284–316.

National Council of Teachers of Mathematics [NCTM]. (1989). *Curriculum and evaluation standards for school mathematics.* Reston, VA: National Council of Teachers of Mathematics.

National Council of Teachers of Mathematics. (2000). *Principles and standards for school mathematics.* Reston, VA: NCTM.

Newton, P. (1999). The place of argumentation in the pedagogy of school science. *International Journal of Science Education, 21,* 553–576.

Parker, G. E. (1992). Getting more from Moore. *PRIMUS, 2,* 235–246.

Pekrun, R. (1992). The impact of emotions on learning and achievement: Towards a theory of cognitive/motivational mediators. *Applied Psychology: An International Review, 41,* 359–376.

Peterson, S. (1955). Benjamin Peirce: Mathematician and philosopher. *Journal of the History of Ideas, 16,* 89–112.

Pillow, B. H. (1999). Children's understanding of inferential knowledge. *Journal of Genetic Psychology, 160,* 419–428.

Pinto, M., & Tall, D. (1999). Student constructions of formal theories: Giving and extracting meaning. In O. Zaslavsky (Ed.), *Proceedings of the 23rd Conference of the International Group for the Psychology of Mathematics Education,* Vol. 1. Haifa, Israel: Israel Institute of Technology.

Pirie, S., & Kieren, T. (1994). Growth in mathematical understanding: How can we characterize it and how can we represent it? *Educational Studies in Mathematics, 26,* 165–190.

Pólya, G. (1954a). *Induction and analogy in mathematics.* Princeton, NJ: Princeton University Press.

Pólya, G. (1954b). *Mathematics and plausible reasoning* (2 vols). Princeton, NJ: Princeton University Press.

Pólya, G. (1962, 1965/1981). *Mathematical discovery* (Volume 1, 1962; Volume 2, 1965). Princeton: Princeton University Press. (Combined paperback edition, 1981. New York: Wiley.)

Pólya, G. (1965). *Let us teach guessing.* Washington, DC: Mathematical Association of America.

Porteous, K. (1986). Children's appreciation of the significance of proof. *Proceedings of the 10th Conference of the International Group for the Psychology of Mathematics Education,* Vol. 1. London: University of London, Institute of Education.

Porteous, K. (1990). What do children really believe? *Educational Studies in Mathematics, 21,* 589–598.

Raman, M. (2003). Key ideas: What are they and how can they help us understand how people view proof? *Educational Studies in Mathematics, 52,* 319–325.

RAND Mathematics Study Panel. (2002). *Mathematical proficiency for all students: Toward a strategic research and development program in mathematics education.* Santa Monica, CA: RAND.

Reid, D. A. (2001). Proof, proofs, proving and probing: Research related to proof. Short Oral presentation. In M. van den Heuvel-Panhuizen (Ed.), *Proceedings of the 25th Annual Conference of the International Group for the Psychology of Mathematics Education*, Vol. I. Utrecht, The Netherlands: Program Committee.

Reid, D. A. (2002a). Elements in accepting an explanation. *Journal of Mathematical Behavior, 20*, 527–547.

Reid, D. A. (2002b). Conjectures and refutations in grade 5 mathematics. *Journal for Research in Mathematics Education, 33*, 5–29.

Reid, D. (2005). *The meaning of proof in mathematics education.* Paper presented at the Fourth Congress of the European Society for Research in Mathematics Education. Sant Feliu de Guixols, Spain.

Reiss, K., Hellmich, F., & Reiss, M. (2002). Reasoning and proof in geometry: Prerequisites of knowledge acquisition in secondary school students. In A. D. Cockburn & E. Nardi (Eds.), *Proceedings of the 26th Conference of the International Group for the Psychology of Mathematics Education*, Vol. 4. Norwich: Program Committee.

Reiss, K., & Renkl, A. (2002). Learning to prove: The idea of heuristic examples. *Zentralblatt für Didaktik der Mathematik, 34*, 29–35.

Renz, P. (1999). The Moore Method: What discovery learning is and how it works. *Newsletter of the Mathematical Association of America, 8/9*, 1–2.

Rin, H. (1983). Linguistic barriers to students' understanding of definitions. In R. Hershkowitz (Ed.), *Proceedings of the 7th International Conference for the Psychology of Mathematics Education.* Rehovot, Israel: Weizmann Institute of Science.

Rittenhouse, P. S. (1998). The teacher's role in mathematical conversation: Stepping in and stepping out. In M. Lampert & M. Blunk (Eds.), *Talking mathematics in school.* Cambridge: Cambridge University Press.

Ross, K. (1998). Doing and proving: The place of algorithms and proof in school mathematics. *American Mathematical Monthly, 3*, 252–255.

Ruthven, K. (1995). The unbearable mightness of being. In L. Healy (Ed.), *Justifying and proving in mathematics: Draft activities.* Unpublished manuscript, London: Institute of Education.

Schiefele, U., Krapp, A., & Schreyer, I. (1993). Metaanalyse des Zusammenhangs von Interesse und schulischer Leistung. *Zeitschrift für Entwicklungspsychologie und Pädagogische Psychologie, 25*, 120–148.

Schifter, D. (1999). Reasoning about operations: Early algebraic thinking, grades K through 6. In L. Stiff and F. Curio, (Eds.), *Mathematical reasoning, K–12: 1999 NCTM Yearbook.* Reston, VA: National Council of Teachers of Mathematics.

Schifter, D., Bastable, V., & Russell, S. J. (2008a). *Developing mathematics ideas casebook, facilitator's guide, and videotape for reasoning algebraically about operations.* Parsippany, NJ: Pearson Learning Group.

Schifter, D., Bastable, V., & Russell, S. J. (2008b). *Reasoning algebraically about operations: Casebook.* Parsippany, NJ: Pearson Learning Group.

Schifter, D., & Fosnot, C. T. (1993). *Reconstructing mathematics education: Stories of teachers meeting the challenge of reform.* Portsmouth, NH: Heinemann.

Schifter, D., Monk, G. S., Russell, S. J., & Bastable, V. (2007). Early algebra: What does understanding the laws of arithmetic mean in the elementary grades? In J. Kaput, M. Blanton, & D. Carraher (Eds.), *Algebra in the early grades.* Mahwah, NJ: Lawrence Erlbaum Associates, Inc.

Schifter, D., Russell, S. J., & Bastable, V. (1999). Teaching to the big ideas. In M. Z. Solomon (Ed.), *The diagnostic teacher: Constructing new approaches to professional development.* New York: Teachers College Press.

Schmittau, J., & Morris, A. (2004). The development of algebra in the elementary mathematics curriculum of V. V. Davydov. *The Mathematics Educator, 8*, 60–87.

Schoenfeld, A. H. (Ed.) (1983). *Problem solving in the mathematics curriculum: A report, recommendations, and an annotated bibliography.* Washington, DC: Mathematical Association of America.

Schoenfeld, A. H. (1985). *Mathematical problem solving.* New York: Academic Press.

Schoenfeld, A. H. (1986). On having and using geometric knowledge. In J. Hiebert (Ed.), *Conceptual and procedural knowledge: The case of mathematics.* Hillsdale, NJ: Lawrence Erlbaum Associates, Inc.

Schoenfeld, A. H. (1988). When good teaching leads to bad results: The disasters of "well-taught" mathematics courses. *Educational Psychologist, 23*, 145–166.

Schoenfeld, A. H. (1989). Explorations of students' mathematical beliefs and behavior. *Journal for Research in Mathematics Education, 20*, 338–355.

Schoenfeld, A. H. (1991). On mathematics as sense-making: An informal attack on the unfortunate divorce of formal and informal mathematics. In J. Voss, D. Perkins & J. Segal (Ed.), *Informal reasoning and education.* Hillsdale, NJ: Lawrence Erlbaum Associates, Inc.

Schoenfeld, A. H. (1992). Learning to think mathematically, problem solving, metacognition, and sense making in mathematics. In D. Grouws (Ed.), *Handbook of research on mathematics teaching and learning.* Reston, VA: National Council of Teachers of Mathematics.

Schoenfeld, A. H. (1994). What do we know about mathematics curricula? *Journal of Mathematical Behavior, 13,* 55–80.

Schoenfeld, A. H. (1999). Looking toward the 21st century: Challenges of educational theory and practice. *Educational Researcher, 28,* 4–14.

Schwab, J. J. (1978). Education and the structure of the disciplines. In J. Westbury & N. J. Wilkof (Eds.), *Science, curriculum, and liberal education: Selected essays.* Chicago, IL, and London: University of Chicago Press.

Schweitzer, K. (2006). "Teacher as researcher: Research as a partnership." In S. Z. Smith, M. Smith, & D. S. Mewborn (Eds.), *Teachers engaged in research preK-2.* Reston, VA: National Council of Teachers of Mathematics.

Sekiguchi, H. (1991). *An investigation on proofs and refutations in the mathematics classroom.* Unpublished doctoral dissertation, University of Georgia, Athens.

Selden A., & Selden J. (2003) Validations of proofs considered as texts: Can undergraduates tell whether an argument proves a theorem? *Journal for Research in Mathematics Education, 34,* 4–36.

Selden, A., Selden, J., Hauk, S., & Mason, A. (2000). Why can't calculus students access their knowledge to solve non-routine problems? In A. H. Schoenfeld, J. Kaput, & E. Dubinsky, (Eds.), *Research in collegiate mathematics education IV.* Providence, RI: American Mathematical Society.

Selden, J., & Selden, A. (1995). Unpacking the logic of mathematical statements. *Educational Studies in Mathematics, 29,* 123–151.

Senk, S. L. (1985). How well do students write geometry proofs? *Mathematics Teacher, 78,* 448–456.

Senk, S. L. (1989). Van Hiele levels and achievement in writing geometry proofs. *Journal for Research in Mathematics Education, 20,* 309–321.

Sfard, A. (1991). On the dual nature of mathematical conceptions: Reflections on processes and objects as different sides of the same coin. *Educational Studies in Mathematics, 22,* 1–36.

Sfard, A. (1992). Operational origins of mathematical objects and the quandary of reification—the case of function. In G. Harel & E. Dubinsky (Eds.), *The concept of function: Aspects of epistemology and pedagogy.* Washington, DC: Mathematical Association of America.

Sfard, A. (2000). On reform movement and the limits of mathematical discourse. *Mathematical Thinking and Learning, 2,* 157–189.

Sfard, A. (2001). There is more to discourse than meets the ears: Looking at thinking as communicating to learn more about mathematical learning. *Educational Studies in Mathematics, 46,* 13–57.

Sfard, A. (2003). Balancing the unbalancable: The NCTM standards in light of theories of learning mathematics. In J. Kilpatrick, G.W. Martin, & D. Schifter (Eds.), *A research companion to principles and standards for school mathematics.* Reston, VA: NCTM.

Shulman, S. (1986). Those who understand: Knowledge growth in teaching. *Educational Researcher, 15,* 4–14.

Shulman, S. (1987). Knowledge and teaching: Foundations of the new reform. *Harvard Educational Review, 57,* 1–22.

Silver, E., & Carpenter, T. (1989). Mathematical methods. In M. Lindquist (Ed.), *Results from the fourth mathematics assessment of the National Assessment of Educational Progress.* Reston, VA: NCTM.

Silver, E. A., Mills, V., Castro, A., & Ghousseini, H. (2006). Blending elements of lesson study with case analysis and discussion: A promising professional development synergy. In K. Lynch-Davis & R. L. Ryder (Eds.), *The work of mathematics teacher educators: Continuing the conversation.* San Diego, CA: Association of Mathematics Teacher Educators.

Silver, E. A., & Stein, M. K. (1996). The QUASAR project: The "revolution of the possible" in mathematics instructional reform in urban middle schools. *Urban Education, 30,* 476–521.

Simon, M. A. (1997). Developing new models of mathematics teaching: An imperative for research on mathematics teacher development. In E. Fennema & B. Nelson. (Eds.), *Mathematics teachers in transition.* Hillsdale, NJ: Lawrence Erlbaum Associates, Inc.

Simon, M., & Blume, G. (1996). Justification in the mathematics classroom: A study of prospective elementary teachers. *Journal of Mathematical Behavior, 15,* 3–31.

Smith, J. (2006). A sense-making approach to proof: Strategies of students in traditional and problem-based number theory courses. *Journal of Mathematical Behavior, 25,* 73–90.

Smith, M. S., & Silver, E. A. (2000). Research-based cases for mathematics teacher education: The COMET project. In M. L. Fernandez (Ed.), *Proceedings of the 22nd Annual Meeting of the North American Chapter of the International Group for the Psychology of Mathematics Education.* Columbus, OH: ERIC Clearinghouse for Science, Mathematics, and Environmental Education.

Smith, M. S., Silver, E. A., & Stein, M. K. (2005a). *Improving instruction in algebra: Using cases to transform mathematics teaching and learning,* Volume 2. New York: Teachers College Press.

Smith, M. S., Silver, E. A., & Stein, M. K. (2005b). *Improving instruction in rational numbers and proportionality: Using cases to transform mathematics teaching and learning,* Volume 1. New York: Teachers College Press.

Sodian, B., & Wimmer, H. (1987). Children's understanding of inference as a source of knowledge. *Child Development, 58,* 424–433.

Sowder, L., & Harel, G. (2003). Case studies of mathematics majors' proof understanding, production, and appreciation. *Canadian Journal of Science, Mathematics and Technology Education, 3,* 251–267.

Stacey, K. (1989). Finding and using patterns in linear generalizing problems. *Educational Studies in Mathematics, 20,* 147–164.

Steen, L. (1988). The science of patterns. *Science, 240,* 611–616.

Steffe, L. & Kieren, T. (1994). Radical constructivism and mathematics education. *Journal for Research in Mathematics Education, 25,* 711–733.

Stein, M. K., Grover, B., & Henningsen, M. (1996). Building student capacity for mathematical thinking and reasoning: An analysis of mathematical tasks used in reform classrooms. *American Educational Research Journal, 33,* 455–488.

Stein, M. K., Smith, M. S., Henningsen, M. A., & Silver, E. A. (2000). *Implementing standards-based mathematics instruction: A casebook for professional development.* New York: Teachers College Press.

Stephan, M., & Rasmussen, C. (2002). Classroom mathematical practices in differential equations. *Journal of Mathematical Behavior, 21,* 459–490.

Stone, C. A. (1993). What is missing in the metaphor of scaffolding? In E. Forman, N. Minick, & C. A. Stone (Eds.), *Contexts for learning: Sociocultural dynamics in children's learning.* Oxford: Oxford University Press.

Stylianides, A. J. (2007). The notion of proof in the context of elementary school mathematics. *Educational Studies in Mathematics, 65,* 1–20.

Stylianides, A. J., Stylianides, G. J., & Philippou, G. N. (2004). Undergraduate students' understanding of the contraposition equivalence rule in symbolic and verbal contexts. *Educational Studies in Mathematics, 55,* 133–162.

Stylianides, G. J. (2005). *Investigating students' opportunities to develop proficiency in reasoning and proving: A curricular perspective.* Unpublished doctoral dissertation, University of Michigan, Ann Arbor.

Stylianides, G. J. (2008). An analytic framework of reasoning-and-proving. *For the Learning of Mathematics, 28,* 9–16.

Stylianides, G., & Stylianides, A. (2008). Proof in school mathematics: Insights from psychological research into students' ability for deductive reasoning. *Mathematics Thinking and Learning, 10,* 103–133.

Stylianides, G. J., Stylianides, A. J., & Philippou, G. N. (2007). Preservice teachers' knowledge of proof by mathematical induction. *Journal of Mathematics Teacher Education, 10,* 145–166.

Stylianou, D., & Blanton, M. (2006). *Undergraduate students' conceptions of proof.* Poster presentation at the AERA conference. San Francisco, CA.

Tall, D. (1989) The nature of mathematical proof. *Mathematics Thinking, 127,* 28–32.

Tall, D. (1991). The psychology of advanced mathematical thinking. In D. Tall (Ed.), *Advanced mathematical thinking.* Dordrecht, The Netherlands: Kluwer Academic.

Tall, D. (1995). Cognitive development, representations and proof. In *Proceedings of Justifying and Proving in School Mathematics.* London: Institute of Education.

Tall, D., & Vinner, S. (1981). Concept image and concept definition in mathematics with ref. to limits. *Educational Studies in Mathematics, 12,* 151–169.

Thurston, W. (1994). On proof and progress in mathematics. *Bulletin of the American Mathematical Society, 30,* 161–177.

Toulmin, S. (1958). *The uses of argument.* Cambridge University Press

Uptegrove, E. B., & Maher, C. A. (2004) Students building isomorphisms. In M. J. Hoines & A. B. Fuglestad (Eds.), *Proceedings of the 28th Conference of the International Group for the Psychology of Mathematics Education,* Vol. 4. Bergen: Bergen University College.

Usiskin, Z. (1980). What should not be in the algebra and geometry curricula of average college-bound students. *Mathematics Teacher, 73*, 413–424.

Usiskin, Z. (1987). Resolving the continuing dilemmas in school geometry. In M. M. Lindquist & A.P. Shulte (Eds.), *Learning and teaching geometry, K–12, 1987 yearbook*. Reston, VA: National Council of Teachers of Mathematics.

van Dormolen, J. (1977). Learning to understand what giving a proof really means. *Educational Studies in Mathematics, 8*, 27–34.

Velleman, D. J. (1994). *How to prove it: A structured approach*. Cambridge: Cambridge University Press.

Vergnaud, G. (1982). A classification of cognitive tasks and operations of thought involved in addition and subtraction problems. In T. P. Carpenter, J. M. Moser, & T. A. Romberg (Eds.), *Addition and subtraction: A cognitive perspective*. Hillsdale, NJ: Lawrence Erlbaum Associates, Inc.

Vergnaud, G. (1983). Multiplicative structures. In R. Lesh & M. Landau (Eds.), *Acquisition of mathematics concepts and processes*. London: Academic Press.

Vinner, S. (1983). The notion of proof—some aspects of students' views at the senior high level. *Proceedings of the 7th International Conference for the Psychology of Mathematics Education*. Rehovot, Israel: Weizmann Institute of Science.

Vinner, S. (1992). The function concept as a prototype for problems in mathematics learning. In G. Harel & E. Dubinsky (Eds.), *The concept of function: Aspects of epistemology and pedagogy*. Washington, DC: Mathematical Association of America.

Vinner, S., & Dreyfus, T. (1989). Images and definitions for the concept of function. *Journal for Research in Mathematics Education, 20*, 356–366.

Voigt, J. (1995). Thematic patterns of interaction and sociomathematical norms. In P. Cobb & H. Bauersfeld (Eds.), *The emergence of mathematical meaning: Interaction in classroom cultures*. Hillsdale, NJ: Lawrence Erlbaum Associates, Inc.

Vygotsky, L. (1962). *Thought and language* (E. Hanfmann & G. Vakar, Trans.). Cambridge, MA: Massachusetts Institute of Technology. (Original work published in 1934.)

Waring, S. (2000). *Can you prove it? Developing concepts of proof in primary and secondary schools*. Leicester: Mathematical Association.

Weber, K. (2001). Student difficulty in constructing proofs: The need for strategic knowledge. *Educational Studies in Mathematics, 48*, 101–119.

Weber, K. (2004). Traditional instruction in advanced mathematics classrooms: A case study of one professor's lectures and proofs in an introductory real analysis course. *Journal of Mathematical Behavior, 23*, 115–133.

Weber, K. (2005). A procedural route toward understanding aspects of proof: Case studies from real analysis. *Canadian Journal of Science, Mathematics, and Technology Education, 5*, 469–483.

Weber, K., & Alcock, L. (2004) Semantic and syntactic proof productions. *Educational Studies in Mathematics, 56*, 209–234.

Weber, K., & Alcock, L. (2005). Using warranted implications to understand and validate proofs. *For the Learning of Mathematics, 25*, 34–38.

Weber, K., Alcock, L., & Radu, I. (2005). Undergraduates' use of examples in a transition-to-proof course. In S. Wilson (Ed.), *Proceedings of the 26th Conference for the North American Chapter of the Psychology of Mathematics Education*. Roanoke, VA.

Weinert, F. E. (2001). Vergleichende Leistungsmessung in Schulen: Eine umstrittene Selbstverständlichkeit. In F. E. Weinert (Ed.), *Leistungsmessungen in Schulen*. Weinheim: Beltz.

Weiss, M., and Herbst, P. (2007, April). *"Every single little proof they do, you could call it a theorem": Translation between abstract concepts and concrete objects in the geometry classroom*. Paper presented at AERA. Chicago, IL.

Wells, G. (1999). *Dialogic inquiry: Toward a sociocultural practice and theory of education*. Cambridge: Cambridge University Press.

Wertsch, J. V. (1991). *Voices of the mind: A sociocultural approach to mediated action*. Cambridge, MA: Harvard University Press.

Werstch, J. V. (1999). The voice of rationality in a sociocultural approach to mind. In L. C. Moll (Ed.), *Vygotsky and education: Instructional implications and applications of sociohistorical psychology*. New York: Cambridge University Press.

Wertsch, J., & Toma, C. (1995). Discourse and learning in the classroom: A sociocultural approach. In L. Steffe & J. Gale (Eds.), *Constructivism in education*. Hillsdale, NJ: Lawrence Erlbaum Associates, Inc.

Weyl, A. (1940). The mathematical way of thinking. *Science, 92*, 437–446.

Wheeler, D. (1990). Aspects of mathematical proof. *Interchange, 21*, 1–5.

Williams, E. (1980). An investigation of senior high school students' understanding of the nature of mathematical proof. *Journal for Research in Mathematics Education, 11*, 165–166.

Wood, T. (1999). Creating a context for argument in mathematics class. *Journal for Research in Mathematics Education, 30*, 171–191.

Wu, H. (1996). The role of Euclidean geometry in high school. *Journal of Mathematical Behavior, 15*, 221–237.

Yackel, E. (2002). What we can learn from analyzing the teacher's role in collective argumentation. *Journal of Mathematical Behavior, 21*, 423–440.

Yackel, E., & Cobb, P. (1996). Sociomathematical norms, argumentations and autonomy in mathematics. *Journal for Research in Mathematics Education, 27*, 458–477.

Yackel, E., & Hanna, G. (2003). Reasoning and proof. In J. Kilpatrick, W.G. Martin, & D. Schifter (Eds.), *A research companion to the principles and standards for school mathematics*. Reston, VA: National Council of Teachers of Mathematics.

Yackel, E., Rasmussen, C., & King, K. (2000). Social and sociomathematical norms in an advanced undergraduate course. *Journal of Mathematical Behavior, 19*, 275–287.

Yandell, B. (2002). *The honors class: Hilbert's problems and their solvers*. Natick, MA: A. K. Peters.

Zack, V. (1991). It was the worst of times: Learning about the Holocaust through literature. *Language Arts, 68*, 42–48.

Zack, V. (1995). Algebraic thinking in the upper elementary school: The role of collaboration in making meaning of "generalization". In L. Meira & D. Carraher (Eds.), *Proceedings of the 19th Annual Conference of the International Group for the Psychology of Mathematics Education*, Vol. 2. Recife, Brazil: Program Committee.

Zack, V. (1999a). Everyday and mathematical language in children's argumentation about proof. In L. Burton (Ed.), The culture of the mathematics classroom. Special Issue, *Educational Review, 51*, 129–146.

Zack, V. (1999b). Nightmare issues: Children's responses to racism and genocide in literature. In J. P. Robertson (Ed.), *Teaching for a tolerant world*. Urbana, IL: National Council of Teachers of English.

Zack, V. (2002). Learning from learners: Robust counterarguments in fifth graders' talk about reasoning and proving. In A. D. Cockburn & E. Nardi (Eds.), *Proceedings of the 26th International Conference for the Psychology of Mathematics Education*, Vol. 4. Norwich: Program Committee.

Zack, V., & Graves, B. (2001). Making mathematical meaning through dialogue: "Once you think of it, the z minus three seems pretty weird." In C. Kieran, E. Forman, & A. Sfard (Eds.), Bridging the individual and the social: Discursive approaches to research in mathematics education. Special Issue, *Educational Studies in Mathematics, 46*, 229–271.

Zack, V., & Reid, D. A. (2003). Good-enough understanding: Theorizing about the learning of complex ideas (Part 1). *For the Learning of Mathematics, 23*, 43–50.

Zack, V., & Reid, D. A. (2004). Good-enough understanding: Theorizing about the learning of complex ideas (Part 2). *For the Learning of Mathematics, 24*, 25–28.

Index

E

Elementary grades, 6, 65
Emotions (and proof), 199
Euclid (Euclidian), 2, 281
Expert (mathematicians), 313
Examples 323 (generic), 286 (prototype)

F

Formal rules (for proof), 290
Functions, 285, 327, 341

G

Geometry, 2, 19, 250
Greek (mathematics), 2, 279

H

Habits of mind, 8
Heuristic (training and proof), 5, 193
Hierarchical structure (proof), 340
Hilbert's fifth problem, 280
History (of proof and mathematics), 2, 280

I

Inquiry-based learning, 309
Instructional
    situation, 28
    scaffolding, 292
Intellectual need, 272, 279
Intermediate value theorem, 324
Intuition (intuitive understanding), 286, 329, 348

K

Key ideas, 333
Knowledge (communally developed vs. individual construction), 308
K-proofs (see Proofs, types of)

L

Language (and proof), 5, 270
Linear algebra, 284
Logical
    equivalence, 205

ordering in proof, 212
notation, 323

M

Mathematical
    Authority, 313
    competency, 191
    disposition, 309
    proof (see Proof, mathematical)
Mathematics as social activity, 272, 307
Metacognition, 299
Misconceptions (on proof), 3–5
Models, 102
Motivation, 194
Moore Method, 309, 345

N

Negotiation of ideas, 10, 272, 303, 308
New Math, 3
Notational systems, 7
Number theory, 309

O

Open number line, 7, 106
Obstacles (epistemological), 279

P

Participation (in learning community), 271
Pattern identification, 235
Pattern spotting, 9, 177
Problem-based instruction, 307
Problem-centered reasoning, 347
Problem solving, x, 136, 289, 308
Professional development, 241
Proof
    and disciplinary practice, 22, 37–38
    and proving, 46–47
    as a generalized argument, 24
    as aid to understanding, 77
    as aid to conviction, 80
    as obstacle, 283
    as public performance, 40–42
        timescales of proof in public performances, 45